Paul Martin studied biology at Cambrid........ where he graduated in Natural Sciences and t........ ...... behavioural biology, and at Stanford University .... Fellow in the School of Medicine. H........ ....searched at Cambridge, before leaving acad........ .... interests. He spent several years as a Whitehall-........ ...... civil servant, latterly as the Director of Communication at ....e Cabinet Office, before becoming a science writer. He is currently a Fellow of Wolfson College, Cambridge. He is co-author with Patrick Bateson of *Measuring Behaviour* and *Design For A Life*. His first book *The Sickening Mind* was described as a 'masterpiece of popularization'. It was shortlisted for the NCR Prize.

More from the reviews:

'Last frontiers are hard to find these days, but Paul Martin has found one: the one third of all human experience that has not been turned into lifestyle. It is something of an embarrassment that in the time of the human genome we still don't really know what sleep is for ... Martin makes a powerful case for the need to take sleep seriously. *Counting Sheep* is an antidote to the symptoms of the frenetic society delineated by James Gleick in *Faster*. I hope it does as well, either as an instant hit or as a sleeper.' PETER FORBES, *Guardian*

'Far-ranging while remaining lucid, fascinating and enjoyable.' THOMAS HODGKINSON, *Literary Review*

'Darwin shaking plants at night to keep them awake. How to spot a sleeping Mediterranean flour moth. R.E.M. (not the band). The Clock gene. The death of Greek hero Perseus. Dickens composing poetry, half-sleeping, on long nocturnal walks. The battle of Stalingrad. The origins of the expressions "nodding off" and "pipe dreams" (the latter to do with opium). Jouvet preventing natural sleep paralysis in cats – "when the sleeping cats entered R.E.M. sleep they would get up and move around, as though acting out a dream." And so on and so on. A pop science Pandora's box of all things sleep-related, made rich and wonderfully well rounded by the author's impressive knowledge – not only scientific, but literary, historical and cultural ... This is a fascinating and comprehensive book.' STEVEN HALL, *City Life*

'Everybody sleeps, everybody dreams and so everybody should have an interest in *Counting Sheep*. Snappily written and cleverly structured ... it is strangely comforting.'

JOANNE HAYDEN, *Sunday Business Post*

'A wide-ranging and freewheeling study of sleep.' *The Week*

'I adored this – reading it kept me awake well past my bedtime. It's chock-full with fascinating stuff – both scientific and cultural – on that most ordinary but oh so mysterious of human activities. The chapter headings alone just make you want to get stuck in – "Strange Tales of Erections and Yawning", "A Brief History of Beds", "Soggy Sheets" and "The Wonderful World of Snoring".' *The Bookseller*

'For those that sometimes have problems drifting off to sleep, this book is on hand to dispel the myths relating to the land of nod. Witty and illuminating, *Counting Sheep* also covers the more in-depth and scientific areas of sleep. A highly recommended read.' *Health Matters*

'This book covers everything you could possibly want to know about sleep and dreams. . . . Paul Martin weaves literary references to sleep, up-to-date science and his own philosophical musings and creates an engaging and fascinating read.' LIZ HOLLIS, *Health & Fitness*

'Martin deals with that third of our life we spend in bed – everything to do with the bedtime hours is here.' TOM O'DEA *Irish Independent*

COUNTING SHEEP

The Science and Pleasures of Sleep and Dreams

PAUL MARTIN

Flamingo
An Imprint of HarperCollinsPublishers

Flamingo
An Imprint of HarperCollins*Publishers*
77–85 Fulham Palace Road,
Hammersmith, London w6 8jb

Flamingo is a registered trade mark of
HarperCollins*Publishers* Ltd

www.**fire**and**water**.com

Published by Flamingo 2003
4

First published in Great Britain by
HarperCollins*Publishers* 2002

Author photograph by Simon James

ISBN 0 00 655172 6

Set in Palatino by Rowland Phototypesetting Limited,
Bury St Edmunds, Suffolk

Printed and bound in Great Britain by
Clays Ltd, St Ives plc

Contents

CONTENTS

PART I

Preliminaries

1

A Third of Life

Man ... consumes more than one third of his life in this his irrational situation.

Erasmus Darwin, *Zoonomia* (1801)

Sleep: a state so familiar yet so strange. It is the single most common form of human behaviour and you will spend a third of your life doing it – 25 years or more, all being well. When you die, a bigger slice of your existence will have passed in that state than in making love, raising children, eating, playing games, listening to music, or any of those other activities that humanity values so highly.

Sleep *is* a form of behaviour, just as eating or socialising or fighting or copulating are forms of behaviour, even if it is not the most gripping to observe. Most of the action goes on inside the brain. It is also a uniquely private experience, even when sharing a bed. When we are awake we all inhabit a common world, but when we sleep each of us occupies a world of our own. Most of us, however, have precious little awareness of what we experience in that state. Our memories of sleeping and dreaming mostly evaporate when we awake, erasing the record every morning.

Many of us do not get enough sleep and we suffer the consequences, often without realising what we are doing to ourselves. The demands of the 24-hour society are marginalising sleep, yet it is not an optional activity. Nature imposes it upon us. We can survive for longer without food. When our sleep falls short in quantity or quality we pay a heavy price in depressed mood, impaired performance, damaged social relationships and poorer health. But we usually blame something else.

Sleep is an active state, generated within the brain, not a mere absence of consciousness. You are physiologically capable of sleeping

3

with your eyelids held open by sticking plaster, bright lights flashing in your eyes and loud music playing in your ears. We shall later see how science has revealed the ferment of electrical and chemical activity that goes on inside the brain during sleep, and how the sleeping brain operates in a quite different mode from waking consciousness. We shall see too how lack of sleep erodes our quality of life and performance while simultaneously making us more vulnerable to injuries and illness. Science amply supports William Shakespeare's view that sleep is the 'chief nourisher in life's feast'.

What is sleep and what is it for? Why do so many people have such problems with it? Why do we dream? Although sleep forms a central strand of human and animal life it is still poorly understood and widely neglected. It is an inglorious example of familiarity breeding contempt. Sleep is so much a part of our everyday existence that we take it for granted. We are ignorant even of our ignorance. In 1758, Doctor Samuel Johnson summed it up like this:

> Among the innumerable mortifications that waylay human arrogance on every side may well be reckoned our ignorance of the most common objects and effects ... Vulgar and inactive minds confound familiarity with knowledge, and conceive themselves informed of the whole nature of things when they are shown their form or told their use ... Sleep is a state in which a great part of every life is passed. No animal has been yet discovered whose existence is not varied with intervals of insensibility. Yet of this change so frequent, so great, so general, and so necessary, no searcher has yet found either the efficient or final cause; or can tell by what power the mind and body are thus chained down in irresistible stupefaction; or what benefits the animal receives from this alternate suspension of its active powers.

The scientists who do know something about sleep often bemoan society's ignorance of it. They point to the vast gap between current scientific understanding of sleep, patchy though it is, and the practical benefits it could bring if that knowledge were absorbed and acted upon by society. Our collective indifference towards sleep has enormous and largely avoidable costs.

A sleep-sick society?

> The mere presence of an alarm clock implies sleep deprivation, and what bedroom lacks an alarm clock?
>
> James Gleick, *Faster* (1999)

All is not well with the state of sleep. Many of us depend on an alarm clock to prise us out of bed each morning, and children's bedrooms increasingly resemble places of entertainment rather than places of sleep. When given the opportunity, we sleep in at the weekends and feel only half awake when we do get up. On that long-awaited holiday we find the change of scenery (or is it the air?) makes us even sleepier. We are told that lying around and sleeping too much will only make us sleepier. But in truth we feel sleepy at weekends and on holidays not because we are sleeping too much, but because we have slept too little the rest of the time.

A century ago the majority toiled long hours while the affluent few idled away their time. Today, however, the more conventionally successful you are, the less free time you will probably have. Having nothing to do is seen as a sign of worthlessness, while ceaseless activity signifies status and success. Supposedly unproductive activities are deprioritised or delegated. And according to prevailing cultural attitudes, sleeping is one of the least productive of all human activities – more worthwhile perhaps than sitting around picking your nose, but not much. In their ceaseless pursuit of work and pleasure the cash-rich buy time from others, hiring them to clean their houses, look after their children and cook their food. But one of the activities you simply cannot delegate to anyone else is sleeping.

Evolution equipped humans, in common with all other animals, with biological mechanisms to make us sleep at roughly the same time every day. However, those mechanisms evolved to cope with a pre-industrial world that was vastly different from the one we now inhabit.

Our daily cycles of sleep and activity are no longer driven by dawn and dusk, but by clocks, electric lighting and work schedules. Sleep has become increasingly devalued in the 24-hour society. Many regard sleep as wasted time and would prefer to sacrifice less of their busy lives to it. We live in a world where there are too many tired, sleep-deprived people. Think of those pinched, yawning faces you can see every day on the trains and in buses and in cars crawling through jams. They look as if they have been brainwashed, but they are just tired.

We pay a steep price for neglecting sleep, in our ignorance and indifference. The scientific evidence tells us that far too many people in industrialised societies are chronically sleep-deprived, with damaging consequences for their mental and physical health, performance at work, quality of life and personal relationships. William Dement, a pioneering scientist in the field, believes that we now live in a 'sleep-sick society'. Scientists have not yet reached a consensus about the precise extent of sleep deprivation in society, but they do all agree that sleepiness is a major cause of accidents and injuries. In fact, sleepiness is responsible for far more deaths on the roads than alcohol or drugs.

Everyone has heard about the need for a balanced diet and physical exercise, even if many of us fail to follow the advice. But sleep is lost in a deep well of ignorance and apathy. Even the medical profession pays it scant regard. Sleep and its disorders barely feature in the teaching of medicine, and few physicians are fully equipped to deal with the sleep problems they regularly encounter. When researchers from Oxford University investigated British medical education in the late 1990s, they discovered that the average amount of time devoted to sleep and sleep disorders in undergraduate teaching was five minutes, rising to a princely peak of 15 minutes in preclinical training. Your doctor is therefore unlikely to be an expert on the subject.

The general public and the medical profession are not the only ones to display a remarkable indifference to sleep. So too do most contemporary writers. Considering that sleep accounts for a third of human existence, it features remarkably rarely in novels, biographies, social histories or learned texts on neurobiology, psychology and medicine. And the few accounts that have made it into print are mostly concerned with what happens when it goes wrong. Insomnia and nightmares loom large in the tiny literature of sleep.

Few biographies mention the sleep behaviour or dreams of their subjects. That part of their story is almost invariably missing, as if somehow we all cease to exist at night. And most of those scholarly books that set out to explain how the human mind works say little or nothing about what goes on during the several hours of every day when the mind is sleeping and dreaming. They are really just books about how the brain works when it is awake. Our neglect of sleep is underlined by its absence from our literature.

Vladimir Nabokov once said that all the great writers have good eyes. What has happened to the eyes of writers as far as sleep and dreams are concerned? It was not always so. Older literature is distinctly richer in references to sleeping and dreaming, perhaps because darkness and sleep and dreams were much more prominent aspects of everyday life before the invention of the electric light bulb and the advent of the 24-hour society. Shakespeare's works are thick with allusions to sleep and dreams, as are Dickens's. We shall encounter some of them later. Meanwhile, to set the right tone, here is Sancho Panza's eulogy to sleep from *Don Quixote*:

> God bless the inventor of sleep, the cloak that covers all man's thoughts, the food that cures all hunger, the water that quenches all thirst, the fire that warms the cold, the cold that cools the heat; the common coin, in short, that can purchase all things, the balancing weight that levels the shepherd with the king and the simple with the wise.

The universal imperative

> Almost all other animals are observed to partake of sleep, aquatic, winged, and terrestrial creatures alike. For every kind of fish and the soft-shelled species have been seen sleeping, as has every other creature that has eyes.
>
> Aristotle (384–322 BC), *On Sleep and Waking*

Sleep is a universal human characteristic, like eating and drinking. Absolutely everybody does it. Sleep occupies about one third of each human life, and up to two thirds of a baby's time. (According to Groucho Marx, the proportion rises to three thirds if you live in Peoria.) It is a common bond that ties us all together. We have no choice: the longer we go without sleep, the stronger our desire for it grows. Tiredness, like hunger and thirst, will eventually force us to do the right thing whether we want to or not.

The dreams that accompany sleep are equally ubiquitous features of human life, even if many of us retain little memory of them after we awake. Dreaming is a classless activity that unites monarchs and paupers, a thought that Charles Dickens mused upon in one of his essays:

7

Here, for example, is her Majesty Queen Victoria in her palace, this present blessed night, and here is Winking Charley, a sturdy vagrant, in one of her Majesty's jails ... It is probable that we have all three committed murders and hidden bodies. It is pretty certain that we have all desperately wanted to cry out, and have had no voice; that we have all gone to the play and not been able to get in; that we have all dreamed much more of our youth than of our later lives.

Sleep is not a specifically human trait, of course. On the contrary, it is a universal characteristic of complex living organisms, as Aristotle deduced more than 23 centuries ago. Sleep is observed in animals of every sort, including insects, molluscs, fish, amphibians, birds and mammals. Within the animal world, sleep does vary enormously in quantity, quality and timing, accounting for anything up to 80 per cent of some animals' lifespans. But they all do it, one way or another. Some species, especially predators, spend more of their lives asleep than they do awake, a fact that TV documentaries and natural-history books seldom mention.

How do we know that an animal is sleeping? It is hard enough sometimes to be sure that a human is asleep, let alone a fish or a fly. The ultimate indicator of whether an animal or person is asleep is the distinctive pattern of electrical activity in its brain. During deep sleep the billions of individual nerve cells in the brain synchronise their electrical activity to some extent, generating characteristic waves of tiny voltage changes that can be detected by electrodes placed on the scalp. We shall be exploring the nature and internal structure of sleep later. The easiest way to recognise sleep, however, is from overt behaviour.

Sleep has several rather obvious distinguishing characteristics. A sleeping person or animal will generally remain in the same place for a prolonged period, perhaps several hours. There will be a certain amount of twitching, shifting of posture and fidgeting. Young animals will suckle while they sleep and ruminants will carry on chewing the cud. But sleepers normally do not get up and change their location. (When they do, we recognise it as a curious phenomenon and call it sleepwalking.)

Sleeping organisms also adopt a characteristic posture. Sloths and bats, for example, sleep hanging upside down from a branch. The Mediterranean flour moth sleeps with its antennae swivelled backwards and the tips tucked under its wings. If you are careful, you can gently lift the sleeping moth's wing without disturbing it – a trick that will definitely not work when it is awake. A lizard will settle on a branch during the hours before sunset, curl up its tail, close its eyelids, retract its eyeballs and remain in that distinctly sleep-like posture all night unless it is disturbed. A partridge, like many birds, will rest its weight on one leg while it sleeps. It is said that some gourmets can tell *which* leg, from its taste.

Monkeys and apes, including humans, usually sleep lying down. Indeed, we are built in such a way that we find it difficult to sleep properly unless we are lying down. People can and sometimes do sleep after a fashion while sitting, notably in aeroplanes, business meetings and school classrooms. If you are really exhausted, you might even manage to snatch some sleep standing up. But sleep taken while standing or sitting upright is generally fitful, shallow and unrefreshing. The non-horizontal sleeper may repeatedly nod off, but as soon as they descend beyond the shallowest stages of sleep their muscles relax, they begin to sway and their brain wakes them up again. That is why we 'nod off'. If you travel frequently on trains or buses, you might have had the dubious pleasure of sitting next to a weary commuter who has nodded off all over your shoulder. Recordings of brain-wave patterns show that people sleeping in an upright sitting position achieve only the initial stages of light sleep, not the sustained, deep sleep we require to wake up feeling truly refreshed. The reason is simple. Our muscles relax when we are fully asleep and we would fall over if we were not already lying down. Our brains therefore do not permit us to enter sustained, deep sleep unless we are in a physically stable, horizontal (or near-horizontal) posture.

Despite the virtual impossibility of sleeping deeply while sitting upright, we are sometimes forced to try. In *Down and Out in Paris and London*, George Orwell describes a particularly unwelcoming form of overnight accommodation that was known to the homeless of prewar London as the Twopenny Hangover. At the Twopenny Hangover the night's residents would sit in a row along a bench. In front of them was a rope, and the would-be sleepers would lean on this rope as though leaning over a fence. In that posture they were supposed to sleep. At five o'clock the next morning an official, wittily known as the valet, would cut the rope so that the residents could begin another day of wandering the streets.

Nowadays, tourist-class airline passengers travelling long distances can enjoy an experience similar to the Twopenny Hangover, albeit at vastly greater expense. George Orwell's autobiographical account of grinding poverty in the late 1920s is also a sharp reminder that lack of money is often accompanied by lack of decent sleep. Rough sleepers rarely get a good night's sleep.

Sleep has several other distinctive characteristics besides immobility and posture. In many species, including humans, individuals return to the same place each night (or each day, if they are nocturnal) in order to sleep. More generally, all members of a given species will tend to choose the same sorts of sleeping places. The distinctive feature of those places is often their security. Birds usually sleep on inaccessible branches or ledges. Many small mammals sleep in underground burrows where they are safer from predators. Fishes lie on the bottom, or wedge themselves into a crevice or against the underside of a rock. We humans prefer to sleep in relatively private and secure places. Given the choice, we rarely opt to sleep on busy streets or in crowded restaurants.

One obvious feature of sleep is a marked reduction in responsiveness to sights, sounds and other sensory stimuli. To provoke a response from a sleeping organism, stimuli have to be more intense or more relevant to the individual. For example, the reef fish known as the slippery dick sleeps during the hours of darkness, partly buried in the sand. While it is in this state, the sleeping slippery dick can be gently lifted to the surface by hand without it waking up and swimming off.

A sort of perceptual wall is erected during sleep, insulating the mind from the outside world. You would still be able to sleep if you had no eyelids, because your sleeping brain would not register what your eyes could see. This sensory isolation is highly selective, however. You can sleep through relatively loud noises from traffic or a radio, but a quiet mention of your name can rouse you immediately. Your brain is not simply blocked off during sleep. Moreover, this reduced responsiveness is rapidly reversible – a characteristic that distinguishes sleep from states such as unconsciousness, coma, anaesthesia and hibernation. A suitable stimulus, particularly one signifying immediate danger, can snap a sleeping person into staring-eyed alertness in an instant.

Another diagnostic feature of sleep is its regular cycle of waxing and waning. Living organisms sleep and wake according to a regular 24-hour cycle, or circadian rhythm. All members of a given species tend to sleep during the same part of the 24-hour cycle, when their environment is least favourable for other activities such as looking for food. For most species this means sleeping during the hours of darkness, but some species do the reverse. Many small mammals, which would be more vulnerable to predators during daylight, sleep by day and forage at night. Aside from a few nocturnal specialists such as owls, birds cannot easily fly in the dark, and most reptiles find it hard to maintain a sufficiently high body temperature to be active during the cool of night. Most birds and reptiles therefore sleep at night. Predators tend to sleep when their prey are asleep and hunt when their prey are up and about.

Sleep, then, is characterised by a special sleeping place and posture, prolonged immobility, a selective and rapidly reversible reduction in responsiveness to stimuli, and a 24-hour cycle. According to these and other criteria, all mammals, birds, fish, amphibians, reptiles and insects that have been inspected have been found to sleep.

Take the humble fruit fly, *Drosophila melanogaster*, for example. This small insect displays all the key features of sleep, and more. Fruit flies alternate between periods of activity and rest according to a 24-hour cycle. They go to sleep in a preferred location where they remain immobile for two hours or more at a time, usually at around the same time of day. While they are asleep, they are much less reactive to sights, sounds and other sensory stimuli. If they are prevented from sleeping they show an increasing propensity to sleep. And when they do sleep after a period of deprivation, they are harder to wake up and they sleep for longer, as if catching up. Young fruit flies sleep more than old fruit flies, just as babies sleep more than adult humans. And, also like humans, their sleep is assisted by sleep-inducing drugs and disrupted by stimulants such as caffeine.

Many small mammals spend more than half their lives asleep. Think of the dormouse, for example. Lewis Carroll's description in *Alice's Adventures in Wonderland* is biologically authentic, apart from the animals' ability to talk grammatical English:

> There was a table set out under a tree in front of the house, and the March Hare and the Hatter were having tea at it: a Dormouse was sitting between them, fast asleep, and the other two were resting their elbows on it, and talking over its head. 'Very uncomfortable for the Dormouse,' thought Alice; 'only, as it's asleep, I suppose it doesn't mind' ...

'You might just as well say that "I see what I eat" is the same thing as "I eat what I see"!'

'You might just as well say,' added the March Hare, that "I like what I get" is the same thing as "I get what I like"!'

'You might just as well say,' added the Dormouse, who seemed to be talking in his sleep, 'that "I breathe when I sleep" is the same thing as "I sleep when I breathe"!'

'It *is* the same thing with you,' said the Hatter.

Dormice (which actually comprise 20 different species of nocturnal rodents) really do spend most of their time asleep, as do many other species of small mammals. The volcano mouse spends more than 17 hours a day asleep. Even the naked mole rat devotes 12 hours a day to sleep, despite the fact that it lives underground and has become so adapted to subterranean life that it has lost the power of sight.

The champion sleepers are two-toed sloths, which dedicate an average of 20 hours a day, or more than 80 per cent of their entire lives, to sleep. (Whereas three-toed sloths are much livelier, sleeping for a mere 17 hours a day.) Close behind come armadillos, opossums and some species of bats, which sleep for 18–19 hours a day. Many lizards spend more than 16 hours a day in sleep. Nearer to home, cats, rats, mice and hamsters sleep for 13–14 hours a day. Birds too can be great sleepers, although their sleep is less obvious to the casual observer because it is fragmented into short episodes. Starlings, which are fairly typical, spend a total of more than nine hours a day asleep, but this is split into many short bouts, each lasting on average only seven minutes. At the other end of the sleep spectrum lurk the wakeful grazers – cows, goats, elephants, donkeys, horses, sheep, deer and giraffes – all surviving on a meagre ration of three or four hours a day. But that is the minimum.

We humans occupy the low-to-middle ground of the zoological sleep spectrum, along with moles – or, at least, we do if we assign ourselves the proverbial eight hours a night. In reality, most people get substantially less than eight hours – a theme we shall be exploring in the next three chapters. On that same theme, it is notable that our closest biological relatives, the apes and monkeys, sleep more than us. Chimpanzees, rhesus monkeys, squirrel monkeys, vervets, patas monkeys and baboons sleep nine or ten hours a night, while the gorilla averages 12 hours. The sleepiest primate is the owl monkey, which clocks up 17 hours a day of sleep, accounting for more than 70 per cent of its life.

Do plants sleep? This is almost, but not quite, as stupid a question as it may seem. In one loose and misleading sense plants do display some behavioural characteristics of sleep. Many plants alter their shape each night, as though curling up to go to sleep. Some species furl their leaves like an umbrella, some allow their leaves to droop as if they need watering, while others, including lettuces and radishes, point their leaves vertically upwards. They all display a distinct 24-hour circadian rhythm.

Pliny the Elder noted this 'sleep of plants' in the first century AD. Eighteen centuries later, Charles Darwin investigated the 'sleep of plants' with a series of ingenious experiments in the garden of his home at Down House in Kent. Darwin was confident that the phenomenon was not true sleep. 'Hardly any one,' he wrote, 'supposes that there is really any analogy between the sleep of animals and that of plants.' As usual, Darwin was right. His experiments demonstrated that plants alter their shape at night to protect themselves from their physical environment. He found, for example, that if he left a plant outside at night, with its leaves tied up to prevent them from drooping, the result was a blackened, shrivelled and dead plant the next morning. Darwin also discovered that a plant will not 'sleep' if it is shaken violently.

Half asleep

> And the small fowl are making melody
> That sleep away the night with open eye
> > Geoffrey Chaucer, Prologue to *The Canterbury Tales* (c. 1387)

Sleep is such an overriding biological imperative that evolution has found ingenious ways of enabling animals to do it in the face of formidable obstacles. Nature, it seems, will do almost anything to ensure that animals sleep.

Consider dolphins, for example. They are air-breathing mammals like us, so they must swim to the surface each time they want to take a breath. They would drown if they fell into deep sleep while deep underwater. One possible solution to this biological design conundrum would be to wake up each time a breath of air was required.

However, evolution has produced a more elegant solution: only one half of the dolphin's brain goes to sleep at a time.

Dolphins are capable of what is known as unihemispheric sleep, in which one hemisphere of the brain submerges into deep sleep while the other hemisphere remains awake. The two halves of the brain take it in turns to sleep, swapping at intervals of between one and three hours. This cerebral juggling trick enables dolphins to sleep underwater without drowning, which is just as well considering that they spend a good third of their lives asleep. Unihemispheric sleep has been recorded in several species of dolphins, porpoises and whales, including bottlenosed and Amazonian dolphins, Black Sea porpoises and white whales.

Despite the apparent convenience of being able to sleep and stay awake simultaneously, very few mammals are capable of unihemispheric sleep. The biological benefits of sleeping with only half of the brain at a time presumably outweigh the disadvantages only under unusual conditions, such as those encountered by air-breathing mammals living in the deep oceans.

Unihemispheric sleep is widespread in birds, however. They do it for a different biological reason. Sleeping with half the brain awake and one eye open allows them to sleep while simultaneously remaining vigilant for predators. In birds, each eye exclusively feeds the visual processing areas in the opposite half of the brain: thus, all the nerve fibres coming from the right eye connect to the left hemisphere of the brain and vice versa. When a bird is in unihemispheric sleep its open eye is the one corresponding to the waking half of the brain, while the closed eye is connected to the sleeping half. If a bird feels relatively safe, it closes both eyes, and both sides of its brain go to sleep.

An experiment with mallard ducks demonstrated how unihemispheric sleep helps birds to stay safe from predators. Four ducks were placed in a row along a perch, the idea being that the ducks at either end of the row would feel more vulnerable to predators than the two in the middle. In the natural world it is generally a bad idea to be on the edge of a group if you might end up as some other animal's dinner. As predicted, video recordings showed that the outer two birds were much more likely to sleep with one eye open than the two on the inside; their unihemispheric sleep increased by 150 per cent. The amount of unihemispheric sleep rose further when the ducks were shown frightening video images of an approaching predator.

The relationship between unihemispheric sleep and vigilance was finely controlled. The exposed birds on the ends of the row preferentially opened their outward-facing eye – the one directed towards potential danger. From time to time, a bird would turn round and switch eyes, so that the open eye was still the one facing out. Simultaneous recordings of brain activity confirmed that the brain hemisphere corresponding to the open eye was always awake, while the hemisphere corresponding to the closed eye was the one in deep sleep.

The one-eyed tactic was effective: when an attacking predator was simulated on a video screen, the birds sleeping with one eye open were able to react in a fraction of a second – far faster than if they had been in deep sleep with both eyes shut.

Humans are not capable of unihemispheric sleep, although at least one writer has played with the fantasy. Damon Runyon wrote of how he once played cards with a fading champion card player who now lacked the stamina to stay awake during marathon games of gin rummy lasting eight or ten hours. When the man lost a game after making a bad play, the punters betting on him to win clamoured to remove their bets from the next game, on the grounds that he was asleep. Then someone pointed out that the allegedly sleeping player's eyes were open, so he must be awake. 'The one on your side is', retorted one of the backers, 'but the one on the other side is closed. He is sleeping one-eye.'

Nice, not naughty

> Like all other forms of pleasure, sleep may become a passion.
> Jean-Anthelme Brillat-Savarin, *The Physiology of Taste* (1825)

Sleep is far more than just a biological necessity. It is also a neglected source of pleasure. Consider this. Activities that are biologically important for survival and reproduction tend to be enjoyable: think of sex, or eating, or drinking, or being successful. Pleasure is one of nature's ways of ensuring that animals do enough of the right things. Whatever happened to sleep? It is clearly essential for survival, and yet for many people it is merely a maintenance activity that brings little positive enjoyment. They sleep because they have to rather than because they want to.

In some respects, sleep has acquired the dismal status that eating had in post-war Britain, where austerity and a cultural blind spot reduced the culinary arts to a joyless act of refuelling. Bland, fatty food was daily shovelled in to keep the boilers stoked, with scant attention paid to its preparation or enjoyment. Fortunately, Britain has since developed more enlightened (if not self-indulgent) attitudes to food, and recent decades have witnessed cooking and eating emerge as pleasurable activities in their own right. For some people in wealthy nations, cooking and eating have become more a form of entertainment than a biological function.

Meanwhile, sleep is mired in the cultural equivalent of a 1950s British canteen meal: an inadequate and faintly unhealthy affair, indifferently concocted and consumed with more haste than enjoyment. Too many people regard sleep as the brain's equivalent of fast food or overboiled cabbage. If gastronomy is 'the reasoned comprehension of everything connected with the nourishment of man', as it was originally defined, then should we not start thinking about sleep in the same way? I hope that by the end of this book you will be pondering the gourmet delights of sleeping, napping and dreaming, and starting to savour more of their lost pleasures for yourself.

PART II

Insufficiencies

2

Sleepy People

She looked a little worn out, a little tired, but, then again, didn't everybody? We all look a bit tired, these days, some more than others.

William Boyd, *Armadillo* (1998)

Many of us in our everyday lives are getting sleep of inadequate quantity and quality, and this is bad for our mental and physical health. Lack of sleep makes us inefficient at work, dangerous behind the wheel of a car and unattractive to be with; it lowers the quality of our lives, causes accidents and makes us more vulnerable to illness. And it is unpleasant.

I am not referring here to the acute sleep deprivation that comes from occasionally staying up all night, although that is common enough in professions such as medicine, the military and politics. Rather, I am talking about the chronic sleep deprivation that accumulates as we continually stint ourselves of sleep, day in and day out, because of the conflicting demands of work and leisure, or because of a sleep disorder, or just because we do not think sleep is important. So, what are the reasons for believing that sleep deprivation is a real problem?

Are we sleep-deprived?

Eight hours they give to sleep.

Sir Thomas More, *Utopia* (1551)

The evidence that chronic sleep deprivation is a common feature of contemporary life comes in several interlinked strands. We will start with perhaps the most obvious one of all: the observation that many people feel sleepy when they are awake.

Assessing the extent of daytime tiredness in society is tricky, not least because so many people have come to regard feeling tired as normal. Nonetheless, numerous scientific studies have unearthed evidence that the problem is real and widespread. For example, a 2001 poll of Americans' sleeping habits found that 22 per cent of adults felt so sleepy during the day that it interfered with their activities. A 1994 survey found that 5 per cent of the British population were experiencing severe daytime sleepiness, while a further 15 per cent felt moderately sleepy during the day. It also found that the people suffering from daytime sleepiness were twice as likely to have a vehicle accident.

A similar picture has emerged from other countries. For instance, a recent study discovered that 10 per cent of middle-aged Finns were excessively tired and tended to fall asleep unintentionally during the day. Their sleepiness was statistically associated with an array of nasty things, including a heightened risk of traffic accidents, premature retirement, depression and anxiety. In Sweden, 9 per cent of adults were found to be suffering from daytime sleepiness, while a survey in Warsaw recorded that 21 per cent of adults felt moderately sleepy during the day. Australian researchers detected excessive daytime sleepiness in 11 per cent of adults. You get the picture. Not even the youngest and healthiest are immune. A French investigation of 58,000 army conscripts discovered that 5 per cent of these fit young men were affected by excessive daytime sleepiness and 14 per cent of them were sufficiently tired to sleep during the day.

Overall, it is safe to conclude that at least one in ten adults in the general population (you, me and the people next door) are currently affected by moderate or severe daytime sleepiness. Some scientists believe the situation is much worse, with up to one in three adults suffering from significant sleepiness. A major review undertaken by the US National Commission on Sleep Disorders Research estimated that as many as 70 million Americans – more than a quarter of the population – were suffering from sleep deprivation or some form of sleep problem, at a direct cost to the national health care bill of about 16 billion dollars a year. The Commission's report concluded that:

A convincing body of scientific evidence and witness testimony indicates that many Americans are severely sleep-deprived and, therefore, dangerously sleepy during the day . . . By any measuring stick, the deaths, illness, and damage due to sleep deprivation and sleep disorders represents a substantial problem for American society.

The problem appears to have deepened over time. Objective evidence about historical changes in sleepiness is hard to find, but there is some. A standard psychological test of personality, which has been regularly administered to large numbers of Americans since the 1930s, revealed that the proportion of men who felt tired during the day was significantly higher in the 1980s than it had been in the 1930s. And the average amount of sleep American students get has fallen by more than one hour over the past three decades. We seem to be a generally wearier bunch than our forebears.

Tired people are certainly common enough in the doctor's waiting room. Physicians frequently encounter patients complaining of feeling tired all the time. The condition even has its own acronym, TATT. Of course not everyone who feels tired all the time is suffering from a lack of sleep. Chronic fatigue can result from anaemia, diabetes, cancer, depression and a whole host of other medical disorders. The distressing condition known as chronic fatigue syndrome (CFS), also referred to as myalgic encephalomyelitis (ME), is characterised by debilitating fatigue, pain in the muscles and joints, impairments in thinking and a general, horrible malaise. CFS is a distinct medical condition that cannot be attributed simply to lack of sleep. The origins of CFS remain controversial, but current theories focus on a combination of malfunctioning immune reactions and psychological factors.

CFS and other medical conditions undoubtedly account for some of the tiredness in society, but not much. The fact is that most people who feel tired during the day *are* just that – tired. They are tired because they have not been getting enough sleep. This explanation is so simple and so blindingly obvious that it is frequently overlooked.

Ordinary, everyday tiredness does not always require a medical explanation, but one is often sought nonetheless. 'Exhaustion' has become a voguish affliction of actors, pop stars and other celebrities. From time to time, a distraught and haggard celeb will flee to a clinic feeling, well, exhausted. The clinic duly subjects them to a battery of medical tests to establish whether they have diabetes, anaemia, ME or a thyroid disorder, while the media throb with stories about personal relationship problems, nervous breakdowns, exotic diets, exotic drugs, professional insecurities and emotional crises – just about everything, in fact, apart from plain tiredness. But given the long hours they are sometimes expected to work, plus the jet lag-inducing international travelling, it would be surprising if our stars of stage and screen did not occasionally feel very tired.

Another reason for believing that sleep deprivation is a common problem is that many of us get less sleep than we want or need, and considerably less than the proverbial eight hours we all supposedly aspire to. The evidence on this point is clear. For example, research found that in the 1990s young American adults were sleeping for an average of 7.3 hours a night. However, even this modest figure was inflated by weekend lie-ins: on weekday nights, Americans slept for an average of only 6.7 hours. A more recent study, which assessed the sleep patterns of middle-aged adults, found an average of only 6.2 hours. Similar conclusions have emerged from other countries. To give just a few illustrations, a survey in Poland found that adults there slept for 7.1 hours on workday nights, while Korean university students averaged 6.7 hours. When the Japanese Ministry of Health conducted a large survey they discovered that almost two thirds normally slept less than seven hours a night, and more than a quarter slept less than six hours. Worse still, half of Japanese high school students reported sleeping six hours or less on weekdays.

The research leaves little doubt that most adults in the USA, UK and other industrialised nations get substantially less than eight hours' sleep most nights of the week, and many get less than seven. But does that matter? How much *should* we sleep?

The mythical inhabitants of Sir Thomas More's idyllic island state of Utopia accorded sleep the priority it truly deserves. They slept for eight solid hours each night. Of the remaining 16 hours, work accounted for only six. (Sensible people.) The Utopians worked for three hours before noon then ate lunch; after lunch they rested for two hours, worked for another three hours then ate supper. They went to bed at about eight in the evening and slept for eight hours. The rest of each day they did as they pleased. Not quite the urban chic lifestyle, perhaps, but a refreshingly different perspective on life.

Conventional wisdom still holds that we need about eight hours of sleep a night. For most of us, however, the reality falls far short of the Utopian ideal, except perhaps at weekends and when we are on holiday. But is an average of, say, six or seven hours a night enough? Obviously, we cannot use Thomas More's sixteenth-century fantasy as a scientific yardstick. So how do we judge the adequacy of sleep?

One approach is to ask people whether they think they are getting enough sleep. A recent study of more than twelve thousand adults did just that, and found that 20 per cent of them felt they were not getting sufficient sleep. Unsurprisingly, one of the factors most strongly associated with insufficient sleep was working long hours. Further evidence came from the Planet Project, said to be the largest opinion poll ever carried out. In 2000, an Internet-based survey was conducted with 1.26 million people in 251 countries. When asked how much sleep they needed in order to feel rested, 47 per cent of people replied eight hours or more. But when asked how much sleep they actually got, only 15 per cent reported sleeping eight hours or more, while 8 per cent said they got less than five hours a night. Many respondents reported having gone without sleep for long periods, whereas a mere 6 per cent said they had never missed a night's sleep.

Six or seven hours of sleep a night is probably not enough for many people on a long-term basis. The experimental evidence suggests that the underlying sleep tendency for a typical healthy adult – that is, the amount of sleep they would take if completely liberated from work schedules and other constraints – is more like eight or eight and a half hours a night. This implies a shortfall of around an hour and a half each night. A shortfall of this size almost certainly matters. Restricting someone's sleep by an hour and a half for just one night will measurably reduce their daytime alertness. The cumulative effects, when sleep is short-changed night after night, are far more pervasive. We shall be looking at the psychological and physical consequences of chronic sleep deprivation in the next two chapters.

A substantial nightly shortfall in sleep is difficult to sustain for more than five or six days in succession without sleepiness seriously impairing daytime alertness, mood and performance. This aspect of human biology might conceivably have contributed to the establishment of the seven-day week, comprising five or six days of work followed by one or two days of legitimised rest, as a standard unit of time. Most humans have been following the seven-day pattern for thousands of years. The seven-day week dates back to the pre-Biblical Sumerian and Babylonian civilisations. Right from the outset, one day of the week was deemed to be a day of rest and recreation. The Babylonians originally named their seven days after the five visible planets plus the Sun and the Moon, but the choice of seven otherwise has no objective basis in astronomy or any obvious feature of the physical environment. The Romans later adopted the seven-day week, dabbled with eight, then reverted to seven. The seven-day week appeared in the Bible, again with the inherent concept of a day of rest. According to the Biblical account of the Creation, God laboured for six days and rested on the seventh. Nowadays, people who work long hours and sleep for only six or seven hours a night really do need that extra time in bed at weekends to stave off sleep deprivation.

The thesis that many people are chronically sleep-deprived is not without its sceptics, however. Yvonne Harrison and Jim Horne at Loughborough University have argued that we all have the capacity to sleep more than we usually do, but only in the way that we can carry on eating after our physiological need for food has been satisfied. In support of their sceptical position, Harrison and Horne have cited, for example, an experiment in which healthy young adults slept for up to ten hours a night for two weeks, getting an extra hour or so of sleep each night. This additional sleep produced some improvements in their reaction times and a slight reduction in daytime sleepiness. However, there were no significant improvements in the volunteers' subjective ratings of their own mood or sleepiness.

Scientists are paid to be sceptical, and counterblasts are an essential part of scientific debate. Nonetheless, such arguments must be set against all the other strands of evidence showing that chronic sleep deprivation is a real and widespread phenomenon. We have considered some of that evidence and there is more to come. First, though, what about you?

Are *you* sleep-deprived?

'... Are you going to bed, Holmes?'

'No: I am not tired. I have a curious constitution. I never remember feeling tired by work, though idleness exhausts me completely.'

Sir Arthur Conan Doyle, *The Sign of Four* (1890)

There is no universal, one-size-fits-all figure for the right amount of sleep. Individuals differ considerably in their sleep requirements. The conventional standard of eight hours a night is somewhat arbitrary, though not far off the mark for most people. But the best yardstick for you is your own preferred sleep duration. You can measure this by conducting a simple, if time-consuming, experiment on yourself. You will need two weeks of complete freedom from the tyranny of work schedules and alarm clocks, which means you will probably have to wait until your next long holiday (and possibly much longer if you have small children).

What you do is this. Every night for about two weeks, go to bed at approximately the same time. Make a note of the exact time just before you turn off the light to go to sleep. Then sleep to your heart's content until you wake spontaneously the next morning – not with the aid of an alarm clock. Note down the time when you woke up and, ideally, a more precise time for when you think you fell asleep the night before. (All being well, this should have been within 10–20 minutes of turning out the light.) Your time of waking should be when you first became fully conscious, not when you eventually stumbled out of bed. Then calculate how long you spent sleeping. The experiment will work much better if you do not roll over after waking and doze for a further hour or two before getting up.

Ignore the figures for the first few days, because you will probably be sleeping longer to compensate for your prior sleep deficit. We tend to feel sleepy on relaxing holidays and at weekends because reality has finally caught up with us; when the usual pressures and stimulation that keep us going through the working week are suddenly removed, we sleep more to catch up on the backlog. This extra sleeping is only transitory, however. Once you have caught up and reached equilibrium you should start to feel livelier and more energetic during the day. Unfortunately, by that time the holiday has usually ended and the normal regime of late nights and early mornings resumes.

After a few days, your nightly sleep duration should settle down to a stable figure. Take the average over the final few days: this represents your preferred sleep duration. It should be somewhere between seven and nine hours, although not everyone fits this pattern. Unless you are elderly or have an unusually relaxed lifestyle, your preferred sleep duration will probably be longer than the time you normally spend sleeping. If the difference is very large – say, two or three hours a night – you could be storing up serious trouble for yourself.

Reasons for not sleeping

> Why all this sleep? – seven, eight, nine, ten hours perhaps –
> with a living to make, work to be done, thoughts to be thought,
> obligations to keep, a soul to save, friends to refrain from losing,
> pleasure to seek, and that prodigious host of activities known
> as life?
>
> Walter de la Mare, *Behold, This Dreamer.* (1939)

Another reason for believing that sleep deprivation is widespread is
that we sleep less nowadays than our ancestors did. There has been
a major shift in sleep patterns in industrialised nations over the past
century, the net result of which is less sleep.

William Dement of Stanford University has argued that humanity
is in the midst of a 'pandemic of fatigue'. Dement estimates that people
in industrialised countries now sleep on average an hour and a half
less each night than they would have done a century ago, and that
most of us consequently walk around with an accumulated sleep
deficit of 25–30 hours. If true, this means we would have to sleep for
an extra two hours a night for two weeks to clear the backlog and
return to equilibrium. Some of us come close to doing just that when
we take a two-week holiday.

Lifestyles in industrialised societies have altered radically over the
past century in ways that have consistently eroded the status of sleep
and dreaming, leaving many of us now probably sleeping less than
at any other period in human history. We live in an era when many
people work long hours, where we have vastly more opportunities
for entertainment and leisure, and where sleep is widely regarded as
the poor relation to other pursuits. Humanity has inadvertently
created lots of reasons for not sleeping.

One reason for not sleeping is that there is always so much work
to do. Many sectors of society now work longer hours, on more days,
than ever before. In early nineteenth-century Britain there were some
40 days in the year when the Bank of England shut its doors to observe
saints' days and anniversaries. By 1830 the number of such holidays
had dropped to 18 days a year, and nowadays there are fewer than
ten. Not only do we spend longer at work, we also spend longer
getting to and from work, as commutes have grown both in distance
and duration. Time spent stuck in a car or public transport is time

that cannot be spent in bed. Ironically, the technological revolution has failed to free us from the shackles of paid work. Quite the reverse, in fact – it has given us the wherewithal to work more productively all day, every day. James Gleick made the point beautifully in *Faster*, his glorious exposé of our no-time-to-lose society:

> Marketers and technologists anticipate your desires with fast ovens, quick playback, quick freezing, and fast credit. We bank the extra minutes that flow from these innovations, yet we feel impoverished and we cut back – on breakfast, on lunch, on sleep, on daydreams.

Cheap mobile communications enable us to stay in touch wherever we are, 24 hours a day. Copious amounts of caffeine, the world's most popular psychoactive drug, help to keep us awake as we squeeze ever more into the day. The puritan work ethic and the cult of time management nag us to do a little bit more at the beginning and end of each day. So we get less sleep. But that is stupid, because people who cut back on their sleep achieve less and feel bad into the bargain. They end up stumbling through the day, fatigued and underperforming, without even realising what they are doing to themselves. They become, to quote one scientific paper, borderline retarded.

The idea of being able to get by on little or no sleep might appeal to some driven souls who would rather use the extra time for other things. (Some people seem to find being at work easier than living a real life.) In J. G. Ballard's short story 'Manhole 69', a scientist who has entirely expunged the need to sleep from three human volunteers sneeringly declares that:

> For the first time Man will be living a full twenty-four hour day, not spending a third of it as an invalid, snoring his way through an eight-hour peepshow of infantile erotica.

(Ballard's sleepless volunteers, needless to say, meet a grisly fate.) If the fantasy of doing with less sleep ever became a reality – which, mercifully, it cannot – those extra hours of wakefulness would just be absorbed by more work. If we could all survive working 20 hours a day then the 20-hour day would become the norm. And we would still feel there were not enough hours in the day.

Fortunately, not everyone aspires to a sleepless world. A few highly successful businessmen have come out of the closet in recent years and openly admitted to sleeping for eight hours or more a night (although some cynics have pointed out that these captains of industry can only afford to have all that sleep because they are amply supported by minions working ridiculously long hours).

Britain is rapidly following the USA in becoming a fully-fledged 24/7 society where the consumer is king and nothing ever closes. Consumers really do want the freedom and flexibility to shop, bank or be entertained at any hour of the day or night, and governments are having to respond to their demands for public services to be continuously on tap, providing 24/7 facilities to the taxpayers who fund them. In 2001 the British government published a report called 'Open All Hours', explaining how public services were raising their game to meet the requirements of the 24-hour society. The report highlighted examples of how public services had responded to demands for extended opening hours. In his foreword, the Prime Minister wrote that 'people living busy working lives ... should be able to access services how and when they want'. The idea of modernising services to suit the needs and convenience of the public is surely laudable and uncontroversial. But there is something crucial missing from the cost-benefit analysis: the impact the 24-hour society is having on the ability of the people who are providing and consuming those services to get enough sleep.

Whatever happened to the technology-enabled revolution in leisure, which the future-watchers so confidently predicted in the 1960s? The main concern in those days was that we would all have too much free time on our hands, not too little. Sebastian de Grazia, one of the more thoughtful advocates from that era, argued for a return to the inner peace that can only come from a capacity for true idleness, combined with an escape from the constant stimulation that prevents people from ever being alone with themselves:

> Perhaps you can judge the inner health of a land by the capacity of its people to do nothing – to lie abed musing, to amble about aimlessly, to sit having coffee – because whoever can do nothing, letting his thoughts go where they may, must be at peace with himself.

If the work ethic does not keep us from our beds, then our insatiable lust for entertainment and amusement surely will. As the Irish poet

Thomas Moore put it, we are inclined to steal a few hours from the night:

> 'Tis never too late for delight, my dear;
> And the best of all ways
> To lengthen our days
> Is to steal a few hours from the night, my dear!

In Shakespeare's *Twelfth Night* the reprobates Sir Toby Belch and Sir Andrew Aguecheek have been up drinking all night. 'I know to be up late is to be up late,' contests Sir Andrew. 'A false conclusion!' avers Sir Toby: 'To be up after midnight and to go to bed then, is early.'

The range of distractions and temptations to seduce us away from our beds has mushroomed since Shakespeare's day. There is so much more to do in developed nations, and so much more wealth to do it with. The only quantity that has remained doggedly constant is the amount of time we have. There are still only 1,440 minutes in a day. So, we opt for the immediate fix of pleasure and stay up late. We know deep down that we will suffer the next day in mood, alertness and performance, but the lures are too appealing and their pleasures are instant.

Psychologists have a technical term – delayed gratification – to describe an individual's ability to forgo an immediate reward in return for a bigger reward later on. It so happens that a capacity for delayed gratification is correlated with intelligence and attainment in life. Most of us, however, display a lamentable lack of delayed gratification when it comes to sleep. William Dement coined another term, 'hedomasochism', to describe the irrational belief that we can do it all, achieving ever more in our work, in our family lives and in our leisure time, all at the expense of sleep. We cannot.

Ancient and modern

> We rise with the lark and go to bed with the lamb.
> Nicholas Breton, *The Court and Country* (1618)

The current predilection for staying awake all hours is very recent in historical terms, let alone when measured against the span of biological evolution. It really took root following the invention of the electric filament light bulb by Thomas Edison in 1879, a negligible fraction of an instant ago in evolutionary terms. Of course, people did stay up after dark in the days before cheap electric lighting – just much less.

There have only ever been two ways for humans to deal with the night: to sleep and doze through it or to light it artificially. Until the nineteenth century the only practical source of artificial light was fire in one form or another. When humans depended on expensive candles or oil for artificial light they went to bed earlier and stayed there longer, unless they were in the wealthy minority. Few people did much work after dark. And when they did use artificial lighting, the fires and candles (and later, the gas mantles) generated light of insufficient intensity to reset their internal biological clocks in the way that much brighter electric lighting can. One electric light bulb produces as much light as a hundred candles and for only a tiny fraction of the cost. Unlike our ancestors, we no longer have to sleep, doze or stay in bed just because it is dark.

To appreciate how different life was for the majority of people living in temperate or northern climates, we need only wind the clock back to the eighteenth century. For most working folk, especially in winter, the sun provided the only serious illumination. *The Natural History of Selborne*, which was written by an English country clergyman called Gilbert White and published in 1788, describes life in a small village in rural England. White's parochial history is said to be the fourth most published book in the English language. In one of his glimpses into the lives of Selborne's human inhabitants, White reminds us that in the days before electric lighting, few people could afford the luxury of routinely staying awake for long during the hours of darkness. The villagers burned rushes to produce light, and even rushes cost money:

> Working people burn no candle in the long days, because they rise and go to bed by daylight. Little farmers use rushes much in the short days, both morning and evening in the dairy and kitchen; but the very poor, who are always the worst economists, and therefore must continue very poor, buy an halfpenny candle every evening, which, in their blowing open rooms, does not burn much more than two hours. Thus have they only two hours' light for their money instead of eleven.

In rural northern Europe of the Middle Ages it was pointless or impossible to work the fields during the dark days of winter, and too costly to heat and light the home all day. Whole families would therefore take to their beds for days at a time. You might not relish the prospect of spending days in bed with nothing to do. (Or perhaps you would?) Boredom would be the big enemy. Boredom, however, is a modern concept. Being alone with our thoughts and dreams is no longer enough for us.

Cold weather was another good reason for staying in bed, as Samuel Pepys recorded in this entry from his diary, written in December 1661:

All the morning at home, lying abed with my wife till 11 a-clock – such a habit we have got this winter, of lying long abed.

But even in warm, sunny climates, our ancestors probably spent more of their time in bed, especially in civilisations that practised the siesta.

In modern industrialised societies we are exposed to an artificial day that is extended by electric lighting and typically lasts for at least 16 hours, regardless of season. The marked seasonal fluctuations in the conception rate, which were once associated with the long winter nights, have almost disappeared now. Moreover, we now pack all of our sleep into a single block of time during the remaining seven or eight hours of darkness. This pattern of sleeping is biologically unusual: in most other species, sleep is split into two or more separate episodes in each 24-hour period. As we shall see, there are reasons for supposing that humans have not always slept in a single, compressed block.

Our daily cycle of sleep and wakefulness is largely determined for us by clocks rather than tiredness. Many of us go to bed when it is time to go to bed, not when we are tired, and wake when we have to wake, not when we choose to. Clocks with minute hands did not become available until the seventeenth century. Until quite recently in history, the majority of people relied on the sun for their timekeeping and lived in a world where the light intensity changed gradually at dawn and dusk, not instantaneously with the flick of a light switch. Moreover, they did not work in offices, factories or shops where they were required to be present at a certain time early every day.

How do humans sleep when left to their own devices in a world where it is dark for more than half the day – as would have been the case in pre-industrial northern countries during winter? To find out, Thomas Wehr at the National Institute of Mental Health in Maryland exposed volunteers to an experimental environment where it was dark for 14 hours a day and they could sleep freely. To begin with, the volunteers slept a lot (some more than 12 hours a day) as they caught up on their backlog. On average, they clocked up an additional 17 hours of sleep during the initial adjustment period. After their sleep deficits had been paid off, they settled down to an average sleep duration of eight and a quarter hours a day. Their mood and energy levels during the day improved consistently over the course of the experiment. When they were awake, they felt more awake and *were* more awake.

As well as sleeping longer, Wehr's subjects also slept differently. Under these conditions of long, dark days and with nothing much to do, their sleep spontaneously divided into two distinct blocks. Typically, they would lie in a state of quiet rest each evening for about two hours before falling asleep. Then they would sleep for about four hours, usually waking at the end of an episode of dreaming. After another couple of hours of quiet rest they slept for a further four hours. On waking in the early morning they would lie in quiet rest for another couple of hours before rising.

Under these pseudoprimitive conditions, then, sleep was preceded, punctuated and terminated by long periods of quiet restfulness. This pattern of sleeping in two distinct blocks of time is known as biphasic sleep. It is typical of many mammals living in the wild and was probably the natural sleep pattern of our ancestors. We all retain the biological capacity for biphasic sleep, despite the profound changes in humanity's environment since the advent of artificial lighting and the 24-hour society. A group of thoroughly modern Americans reverted to biphasic sleep within days of being given the opportunity. The nearest contemporary equivalent is the afternoon sleep of the siesta, a custom that still survives in some countries.

The predominant lifestyles of artificially-lit industrialised societies have led us to compress our sleep into a single block of seven or eight hours, as though we were living permanently in midsummer (but usually without the siesta). We have jettisoned the additional hours of quiet rest and the seasonal variations that once accompanied human sleep. We have also lost the main channel that once existed to our dreams. During the 14-hour nights, as they alternated between sleep and quiet rest, Wehr's volunteers usually awoke from dreaming, giving them ample opportunity to lie quietly in the dark and contemplate their dreams. In later chapters we shall consider why dreaming evolved and what it does for us.

A quite different reason for believing that many people nowadays are chronically sleep-deprived is the mass of evidence that sleepiness is a major cause of accidental injuries and deaths. We shall now look at how sleepiness jeopardises safety-critical activities such as driving a car, flying an aeroplane, being a doctor, running a country and operating a nuclear power plant.

Sleepy drivers

Till o'er their brows death-counterfeiting sleep
With leaden legs and batty wings doth creep.
        William Shakespeare, *A Midsummer Night's Dream* (1595–6)

Accidents are one of the leading causes of death in developed nations, and sleepy people are responsible for many of them. A remarkably large proportion of vehicle accidents are the direct or indirect result of tired drivers losing concentration or falling asleep at the wheel. A few examples may give a flavour of the carnage they cause. In March 1994 near Barstow in California, a pickup truck carrying 20 people veered off the road and crashed into a culvert after the driver apparently fell asleep at the wheel. The driver survived but 12 passengers died. In July 1995 near Roquemaure in France the driver of a bus carrying Spanish students from Amsterdam to Barcelona seemed to nod off then wake abruptly as his bus scraped a passing truck. He lost control and the bus swerved wildly before rolling over several times. The accident caused 22 deaths and 32 injuries. In February 2001 a sleep-deprived driver caused the Selby rail disaster in the UK, after he fell asleep at the wheel and his vehicle crashed onto a railway line. Ten train passengers died. The driver, Gary Hart, admitted getting no sleep the night before the crash, but claimed he could still drive safely. He was sent to prison.

Scientists have judged that sleepiness is a factor in at least 10 per cent of fatal car crashes in the USA and more than 50 per cent of fatal crashes in which a truck driver is killed. A 1994 report by the US National Commission on Sleep Disorders concluded that driver fatigue contributed to 54 per cent of all vehicle accidents in the USA.

A comparable situation applies in the UK and elsewhere. Research concluded that at least 10 per cent of vehicle accidents in the UK are related to sleepiness, though some experts have put the figure much higher. Two large surveys in England found that sleepiness was a causal factor in 16 per cent of accidents to which the police were summoned and at least 20 per cent of accidents on motorways. Driving on a motorway is generally more monotonous than driving on a minor road, and monotony heightens the risk that a tired driver will fall asleep at the wheel. Half the drivers involved in these sleep-related accidents were men under the age of 30 and many of the accidents involved truck drivers, company cars or workers returning home from night shifts.

For every sleepy driver who actually crashes there are uncounted numbers who have had near misses. A large survey by the British Transport Research Laboratory found that 29 per cent of drivers had come close to falling asleep at the wheel within the previous year, while other research established that at least 5 per cent of middle-aged male drivers had actually fallen asleep while driving on several occasions. Not surprisingly, drivers suffering from moderate or severe daytime sleepiness are at least twice as likely to have a vehicle accident.

The official statistics tend to underestimate the true extent of the sleepiness problem, and it is easy to see why. Drivers who survive crashes are naturally reluctant to admit that they dozed off at the wheel, even if they recollect doing it. And it is difficult to prove legally that sleepiness caused a crash (especially if the driver is dead). Unlike alcohol, drugs or mechanical defects in a vehicle, sleepiness leaves few evidential traces. You can easily measure how drunk someone is at the roadside immediately after an accident. But measuring sleepiness is neither quick nor easy, and in practice it is simply not done.

The systematic under-reporting of fatigue was highlighted by accident statistics for Italian highways. Over the period from 1993 to 1997 the Italian authorities officially ascribed the cause as sleep in only 3 per cent of accidents. However, by analysing the data in more depth, researchers were able to estimate that sleepiness had probably contributed to about 22 per cent of accidents. If true, this means that the official statistics had underestimated the hazard of driver sleepiness by a factor of seven.

The time of day is a major element in the relationship between sleepiness and accidents. Thanks to our natural circadian rhythms, we all feel sleepier at certain times in the 24-hour cycle (usually in the early hours of the morning and again in the afternoon) regardless of how much sleep we have had. As expected, sleepiness-related vehicle accidents occur most often in the early hours of the morning and in the afternoon, during these natural peaks in sleepiness. Older drivers are particularly susceptible to afternoon sleepiness, whereas younger drivers are more prone to crashing late at night or in the early hours. When researchers analysed the data for accidents in which the driver had been injured or killed (excluding those involving alcohol) they found that young drivers were between five and ten times more likely to crash late at night than during the morning.

Driving late at night poses a risk for train drivers as well. In one study, scientists monitored train drivers while they drove the same route, both by day and at night. The train drivers felt much sleepier when driving at night, and physiological measurements mirrored their subjective feelings. Their brain waves, heart rates and eye movements at night were all characteristic of sleepy people. Four of the 11 drivers who were monitored admitted to dozing off during the night journey and two of them failed to respond to signals. Sleepiness has almost certainly caused numerous rail crashes over the years, but again the official statistics have systematically underestimated its importance.

The sleepiness experienced by many drivers is partly a product of natural circadian variations in wakefulness. But much of the blame rests with simple lack of sleep. And when lack of sleep is combined with driving at odd hours, the effect can be lethal. Prevailing social attitudes towards this issue are frankly perverse. Many parents think nothing of packing their family into a car and then driving long distances to a holiday destination while they are seriously tired. They would be horrified at the thought of doing this while drunk, but the effects of tiredness and alcohol on their ability to drive safely are strikingly similar, as we shall see in the next chapter. The number of drivers involved is large. In August 1996 French researchers assessed the extent of sleep deprivation among drivers during the holiday season, by randomly stopping two thousand cars at tollbooths and interviewing the drivers. It transpired that half of them had slept less than they would normally have done during the previous 24 hours. On average, these happy holidaymaking drivers had slept for 3.4 hours less than normal.

Many long-haul truck drivers get insufficient sleep, with potentially serious consequences for their performance and safety. When investigators studied truck drivers working in the USA and Canada they found that the drivers spent on average slightly more than five hours a day in bed and got slightly less than five hours' sleep. This was much less than their self-reported ideal of more than seven hours a day. Nearly half the truck drivers augmented their sleep by napping, but the naps were not sufficient to compensate. Video and EEG brain-wave recordings revealed that more than half the drivers had at least one period of drowsiness while they were driving, and two actually fell asleep at the wheel.

Even changing the clocks can be dangerous. The switch to daylight-saving time each spring reduces the length of one night by one hour, which slightly disrupts sleep routines for the next few nights. The extra sleepiness caused by even this apparently trivial disturbance is enough to generate a statistically significant seasonal rise in traffic accidents. Fatal accidents peak on the day immediately following the changeover. Alcohol-related accidents also rise during the week after the clocks change, probably because the effects of alcohol and sleepiness reinforce each other.

You might think that changing clocks in the opposite direction each autumn would have the reverse effect, but you would be wrong. When researchers analysed 21 years of US vehicle accident statistics, they found that the switch back from daylight-saving time each autumn was also accompanied by an increase in fatal accidents – despite the fact that in this case everyone got an extra hour in bed. The likely explanation is that many people anticipated the extra hour in bed by staying up even later the night before. This was borne out by the fact that the rise in fatal accidents around the autumn changeover was most marked just before the clocks changed, and especially in the early hours of the morning.

One of the most alarming aspects of daytime sleepiness is that we can fall asleep briefly without even noticing. You are unlikely to be aware that you have slept unless your sleep has lasted for at least a couple of minutes. Tired people can therefore fall asleep at the wheel of a speeding vehicle for tens of seconds at a time and never even know. Researchers measured this phenomenon by waking volunteers after daytime naps of varying durations and asking them if they had been asleep. (Their sleep was confirmed by objective physiological measures.) After one minute of sleep only 15 per cent of subjects had any awareness that they had been asleep, and only 35 per cent were aware even after five minutes of sleep. The upshot is that so-called microsleeps, lasting anything up to a minute, often go unnoticed. Suppose a sleepy driver lapses into a microsleep for only ten seconds while driving on a motorway at 70 miles an hour. During that brief, unnoticed lapse in waking consciousness the vehicle will cover about 70 car lengths. It is virtually certain that while you are reading these words someone, somewhere is microsleeping at the wheel of a speeding vehicle.

One apparent obstacle to prosecuting drivers who fall asleep at the wheel is proving that they were aware of their dangerous state and are therefore legally responsible for their actions. No one could reasonably claim to have been completely unaware that they were dangerously drunk, but a driver might conceivably claim to have been oblivious of being sleepy before crashing. However, the experimental evidence suggests otherwise. Sleep does not occur spontaneously without prior warning in the form of sleepiness.

Drivers who fall asleep at the wheel may not recall the actual moment of falling asleep, but they will almost certainly remember feeling sleepy beforehand. Scientists established this by monitoring sleep-deprived volunteers while they drove a simulator. The sleepier the drivers felt, the more mistakes they made. Serious errors, of the type that might have caused a crash in real life, were always preceded by prolonged feelings of sleepiness. By the time an 'accident' took place the tired driver had invariably been consciously fighting sleepiness for some time. The strong implication is that drivers who fall asleep at the wheel in real life will almost certainly have felt noticeably sleepy beforehand. The problem is that so many sleepy drivers press on regardless, fighting their sleepiness and risking lives. Many drivers harbour the illusion that they will not fall asleep at the wheel provided they fight hard enough. What they fail to appreciate is that if you are sufficiently sleepy you will eventually fall asleep, no matter how hard you resist.

Not all sleepy drivers are sleepy because of sleep-deprived life-styles. Some are sleepy because they have a medical sleep disorder, often undiagnosed. The most common of these, called sleep apnoea, involves the repeated interruption of breathing during sleep. We shall be taking a closer look at sleep apnoea in chapter 15. Individuals who suffer from this disorder can become severely sleep-deprived, although they rarely know why. The daytime sleepiness caused by the repeated disruption of their sleep every night can severely impair their driving performance.

Sleepy drivers not only have more accidents, they also have worse accidents. The hallmark of an accident caused by a driver falling asleep at the wheel is the absence of skidmarks. Of all the crashes that are attributed to drivers falling asleep, more than three quarters involve the car driving off the road and more than half involve high speeds.

Car and truck manufacturers have done little to tackle the safety hazard created by sleepy drivers. Driver fatigue remains one of the biggest weak spots in vehicle safety, perhaps because it is much easier to modify the design of a vehicle than to modify the behaviour of humans. However, some promising technology is being developed that may show the way. One system uses cameras mounted in the dashboard to track the driver's eye movements. It exploits the fact that people blink in a characteristic way when they are about to fall asleep. The device warns the driver if the blink frequency indicates a risk of nodding off at the wheel. IBM is developing an even more sophisticated system, known as the Artificial Passenger. An intelligent computer, which knows the driver's personal profile and interests, holds a conversation with the driver. It asks questions and even tells jokes (though humour is reportedly not yet one of its strengths). If the driver's responses are slow, flat in intonation and fail to make sense, the Artificial Passenger may judge that the driver is sleepy and urgently needs to be revived. If so, it will automatically open one of the car's windows, sound an alarm or even activate a device that sprays cold water in the dozing driver's face.

Governments are only just beginning to wake up to the carnage caused on our roads by sleepiness, having focused for so long on the dangers of alcohol. And yet sleepiness accounts for far more road deaths than alcohol, let alone drugs.

Sleepy pilots

My spirits grow dull, and fain I would beguile
The tedious day with sleep.

William Shakespeare, *Hamlet* (1601)

Fatigue and chronic sleep deprivation are obviously of crucial relevance to aviation safety. Tired pilots are bad pilots, for all the reasons that tired drivers are bad drivers. How big is the problem in practice?

Historically, severe fatigue among aircrews has sometimes been a major problem during crises where huge demands have been placed on precious personnel. Take, for example, the Berlin airlift of 1948–9. In June 1948 the forces of the former Soviet Union occupying eastern Germany began a blockade of road, rail and other communications between Berlin and the West. An international crisis ensued. The USA and UK mounted a huge airlift operation to supply West Berlin with food and other essential supplies. The airlift continued for 11 months until the Soviets eventually withdrew their blockade. During that time Allied planes delivered more than two million tons of food, fuel and other supplies to the beleaguered residents of West Berlin. To sustain this huge effort, the aircrews worked punishing schedules with grossly inadequate sleep. There were many accidents, some of them the result of fatigue. A special investigation during the crisis led to immediate improvements in the aircrews' working conditions and sleeping quarters, which probably made a material contribution to the ultimate success of the whole operation.

Even in peacetime, tiredness is not unknown on the flight deck. Most airline flight crews experience some sleepiness and impairment in their performance, especially during long-haul and overnight flights. In one recent study, scientists from the British Defence Evaluation and Research Agency monitored 12 airline pilots during routine nine-hour flights between London and Miami. Recordings of their EEG brain-wave activity and eye movements revealed that 10 of the 12 pilots either slept or displayed signs of significant sleepiness during the flights. These episodes were often brief, lasting less than 20 seconds. Microsleeps of this brevity generally go unnoticed, and the pilots would probably have been unaware of drifting off.

Scientists from the NASA Ames Research Center in California also detected fatigue among flight crews on commercial long-haul flights. The crews on these flights, which crossed up to eight time zones, became measurably sleep-deprived. They felt more fatigued than normal, consumed more caffeine, ate more snacks and reported more minor health problems such as headaches, nasal congestion and back pain. Their sleep loss was made worse by the fact that their circadian rhythms did not have time to synchronise with local times. Their natural low points in alertness therefore often occurred while they were on duty, amplifying their sleepiness.

Jet lag is not just unpleasant and stressful – it also has physical effects on the brain. Researchers compared two groups of female flight attendants who had all been working on long-haul flights for at least five years. Half the women were in jobs that allowed them two weeks to recover between long-distance flights, while the other half usually had only a few days' rest in between. The women who had little time between flights performed significantly worse on tests of learning and memory; their reactions were slower and they made more mistakes. More significantly, brain scans revealed distinct physical changes in their brains. A region of the brain known as the right temporal lobe had shrunk significantly. The women with the most shrunken right temporal lobes also had the highest levels of cortisol, a stress hormone that is known to affect the structure of the brain and the functioning of the immune system.

The implication of this research is that people who regularly fly long distances, crossing more than six or seven time zones, should ideally allow at least ten days to recover before doing it again. The research also raises questions about the policies of airlines that require their flight crews to fly long haul without adequate rest periods in between. The thought of sleepy, jet-lagged pilots with wizened right temporal lobes and impaired mental abilities slumped behind the controls of jumbo jets is mildly alarming.

Flying for a living can be tiring even when it does not involve crossing multiple time zones and becoming jet-lagged. Like workers in many other industries, flight crew are often required to start work early in the morning, and this alone can starve them of sleep. Researchers who monitored the sleep of female cabin crew found that when the women worked early mornings their sleep was reduced to an average of just over five hours. This is not enough sleep for the vast majority of people. Early-morning working was also unpleasant and mildly stressful for these women. They reported feeling apprehensive about having to rise early, they felt sleepy during the day and they complained more about their sleep being unrefreshing.

The only cure in situations like these is getting enough sleep, and at the right time of day. But that is not always possible on long-haul flights. Napping can provide a short-term palliative. If all else fails, a British firm has patented a technological aid to keep airline pilots awake. Worn like a wristwatch, it uses a motion sensor to monitor the pilot's movements. A loud alarm sounds if there has been no movement for a few minutes. Personally, I would prefer not to find myself on a plane flown by a pilot who needs one of these devices. But if I do, I hope it works.

Space flight is even less conducive to sleep than air travel. Astronauts can and do sleep in space, but not very well. Space flight confuses the body's internal clock and reduces both the quantity and quality of sleep. Astronauts on the Space Shuttle were typically getting only five or six hours of poor quality sleep a night, and often resorted to sleeping pills. More than 40 per cent of Space Shuttle astronauts took medication for sleep disturbances – about the same proportion as took drugs for motion sickness.

In recent years, NASA has been giving its astronauts doses of the 'sleep hormone' melatonin to help them sleep. (We shall see what melatonin does in chapter 6.) However, research by Charles Czeisler at Harvard Medical School found that melatonin actually had little beneficial effect on astronauts' sleep. What did work, however, was covering the astronauts in electrodes to monitor their sleep. Czeisler discovered that Space Shuttle astronauts slept better when they were festooned with electrodes and physiological monitoring equipment. The most likely explanation is simple and psychological. The astronauts had probably been sleeping badly because they were so focused on performing their many duties. Swathing them in sleep-monitoring electrodes convinced them that sleep was also a legitimate and important part of their duties, and they consequently relaxed and slept better despite the marginal discomfort.

Sleepy doctors

> Think not, is my eleventh commandment; and sleep when you
> can, is my twelfth.
>
> Herman Melville, *Moby-Dick* (1851)

To find the prime example of skilled professionals who routinely perform demanding, safety-critical tasks while severely sleep-deprived we need look no further than medicine. Chronic sleep deprivation is rife among hospital doctors, who are not superhuman enough to be immune from its consequences. Lack of sleep impairs their mood, judgment, decision making, thinking abilities and communication skills just like anyone else. One commentator recently described medical training in the USA as 'a gruelling endurance test in which patients are often those most at risk'. The situation in the UK is no better.

Just how tired are doctors? According to the research data, some of them are very tired indeed. A study of American physicians undergoing postgraduate medical training illustrates the problem. During a typical 36-hour period of on-call duty, the interns spent less than five hours in bed and slept for an average of less than four hours. That is not enough. Another American study found that three out of four residents in obstetrics and gynaecology were working between 61 and 100 hours a week, and more than two thirds reported getting less than three hours' sleep while on night call. Perhaps unsurprisingly, a large majority wanted limits placed on their work hours, despite concerns that this might restrict their professional experience. Much the same is true for doctors in other countries. For instance, house officers in Stockholm hospitals were found to sleep for an average of only four hours when on night call.

Three or four hours' sleep is not enough for most people. Research has shown that doctors who have slept for less than five hours in the previous 24 display significant deteriorations in their memory, intellectual skills, language and numeracy. Doctors working night shifts get less sleep than those on day duty, and their performance is consequently worse. A study at Stanford University found that emergency physicians slept for an average of 6.3 hours after working day shifts, but only 5.2 hours after night shifts. Their performance and mood suffered accordingly: those working nights had slower reaction times and took one third longer to perform a standard medical procedure. Their performance deteriorated during the course of a night shift and they became progressively more likely to make mistakes. They also felt less alert, less motivated, less happy and less clear-thinking than when they were on day shift. Given the choice, any sane patient would want to be treated by a doctor working day shifts.

As we shall see in the next chapter, sleep deprivation has a big impact on tasks requiring sustained concentration and effort. But tired people are often able to perform simple or engaging tasks in short bursts. Sleep-deprived doctors usually cope surprisingly well with brief but invigorating crises. Problems are more likely to arise with routine, repetitive tasks requiring prolonged attention. It might be relevant that the impact of sleep deprivation is found to vary somewhat between the different medical specialities, with surgeons being the least affected.

We will also see in the next chapter that sleep deprivation erodes our mood, motivation, social skills, communication skills, creativity and lateral thinking. Again, doctors are no exception. Psychologists who assessed junior doctors after a night of dealing with emergency admissions found deteriorations in their mood and motivation, as well as the usual impairment in short-term memory. Sleep-deprived doctors also perform significantly worse on measures of creative thinking and originality. They are less capable of solving complex problems that require originality and non-linear thinking, such as diagnosing an unusual condition.

To put icing on the cake, sleep-deprived doctors have an alarming tendency to fall asleep when driving their cars. An American study of paediatricians found that half of them admitted to having fallen asleep while driving, almost always after a night on duty. The on-call doctors notched up substantially more traffic accidents and traffic citations than their faculty colleagues. Their propensity to fall asleep at the wheel was unsurprising, considering they got less than three hours of sleep during on-call nights. So, after unintentionally jeopardising their patients' lives while on duty in the hospital, sleep-deprived doctors put themselves and other road users at risk while driving home.

The madness of politicians

> I have noted as something quite rare the sight of great persons who remain so utterly unmoved when engaged in high enterprises and in affairs of some moment that they do not even cut short their sleep.
>
> Michel de Montaigne, 'On Sleep', *Essays* (1580)

Long hours and inadequate sleep are standard features of political life. Those highly motivated, hardy individuals who survive the fierce competition and reach the top must have an above-average capacity for coping with little sleep. Having got to the top, they then set a bad example to the rest of us by projecting an image of tireless and unceasing industry. To accuse a politician of looking tired is frankly insulting. But they are only human, and inside that aura of sleeplessness there often lurks a tired person who secretly wants to spend more time asleep in their own bed.

Mythology and image-making abound when politics meets sleep.

During Margaret Thatcher's tenure as prime minister an absurd myth was fostered that it is both feasible and admirable for people routinely to sleep for only four hours a night and work hard for the remaining twenty. Hogwash. With the possible exception of a tiny minority of extraordinary individuals, humans simply do not thrive or perform well for long on four hours' sleep a night.

Negotiators sometimes deliberately exploit the debilitating effects of acute sleep deprivation to achieve their aims. People who have hardly slept for two or three days will agree to almost anything at four o'clock in the morning. Dragging out negotiations over several days may be irksome, but it can work if you make sure your side gets more sleep than the opposition. But more often than not, tiredness just gets in the way of rational politics. In 1997, after a sleep-deprived marathon of negotiation, representatives of 160 nations agreed the Kyoto Protocol, aimed at reducing global emissions of greenhouse gases. Three years later, fatigue helped to set back the environmental cause. In November 2000 an international summit convened in The Hague to thrash out unresolved problems left hanging by the Kyoto treaty. The negotiations ground on for 12 long days, leaving the delegates exhausted. In the early hours of the morning on the final day the British deputy prime minister, John Prescott, proposed a deal that he thought would break the logjam. But the deal collapsed, reportedly because the French delegate refused to make a difficult decision. A furious Prescott laid the blame squarely on the French environment minister. He told journalists: 'She got cold feet, felt she could not explain it, said she was exhausted and tired and could not understand the detail and then refused to accept it. That is how the deal fell.' The summit ended without reaching agreement.

The burden of work on the politicians and officials who run our nations has grown inexorably over the years. The number of decisions they must take has mushroomed, as the world has become an ever more complex, law-bound and media-scrutinised place. The insatiable demands of the 24-hour news media add greatly to the load. Political leaders are frequently overloaded, with insufficient time to think and formulate policy, let alone get enough sleep.

Academic observers of the British government scene calculated that between the 1960s and the 1990s the average working day for government ministers grew from 14 hours to 18 hours. The eminent political historian Peter Hennessy described the job of a British cabinet minister as 'a conveyor belt to exhaustion and underachievement all round', while a former senior adviser to the prime minister wrote that 'Ministers are governed by diaries which seem designed to break them in physique or spirit in the shortest possible time.'

The diaries of the late Alan Clark, who served as a government minister in the 1980s and early 1990s, give illuminating glimpses into the sheer grind of ministerial life. Clark observed colleagues who were 'boss-eyed' with fatigue after working past midnight. One diary entry from 1984 describes how the civil servants would always find more work for him to do, no matter how little sleep he had had:

> Today has been vilely full. Went early to Leicester after a late, late vote and impossible to drowse in the train as officials were watching me beadily in case (their excuse) anything in the brief 'needed explaining'. I dropped off, as good as, several times during monologues at the various offices.

To add to the hazards of politics, Clark's life was occasionally jeopardised by an exhausted government driver who had a tendency to nod off while conveying the minister along motorways at antisocial hours. Alan Clark's experiences were by no means unusual. Geoffrey Howe, who was Foreign Secretary in the 1980s and who, like Alan Clark, worked under the notoriously unsleeping eye of Margaret Thatcher, described his gruelling work regime like this:

> During six years at the Foreign Office I took home, to work through overnight while others slept, no less than 24 tonnes of paper ... Six o'clock was my normal time for getting up. My average bedtime was about four hours earlier.

The long-hours culture affects not only the elected politicians but also the officials who serve them (and, some would claim, run the country). Sir John Coles, who was head of the British Diplomatic Service from 1994 to 1997, described the problem like this:

> Long working hours, pressure and flurry were part of Foreign Office culture. We liked to feel busy and under pressure ... But it became necessary to question some of this culture. These things did not necessarily lead to good policy. Tired, pressurised officials were liable to make mistakes.

The demands now are, if anything, even greater. In December 2000 Tony Blair was asked, during an Internet chat forum with members of the public, what he remembered dreaming about on the night after his first general election victory in 1997, and what was the last dream he remembered. The prime minister's answer was revealing about the punishing lifestyle that goes with his job:

> I don't remember getting much sleep at all that night ... After a couple of hours' sleep, we were up early to prepare for going to Buckingham Palace. As for dreams, I've not had much chance for sleep over the past few days, let alone dreams.

Jet lag caused by frequent international travelling adds to the problem. In the immediate aftermath of the terrorist attacks in the USA on September 11, 2001, Tony Blair engaged in a gruelling programme of shuttle diplomacy that saw him travel more than 40,000 miles on 31 international flights in the space of a few weeks. And the fact that he then looked tired made the front pages of the newspapers.

British Members of Parliament work some of the strangest, if not the longest, hours of any legislature in the world. In October 2000 *The Times* published the results of one of the most detailed surveys ever carried out into MPs' lifestyles. The survey revealed a nightmarish world of long hours and chronic sleep deprivation. Most MPs said they worked between 71 and 80 hours a week, with one in six working up to 90 hours a week. One government minister logged a working week of 91 hours in Westminster followed by 20 hours at the weekend, leaving an average of eight hours a day for everything else including travelling, family life and a little sleep.

The response from one MP was illustrative. The constituency he represented, and where his family still lived, was a long way from London. He would leave home at five a.m. on a Monday morning and return in the small hours of the following Friday morning. His weekend at home would be spent on constituency business, and then there were the occasional foreign trips with a parliamentary committee. He and his partner planned to go out together on Saturday nights but he was usually so tired by then he would fall asleep. Once, while driving home from London, he had almost fallen asleep at the wheel and crashed. There has been some modernisation of parliamentary schedules, in response to pressure from MPs, but the long-hours culture remains deeply embedded.

Some politicians cope with the long hours with the assistance of drugs of various kinds. In the USA, coke is a favourite – particularly the diet cola variety. During the US presidential election campaign in 2000, candidate Al Gore engaged in an electioneering programme of awesome intensity, involving 19-hour days and an itinerary that criss-crossed the continent. 'Our campaign consists of a lot of long days and a lot of short nights,' said Gore's spokesman. 'While some candidates may look for their feather pillows, Al Gore is looking for every single undecided voter he can find.' To help him remain awake and vaguely sentient, Gore reportedly drank copious amounts of Diet Coke. One of his aides was explicit about the reason: 'These are high-caffeine days. He needs his fuel to get through them.' Sadly for Gore, the caffeine was not enough.

George W. Bush, who beat Al Gore by the slimmest of slim margins, became notorious during the election campaign for his verbal fluffs and tortured syntax. Whole books have been dedicated to Bush's gaffes, malapropisms and garbled sentences. One American psychologist even suggested that Bush's difficulties with the spoken word resulted from a lack of sleep in someone who apparently needed a lot of it.

Perhaps the defeated Al Gore could draw a minuscule crumb of comfort from an informal survey, which was conducted several months after the presidential election and reported to the International Conference for the Study of Dreams in July 2001. This survey found that conservative Republican supporters were nearly three times more likely to experience nightmares than their less conservative Democrat opponents. Half of the dreams recalled by Republicans were nightmares, compared with fewer than one in five of Democrats' dreams. Moreover, the conservatives' dreams were generally more frightening and more aggressive in content.

When the next big crisis erupts on the world stage, remember this. The politicians and officials who will be handling that crisis will be getting little sleep, perhaps for days at a time, and they will consequently become even more sleep-deprived than they already were. Their reactions, judgment, rationality, mood, memory, creativity and social skills will deteriorate, and they will become more prone to taking inappropriate risks. You might conclude that the world would be a safer and saner place if our leaders and their officials spent more time in bed (asleep).

Truly, madly, sleepily

I have an exposition of sleep come upon me.
William Shakespeare, *A Midsummer Night's Dream* (1595–6)

The human errors caused by tiredness sometimes have truly catastrophic consequences, and there have been plenty of man-made disasters to prove it.

Tiredness lay behind the environmental disaster that occurred when the supertanker *Exxon Valdez* ran aground in Prince William Sound, Alaska, spilling 11 million gallons of crude oil into the pristine waters and polluting thousands of miles of shoreline. The official investigation by the US National Transportation Safety Board concluded that sleep deprivation was a direct cause. The accident took place just after midnight on 24 March 1989, when the *Exxon Valdez* was under the control of the third officer. He had slept for only six hours during the preceding 48 hours and was therefore substantially sleep-deprived. It appears that he fell asleep on duty. Media reporting at the time suggested that alcohol was to blame for the accident, but the real culprit was fatigue. In addition to the appalling environmental damage, one man's tiredness cost his employer more than five billion dollars in punitive damages.

Tiredness contributed to the US Space Shuttle Challenger disaster. On a freezing cold morning in January 1986 the Challenger ascended for ten miles and then exploded, killing the crew of seven astronauts. It later emerged that crucial rubber O-ring seals had failed catastrophically at the low temperatures prevailing that morning. The US Presidential Commission that investigated the disaster concluded that there had been serious flaws in the decision-making processes leading up to the launch. The danger had been foreseeable, but tiredness had contributed to the bad decision to launch despite the icy conditions. Key managers had slept for less than two hours the night before and had been on duty since the early hours of the morning. They were at a dangerously low ebb when the fateful decision to launch was taken. The official report noted that 'fatigue is not what caused the accident, but it didn't help the decision-making process'. It also commented that 'working excessive hours, while admirable, raises serious questions when it jeopardises job performance, particularly when critical management decisions are at stake'. Bear in mind that when official investigations search for the causes of disasters they instinctively focus on tangible, physical causes like rubber O-rings and air temperatures; intangible human factors like fatigue are seldom in the forefronts of investigators' minds.

Lack of sleep has contributed to, if not caused, a string of disasters and near disasters in nuclear power plants. Many of them occurred in the early hours of the morning – a common feature of sleep-related accidents – and stemmed from failures by human operators to make sensible decisions when faced with the relevant information. Research has confirmed that nuclear power-plant operators who work night shifts experience real problems with sleepiness, distractibility and poor alertness. Even if they are not particularly sleep-deprived, they are unlikely to perform well in the early hours of the morning. And that is when the accidents happen.

The worst nuclear incident so far in the USA took place in March 1979, when the reactor at the Three Mile Island power station near Harrisburg in Pennsylvania came close to meltdown. The near-disaster at Three Mile Island arose in the early hours of the morning, after operators failed to recognise what their instruments were plainly telling them – namely, that an automatic valve had closed, cutting off the water supply to the coolant system. The reactor shut itself down automatically, as it was designed to do when a malfunction like that occurred. But a series of errors by the human operators led to a dangerous loss of coolant from the reactor core and almost turned an incident into a catastrophe. Radioactive gases were released from the partially exposed reactor core, but the containment vessel fortunately prevented them from escaping into the environment. Although no one died as a direct result of the Three Mile Island accident, it had a massive impact on the American nuclear industry. The damaged reactor took ten years to decontaminate and remained unusable. Fatigue is believed to have contributed to the operators' repeated failures to handle the incident correctly.

The worst nuclear accident thus far in history occurred at the Chernobyl nuclear power station in April 1986. It too was sleep-related, and started in the early hours of the morning when the operators were at their lowest ebb. The disaster happened when the engineers operating one of the station's four nuclear reactors made a series of irrational judgments. They attempted an ill-conceived experiment that involved shutting down the reactor's regulatory and emergency safety systems and withdrawing most of the control rods from the core, while allowing the reactor to continue running. The operators exhibited the alarming propensity to take inappropriate risks that is characteristic of tired people. As a later report put it, they behaved 'like intelligent idiots'.

The reckless behaviour of Chernobyl's operators caused a chain reaction. At 1:23 a.m. on the morning of 26 April a series of explosions blew the reactor apart. There was a partial meltdown of the reactor's graphite core and it caught fire. Large amounts of radioactive material were released into the environment – several times the amount created by the atom bombs dropped on Japan in World War Two. Some of it was carried by winds and contaminated several western European countries, including France and the UK. Thirty-two people at the Chernobyl plant died at the time of the accident and several more died soon after from severe radiation exposure. The long-term damage to the health of populations living in affected areas remains a matter of controversy, but it is undoubtedly huge. Several thousand people have died, or will die, as a result.

Over and over again, man-made disasters like Chernobyl and *Exxon Valdez* have occurred at night or in the early hours of the morning, when people's reactions and judgment are at their weakest. We saw earlier that drivers are much more likely to have a serious crash late at night than in the middle of the morning. Almost everyone who works night shifts displays signs of sleepiness and impaired performance, and it is not difficult to see why. Working at night forces people to perform at a time when their biological clocks are telling them to sleep, and to sleep when their biological clocks are telling them they should be awake. They perform worse when they are at work, and they are less able to sleep when they go home, as a result of which they become tired and accident-prone. Add chronic sleep deprivation to the brew and you have a potentially lethal concoction. And we all have to live with the consequences.

The price of eternal vigilance is liberty

> Care is heavy, therefore sleep you.
>
> Thomas Dekker, *Patient Grissil* (1603)

If society were to recognise the true importance of sleep, then attitudes towards tiredness on the roads and in the workplace might become more enlightened. In a more sleep-conscious world it would no longer be socially acceptable, let alone admirable, for people to drive or turn up for work suffering from severe fatigue, any more than it is now acceptable to be drunk in the workplace or behind the wheel of a car. Napping during working hours would be tolerated and even encouraged, rather than stigmatised as a sign of sloth, drunkenness or illness. Meanwhile, society continues to turn a blind eye to people driving cars, flying aeroplanes, practising medicine, operating safety-critical machinery and running nations when they are mentally and physically impaired by lack of sleep.

In the next chapter we shall see that sleep-deprived people are bad at making decisions and communicating those decisions to others. Their judgment is impaired, they are easily distracted, they respond poorly to unexpected information, they lack flexibility, they persist with inappropriate solutions to problems and they are prone to taking foolish risks. These are not the characteristics any of us would wish to see in the people who make life-and-death decisions in the corridors of power, hospitals, flight decks or nuclear power stations.

3

Dead Tired

You lack the season of all natures, sleep.

William Shakespeare, *Macbeth* (1606)

Does it really matter that many people in industrialised countries no longer get enough sleep of sufficient quality at the right times? We have seen one reason: the fact that sleepiness causes accidents. But far more than that, inadequate sleep matters because of what it does to our minds and our bodies each and every day.

Sleep is eloquent in its absence. We know that if we miss a night's sleep we will feel bad the next day. But the unpleasant sensations of acute fatigue evaporate after a good night's sleep and we soon forget. Far less obvious are the insidious, cumulative consequences of seldom getting quite enough sleep, night after night, week after week. Chronic sleep deprivation creeps up on us. It has pervasive effects on our mood, social skills and mental abilities – especially judgment, creative thinking and problem solving. It can also impair our physical health and make us more vulnerable to disease, as we shall see in the next chapter. However, the first and most obvious symptom of insufficient sleep is sleepiness, and that is where we shall start.

Sleepiness

Life is one long process of getting tired.

Samuel Butler, *Notebooks* (1912)

The longer you go without sleep, the sleepier you feel. Objective measurements prove that there is indeed a close relationship between sleep deprivation and sleepiness. That relationship is 'dose-dependent', which means that the longer you have been deprived of sleep, the faster you will fall asleep when given the opportunity. Really tired people can fall asleep almost anywhere, as William Shakespeare observed:

> Weariness
> Can snore upon the flint, when resty sloth
> Finds the down pillow hard.

If you are able to lie on a hard floor and go to sleep immediately during the middle of the day, you are probably sleep-deprived. (An obvious point, but no less true for that.)

Sleep will eventually force itself upon you if you go long enough without it. When scientists carry out sleep-deprivation experiments with human volunteers they often have to work quite hard to keep their subjects awake. Someone who has been deprived of sleep for two or three days requires stimulation, activity and variety to stop them falling asleep, no matter how well motivated they are to stay awake. Ian Oswald, a British scientist who conducted sleep-deprivation experiments in the 1960s, recalled walking the streets of Edinburgh flanked by exhausted volunteers whom he was working hard to keep awake. Even then, he would often see their eyelids closing as they walked along in a twilight state somewhere between waking and sleep.

Sleepiness and its flip side wakefulness determine how alive you feel and how well you perform every day of your life. But until the late 1970s there was no standard tool for measuring sleepiness. Then Mary Carskadon of Brown University in the USA invented a simple technique called the multiple sleep latency test (MSLT). It measures sleepiness like this. The subject is invited to go into a dark, quiet room in a sleep laboratory, to relax and fall asleep. Sensors detect the physiological signs of sleep (of which more later). As soon as the subject falls asleep, he or she is woken up. The key measurement is the time taken to fall asleep. The purpose of the MSLT is to measure *daytime* sleepiness, so the tests are generally carried out between ten in the morning and eight in the evening. The procedure is usually repeated four or five times during the course of a day.

The length of time between lying down and falling asleep is called the sleep latency. A daytime sleep latency of 15–25 minutes is generally interpreted as a reassuring sign of normality. Someone who is sleep-deprived will fall asleep faster. A sleep latency of less than ten minutes raises questions; there is a good chance that the person concerned will be experiencing intrusive daytime sleepiness, especially during the normal afternoon dip in wakefulness or when bored. A daytime sleep latency of less than five minutes usually signifies excessive tiredness, implying that the individual is substantially sleep-deprived or suffering from a sleep disorder.

Not everyone who falls asleep quickly in the MSLT is sleep-deprived, however. There are a few healthy people who can fall asleep in a trice, given the opportunity and the will. These individuals have the ability to go out like a light even after they have had unlimited amounts of sleep and without displaying any other signs of sleep deprivation. They just seem to be unusually good at relaxing and going to sleep during the day, regardless of their need for sleep. An unusually short sleep latency should therefore be viewed as strong evidence of sleep deprivation, but not conclusive proof.

The MSLT is conducted in a sleep laboratory, using electronic sensors to determine the onset of sleep objectively. There is, however, a crude, do-it-yourself version of the MSLT. This entails lying on a bed under sleep-friendly conditions – lights off, shoes off, TV off, no noises off, and so on – with one hand dangling over the edge of the bed and holding a metal object such as a spoon. Under the dangling hand rests a plate or some other hard surface. When you fall asleep your muscles relax and the spoon drops onto the plate with a loud clatter. All being well, you should then wake up, check your stopwatch and write down your sleep latency. If, on the other hand, you wake up several hours later to find your spoon lying on the plate, it may be that you are sleep-deprived. An even simpler way of assessing your own subjective sense of sleepiness is with a self-rating method called the Stanford Sleepiness Scale. This is a seven-point scale, ranging from 1 (defined as feeling active, vital, alert, wide awake), through 2 (functioning at a high level, not at peak), 3 (relaxed, not full alertness, responsive), 4 (a little foggy, not at peak, let down), 5 (tired, losing interest, slowed down), and 6 (drowsy, prefer to be lying down), to 7 (almost in a reverie, hard to stay awake).

Sleepiness is affected by many other factors besides the length of time since you last slept. Foremost among these is the time of day. Other things being equal, it is easier to fall asleep at certain times of the day and harder at other times. Wakefulness fluctuates naturally over the 24-hour cycle, with low points occurring in the early hours of the morning and again in the afternoon. This daily rhythm in wakefulness is mirrored by fluctuations in performance, which manifest themselves in many ways. For instance, you are more likely to crash your car or make a mistake when performing a difficult task in the early hours of the morning or in the afternoon. Sleep deprivation amplifies the effects of time of day, so the more sleep-deprived someone becomes, the deeper their daily troughs in wakefulness.

Stimulation and activity can mask sleepiness, at least for a while. Busy people often fail to notice how tired they really are until they relax at the weekend or on holiday. Their true state then becomes apparent. Sleepiness is even influenced to some extent by the nature of your previous activity. Experiments have confirmed, for example, that people feel sleepier and fall asleep more quickly after watching television than after taking a walk.

Fighting the beast

Thou art inclined to sleep; 'tis a good dullness,
And give it way: I know thou canst not choose.
William Shakespeare, *The Tempest* (1611)

What does it feel like to be seriously tired? For most of us real exhaustion is thankfully a rare experience. But for some, it is all too real. One of the most graphic accounts ever written about crushing fatigue can be found in the autobiography of the American aviator Charles Lindbergh. In 1927 Lindbergh made the first solo non-stop flight across the Atlantic in his single-engine plane *The Spirit of St Louis*. It lasted 33.5 hours. In his 1953 Pulitzer Prize-winning autobiography, Lindbergh described how his flight from New York to Paris came close to disaster – not because of mechanical failure or bad weather, but because of simple lack of sleep.

Before his marathon flight Lindbergh was told that it would be impossible for one man to fly an aeroplane alone for 30 or 40 hours without sleep. Lindbergh disagreed: he had worked for more than 40 hours at a time without sleep, so he did not see why he should be incapable of flying for that long sitting down. He was almost proved wrong.

The timing of Lindbergh's departure remained uncertain until the last moment. It all depended on the weather. When an opportunity suddenly arose to take off the following morning at daybreak, Lindbergh went back to his hotel to cram in a few hours' sleep. As a professional pilot he had learned from experience that every little bit of sleep helps. Even so, it was close to midnight before he reached his room, leaving barely three hours if he was to be ready to fly at dawn. As he eventually lay down to sleep, Lindbergh regretted the flawed planning that would force him to set off in a state of sleep deprivation. He knew he would have been better off starting his marathon fully rested – not just because that would make him a better pilot, but also because tiredness would prevent him from fully appreciating the experience. Lindbergh wanted to enjoy his flight. Even then, in those precious few hours, his sleep was further eroded when someone woke him to ask a stupid question. By three o'clock that morning Lindbergh was at the airfield preparing to take off. And so began his epic, sleepless flight across the Atlantic.

After only a few hours in the air, Lindbergh felt tiredness creeping up on him. How pleasant it would be, he mused, to doze off for a few seconds. He shook himself. He could not afford to feel like that so early in the trip. Later that day, and still less than nine hours into the flight, fatigue hit him again:

> My eyes feel dry and hard as stones. The lids pull down with pounds of weight against their muscles. Keeping them open is like holding arms outstretched without support. After a minute or two of effort, I have to let them close . . . My mind clicks on and off, as though attached to an electric switch with which some outside force is tampering. I try letting one eyelid close at a time while I prop the other open with my will. But the effort's too much. Sleep is winning. My whole body argues dully that nothing, nothing life can attain, is quite so desirable as sleep.

If sleepiness weighed so heavily upon him now, how could he get through the night, to say nothing of the dawn and another day and its night and possibly even the dawn after that? Lindbergh was ashamed. How could he let something as trifling as sleep ruin the record-breaking flight he had spent so many months planning? How could he face his sponsors and admit he had failed to reach Paris because he was sleepy? This must be how an exhausted sentry feels, he thought: unable to stay awake, yet knowing he will be shot if he is caught napping. He had no choice but to battle against his fatigue, minute by minute. In the end, it would all come down to sheer will power.

As the first traces of dawn began to appear on the second morning, Lindbergh felt the overwhelming desire to sleep falling over him like a quilt. Dawn was the time he had dreaded most:

> Like salt in wounds, the light of day brings back my pains. Every cell of my being is on strike, sulking in protest, claiming that nothing, nothing in the world, could be worth such an effort; that man's tissue was never made for such abuse. My back is stiff; my shoulders ache; my face burns; my eyes smart. It seems impossible to go on longer. All I want in life is to throw myself down flat, stretch out – and sleep.

Lindbergh searched for some way to stay alert. Shaking his body and stamping his feet no longer did any good. He had no coffee with him, but consoled himself with the thought that he had long since passed the stage when coffee could have helped. He pushed the stick forward and dived down into a ridge of cloud, pulling up sharply again after clipping through its summit. That woke him up a little, but not for long. He was thankful that *The Spirit of St Louis* had not been designed to be a stable aeroplane. The very instability that made it difficult to fly now guarded him against catastrophic errors. The slightest relaxation of pressure on stick or rudder would start a climbing or a diving turn, hauling him back from the borderland of sleep.

In the twentieth hour sleepiness temporarily gained the upper hand. Lindbergh suddenly awoke to find the plane diving and turning: he had been asleep with his eyes open. The realisation that he had lost control of himself and the plane was like an electric shock, and within seconds he was back in command. But as time passed, and no new emergencies occurred, he lapsed back into a dreamlike state, unsure whether he was dreaming through life or living through a dream. Over and over again he fell asleep with his eyes open, knowing he was falling asleep and unable to prevent it. Extreme measures were needed. He struck his face sharply with his hand, but felt hardly any sensation. He hit his face again, this time with all his strength. All he felt was numbness. Not even pain would come to his rescue. He broke open a capsule of ammonia and inhaled, but smelt nothing. Lindbergh realised how deadened his senses had become.

After 24 hours in the air and with more than a thousand miles still to go, Lindbergh seriously doubted whether he could stay awake long enough to avoid crashing into the Atlantic. But just as he felt death and failure staring him in the face, he began to turn the corner. The seriousness of his crisis had at last broken the spell of sleep and summoned up his last reserves of mental strength. He felt as though he was recuperating from a severe illness. And, as the history books relate, Charles A. Lindbergh made it to Paris and became an international hero. His remarkable last-minute rally was almost certainly a reflection of his circadian rhythm. His next peak of alertness came just in time. If his flight had continued much longer, he would inevitably have plunged back down into another circadian trough of fatigue from which he might never have ascended.

Lack of sleep has always been one of the least glamorous aspects of life at sea. In 1938 the 18-year-old Eric Newby signed on as an apprentice aboard one of the last square-rigger sailing ships. Newby's autobiographical account of his voyage, *The Last Grain Race*, charts his experiences as he made the round trip from Belfast to Australia and back again via Cape Horn. The ship had a minimal crew, so each man worked hard and slept little. 'I had never been so tired in my whole life,' wrote Newby, 'far too exhausted to appreciate the beautiful pyramids of sail towering above me.' After coming off watch he would fall into a dreamless sleep, so deep that when he awoke and went up on deck again he felt like a sleepwalker. On the return voyage his ship was hit by an awe-inspiring storm that soon had the crew's compartment awash in six inches of water. But the sailors who were not on duty snored through it all, so great was their appetite for sleep. Like soldiers preparing for the next battle, they lay 'absorbing sleep greedily like medicine'.

The writer C. S. Forester charted the sleepiness of the long-distance mariner in his Hornblower novels, which relate the fictional Royal Navy career and Napoleonic War adventures of Horatio Hornblower, a character partly modelled on Admiral Horatio Nelson. In one story, the gallant Hornblower is exhausted from lack of sleep after a daring escape from France and a prolonged battle with pursuing enemy ships. Like Charles Lindbergh, Hornblower experiences the disturbing sensation that his mind has become disconnected from his body:

> His voice sounded strange and distant in his own ears, like that of a stranger speaking from another room, as he issued his orders;

the very hands with which he held the ropes seemed not to belong to him. It was as if there was a cleavage between the brain with which he was trying to think and the body which condescended to obey him.

Sleep deprivation was a fact of life for Hornblower's real-life counterpart as well. When Horatio Nelson was commanding a Royal Navy warship he seldom had more than two hours of uninterrupted sleep and sometimes stayed on deck all night. However, like many leaders famed for coping with little sleep, Nelson had a well-developed ability to take catnaps during the day. He would nap in his cabin in a black leather armchair, his feet up on a chair. As we shall see later, a faculty for napping has enabled many high achievers to cope with meagre rations of night-time sleep.

Even in the twenty-first century, napping is a crucial skill for sailors – especially when sailing single-handed. In February 2001 the British yachtswoman Ellen MacArthur crossed the finishing line of the Vendée Globe boat race after 94 days alone at sea. She had travelled 24,000 miles across three oceans to become the fastest woman to sail single-handedly around the globe. For 13 weeks she had managed her 18-metre yacht in some of the world's roughest seas by herself. The only way Ellen MacArthur could survive was to become a past master of napping, and divide her sleep into multiple brief naps. During the 94-day voyage she took 891 naps, each lasting on average 36 minutes, giving her a total of about five and a half hours of sleep a day. Even sleeping in such short bursts, MacArthur frequently had to rely on what she called her 'sixth sense' to wake her when something urgently required her attention.

A soil for peevishness

> He that sleeps feels not the toothache.
> William Shakespeare, *Cymbeline* (1609–10)

Lack of sleep does far more than just make us feel sleepy, however: its tentacles reach out and twist our emotional, cognitive and physical states. One of the first casualties is mood. Tired people are emotionally less resilient and more prone to irritation or sadness. Tiredness also impairs our social and emotional skills, with potentially damaging consequences for personal relationships. A tired person can be physically present but psychologically and emotionally absent. In *The Screwtape Letters*, C. S. Lewis imagines an experienced devil instructing his neophyte nephew on how to corrupt a young human. The best way, he advises, is through fatigue:

> The paradoxical thing is that moderate fatigue is a better soil for peevishness than absolute exhaustion ... It is not fatigue simply as such that produces the anger, but unexpected demands on a man already tired.

After a night without sleep, healthy people exhibit clear disturbances in mood, which are characterised by irritability, tension and reduced vigour. These symptoms normally evaporate after a good night's slumber. Sleep-deprived people, like drunks, lose their social inhibitions and behave in inappropriate ways; they are prone to outbursts of childish humour, which others around them do not always find hilarious. (And strangely, for reasons that remain unclear, acute sleep deprivation can also have the counterintuitive effect of stimulating the libido.)

Severe sleep deprivation can induce feelings of persecution and mild paranoia. It is well known among sleep scientists that the volunteers who take part in their sleep-deprivation experiments often become irritable and impatient. Some subjects become slightly paranoid, convinced that the researchers and fellow volunteers are plotting against them. In rare instances, exhaustion can provoke more dramatic changes. In one documented case, a previously healthy man became psychotic after four nights of badly disrupted sleep and believed he was the Messiah.

Chronic sleep deprivation – that state of never getting quite enough sleep, day in day out, over a prolonged period – is far more common in everyday life than the acute deprivation that comes from having no sleep at all for one or two nights. And chronic sleep deprivation can have just as much cumulative impact, leaving even the saintliest person with a shorter fuse. The writer John Seabrook described the debilitating fatigue that goes with having a small baby like this:

> The burning eyes; the band of fatigue that tightens around the skull, a sensation some liken to the feeling that you're always

wearing a hat; the irritation – at each other, at friends, at the cat's water bowl, which I kept kicking by accident . . .

Lack of sleep and tiredness are obviously not wholly responsible for the tetchiness, aggression and petty violence of everyday life, but it is a racing certainty that they contribute towards making the world a nastier place. Conversely, there is little doubt that good sleep makes us feel better. In one study, researchers issued volunteers with pocket computers on which they logged their sleep patterns, moods and social interactions over a two-week period. The results showed that going to sleep earlier in the evening was consistently associated with better mood and better social interactions the following day.

Tired people are stupid and reckless

Fatigue makes women talk more and men talk less.
C. S. Lewis, *The Screwtape Letters* (1942)

Besides making us grumpy and poor company, sleep deprivation impairs our mental abilities in many subtle and not-so-subtle ways. In brief, tired people are stupid and reckless. Sleep deprivation damages our ability to perform tasks that require attention, thought, judgment, memory, social skills or communication skills – which covers the ground fairly comprehensively. Tired people also seem to lose sight of the consequences of their actions, liberating them to do silly and sometimes catastrophic things: the Chernobyl and *Exxon Valdez* disasters were just two examples.

Even modest sleep loss will measurably reduce your mental performance. After one night without sleep, you will have slower reactions, make more mistakes and find it harder to maintain attention. Unsurprisingly, two or three days of sleep deprivation produce even bigger impairments. Young adults are no more resistant than older people. If anything, they are more vulnerable. When researchers compared the consequences of one night of total sleep deprivation on healthy 80-year-olds and 20-year-olds, they recorded larger disturbances in the mood and cognitive performance of the younger subjects.

Reducing someone's sleep for several nights in a row can undermine their performance just as much as completely depriving them of sleep for one or two nights. When scientists limited healthy young adults to an average of only five hours' sleep a night for a week, the subjects became progressively sleepier and their performance deteriorated significantly. Two full nights of catch-up sleep (equivalent to a restful weekend) were needed to reverse the decline.

The armed forces have understandably maintained a long-standing interest in how well people hold up when they are deprived of sleep, as often happens in conflict. Experience shows that soldiers can continue to perform reasonably well after days of sleep deprivation under combat conditions, buoyed up by adrenaline, physical exertion and strong motivation. With sufficient stimulation and will power, military personnel can usually keep going without sleep for three or four days before they keel over. In one study, for example, men were assessed throughout a strenuous combat training course lasting several days. Some of the trainees were allowed no sleep at all, while others were permitted a few hours in the middle of the course. All the men displayed a substantial deterioration on measures of mood, vigilance and reaction time, with those who got no sleep performing even worse than those who got some. By the end, the trainees who had not slept at all were suffering from clinical symptoms including sensory disturbances.

Sleep-deprived people can often perform certain tasks reasonably well in short bursts, despite their fatigue. The real problems arise if they have to sustain their effort for any length of time. This lack of staying power was highlighted by experiments conducted in the 1950s at the Walter Reed Army Institute of Research in the USA. Sleep-deprived volunteers scored well in tests of reaction times when they were only required to react a few times over a period of one minute. But when they had to make lots of rapid responses spread irregularly over a period of 15 minutes, their performance fell apart. Sometimes they reacted quickly and sometimes they reacted very slowly or not at all. The more sleep-deprived they were, the more their responses varied. That experiment illustrates a general pattern that has emerged from studies of sleep deprivation – namely, that the performance of tired people becomes much more variable and inconsistent. Their accuracy and speed can fluctuate wildly over periods of a few minutes. In one study, for example, scientists monitored the performance of volunteers throughout an 88-hour period of total sleep deprivation. The subjects' performance on a task requiring vigilance and co-ordination was much more variable (as well as just plain worse) than a comparison group, and the longer they went without sleep the more variable their performance became. These wild fluctuations in performance probably arose because their attention was repeatedly blotted out by microsleeps. As we saw earlier, someone who is very tired can lurch from wakefulness to sleep and back again in a matter of seconds without even noticing.

Much of the research on sleep deprivation has focused on how it impairs people's ability to perform tasks requiring rapid but relatively simple decisions, involving very basic skills such as the ability to maintain attention under monotonous conditions. These skills are sensitive to sleep deprivation (which is partly why they have been so frequently measured) but they also have only a tenuous bearing on what happens in reality. The sorts of judgments and decisions we all have to make in real life usually demand far more than just the ability to stay awake and press buttons. Real decisions require us to assimilate and process complex, incomplete and often contradictory information, to keep track of our actions, assign priorities, ignore distractions and communicate with other people. The truly frightening aspect of sleep deprivation is the way it erodes all of these capacities.

Sleep-deprived people are bad at making complex decisions that require them to revise their plans in the light of unexpected news, to ignore irrelevant information and to communicate effectively. These are precisely the capabilities that we all need most when dealing with the vagaries of normal life – as do politicians, managers, doctors, military commanders and other key decision makers.

Even a single night without sleep will impair your ability to think flexibly and creatively. Sleep-deprived people perform badly on all aspects of creative thinking, including originality, flexibility, generating unusual ideas, being able to change strategy, word fluency and nonverbal planning. Tired people tend to persist with their current activity regardless of whether it is appropriate – a characteristic that psychologists call perseveration. In one cleverly revealing experiment, sleep-deprived volunteers played a realistic marketing game that required them to make complex decisions and then continually update those decisions in the light of new information. After 36 hours without sleep, the well-motivated subjects were still able to read and absorb written information. But their ability to make sound decisions based on that information deteriorated markedly. As they became more sleep-deprived they found it increasingly difficult to recognise when to change tactics in light of changed circumstances. Their thinking became rigid and they tended to persist with incorrect responses.

Eventually, after 36 hours without sleep (the equivalent of losing just one night's sleep) their ability to play the game collapsed.

Sleep-deprived people reveal their blunted creativity and mental inflexibility through alterations in their spoken language. After 36 hours of sustained wakefulness, people display a marked deterioration in verbal fluency and inventiveness. They are more reliant on routine responses and tend to become fixated with a particular category of words. They also start using inappropriate, monotonous or flattened intonation when reading out loud. Think about how people talk when they are drunk and you will get the picture. Tired people are bad at finding the right words. Their language becomes less spontaneous and expressive, and they are less willing to volunteer information that others might need to know. All in all, they are worse at communicating their thoughts, feelings, decisions and actions. Being a poor communicator is unhelpful, whether you are a head of state trying to deal with a crisis or simply someone who values their social life.

The supposedly robust adolescent is, if anything, even more vulnerable than older people. One study found that when youths aged 10–14 years were restricted to only five hours in bed for one night, their verbal creativity, verbal fluency and ability to learn new abstract concepts were all impaired. Less complex mental functions, such as rote learning, were unaffected by this modest degree of sleep restriction. So the sleep-deprived adolescent would still be able to go through the basic motions at school the next day, but would be unable to perform to anything like their true potential. And it is not entirely unknown for adolescents to go to school after getting less than a full night's sleep.

As well as sapping our ability to perform, sleep deprivation also impairs our motivation. Tired people just can't be bothered. In one experiment, researchers persuaded a group of male students to work continuously for 20 hours without sleep and then observed them in various social settings. Sleep deprivation diminished everyone's performance, but the students who were working in groups performed even worse than those who were working individually. This was because working in a group gave them the opportunity to loaf – and they did.

Another alarming characteristic of sleep-deprived people, and something else they have in common with drunks, is their propensity to take risks. Experiments have shown that as people become more fatigued, they are more attracted by actions that might bring big rewards and less worried by the possible negative consequences of those actions. For instance, French scientists assessed the risk-taking propensities of military pilots who were involved in a maritime counter-terrorism exercise. This required them to make strenuous night flights, depriving them of sleep. The data showed that as the exercise progressed, the pilots became more impulsive. In effect, sleep-deprived people become more reckless and foolhardy. This is yet another characteristic that you would not wish to encounter in someone who is in charge of a hospital ward or a political crisis or a nuclear power station, or the family car for that matter.

Alcohol, beauty and old age

> It provokes the desire, but it takes away the performance.
> William Shakespeare, *Macbeth* (1606)

Tired people resemble drunk people in several respects. Most obviously, tiredness and alcohol both hamper our ability to perform tasks that require judgment, attention, quick reactions and coordination – such as driving a car, for example. Missing a night's sleep has an impact on driving skills comparable to drinking a substantial amount of alcohol. In both cases safe driving is no longer possible; in one case, however, it is both legal and socially acceptable.

To compare the effects of sleep deprivation and alcohol, researchers persuaded healthy young men to drive a simulator while under the influence of varying degrees of sleepiness and with varying blood-alcohol levels of up to 0.08 per cent. (For comparison, the legal limit for driving in most countries is 0.08 per cent blood alcohol and in Scandinavian countries it is 0.05 per cent.) Alcohol undermined their ability to maintain a suitable speed and road position, and so did sleep deprivation. Drivers who had been awake for 21 hours (hardly a remarkable feat) performed as badly as drivers with a blood-alcohol level of 0.08 per cent. So, next time you miss a night's sleep, try to remember (if you can) that your driving ability will be as bad as if you had a blood-alcohol level that would be illegal in most countries.

Recent research has underlined the parallel between tiredness and alcohol. Scientists assessed volunteers driving on a closed course, both after drinking alcohol and after sleep deprivation. Detailed comparisons showed that both acute sleep deprivation (one night with no sleep) and chronic sleep deprivation (two hours less sleep than normal each night for a week) caused impairments in performance and reactions that were almost indistinguishable from those caused by illegally high blood-alcohol levels. The inescapable conclusion is that even relatively modest sleep deprivation, of the sort that many people experience in everyday life, has potentially dangerous consequences. Shaving a couple of hours off your sleep each night for a week can make you as incapable behind the wheel as a drunk driver.

Another striking similarity between sleep deprivation and alcohol is that as well as impairing our performance, sleep deprivation also undermines our ability to realise that our performance has been impaired, as we shall shortly see. (Try saying that sentence quickly when you are tired or drunk.) Tired people, like drunk people, have a misplaced confidence in their own abilities. This dangerous trait was highlighted in an experiment in which students took cognitive tests after they had been deprived of sleep for 24 hours. Predictably, they performed worse than subjects who had slept well the night before. However, when asked to assess their own performance, the sleep-deprived subjects awarded themselves *higher* ratings than did the non-deprived subjects. Tiredness had marred their ability to appreciate their own inability.

Cultural attitudes towards being tired or drunk behind the wheel of a car are very different, despite the fact that the net outcomes are remarkably similar and sometimes fatal. Driving when drunk is illegal and (nowadays) socially unacceptable, whereas driving when tired is neither. People still boast about their feats of sleep-deprived driving, in the way that people once boasted of their ability to drive after downing stupefying quantities of booze. Of course, one big difference between drinking alcohol and going without sleep is that drinking alcohol is enjoyable. Sleep deprivation has only the bad bits to offer, including the hangover.

Being tired and drinking alcohol are obviously not mutually exclusive. Indeed, the two often go hand in hand. A night on the town is often a night of little sleep. What is more, they reinforce each other. Tiredness amplifies the effects of alcohol and vice versa. It is a common experience (except presumably among lifelong teetotallers) that alcohol packs a heftier punch when we are tired – say, at the end of a hectic working week. Even moderate amounts of alcohol produce a big sedative effect after insufficient sleep. Conversely, the effects of alcohol are somewhat blunted if you are well rested. Experiments have found that alcohol reduces the alertness of people who have had less than eight hours' sleep the previous night, whereas the same amount of alcohol makes no difference if they have had unlimited sleep. If you want to enjoy a few drinks without wilting, make sure you sleep properly the night before.

Another common experience is that the effects of alcohol vary according to the time of day when it is drunk. This happens because alcohol reinforces our natural circadian rhythm in sleepiness, exaggerating the troughs in alertness that normally occur in the early hours of the morning and again in the afternoon. Alcohol consumed at lunchtime or in the early afternoon therefore tends to have a bigger impact than the same amount consumed in the early evening, when alertness is normally higher. Lunchtime boozing really is more likely to make you fall asleep at your desk or crash your car than those pre-dinner 'sharpeners'.

Scientists have also uncovered some intriguing parallels between sleep deprivation and the normal ageing process. As we grow older, our performance declines on many psychological and neurological measures. A similar pattern of deterioration is observed in young people after sleep deprivation. One experiment found that after 36 hours without sleep, adults in their twenties had a performance profile similar to that of non-sleep-deprived people aged about 60. So, if you are in your twenties and you want to know how it feels to have the brain of a healthy 60-year-old, just stay up all night. Then you will know. The reason why ageing and sleep deprivation exert similar effects may be because they both impair the functioning of the pre-frontal cortex, a region of your brain that is extremely active when you are awake.

To complete the three-way permutation, alcohol and old age make an unpleasant cocktail. Being old is generally bad for sleep, and so is being an alcoholic. Being old *and* alcoholic is even worse. The elderly are more prone to insomnia, as we shall see in a later chapter. Alcoholism is also accompanied by sleep problems. And when old age and alcoholism combine within the same person they reinforce their malign influences on sleep. Researchers have found that older alcoholics have significantly worse sleep problems than younger alcoholics.

Sleep, or lack of it, can also affect physical appearance – a belief encapsulated in the term 'beauty sleep'. Although sleeping all hours will not necessarily make you more beautiful, prolonged lack of sleep will detract from your physical charms. Animals that have been experimentally deprived of sleep for long periods develop unsightly skin disorders. Lack of sleep in humans, especially adolescents and young adults, might exacerbate skin problems such as acne. Sleep deprivation weakens the ability of the skin to maintain its normal protective functions as a barrier against dirt and microbes. That said, the scientific evidence for a direct causal link between sleep and an unblemished complexion remains sparse, Sleeping Beauty notwithstanding.

Lack of sleep probably does contribute to the depressing tendency of men to become pot-bellied and flabby in middle age. In men (but not women) almost the entire day's production of growth hormone within the body occurs during sleep. The less sleep a man gets, the less growth hormone his body produces. As part of the normal ageing process, the total amount of sleep and the production of growth hormone both decline in parallel. Scientists have suggested that this age-related fall in growth-hormone production could be responsible for the systematic replacement of muscle by flab, better known as middle-aged spread. If so, dwindling sleep might be an important ingredient in the expanding waistline, double chins and spindly legs that help to make male middle age such a joy.

Champion wakers

> I'll wake mine eyeballs out.
>> William Shakespeare, *Cymbeline* (1609–10)

What happens to people if they get no sleep at all for a long time? One of the first scientific experiments on sleep deprivation dates from 1896, when Professor G. T. W. Patrick and his colleague Dr Allen Gilbert of the Iowa University Psychological Laboratory kept three volunteers awake for 90 hours. Patrick and Gilbert charted the now classic signs of prolonged sleep deprivation, including progressive deteriorations in reactions, memory and sensory acuity, together with a decline in body temperature.

Their first experimental subject, an assistant professor at the university, suffered his worst fatigue during the second night. Like Charles Lindbergh, he found that dawn was the cruellest time. He experienced visual hallucinations in which the air seemed full of red, purple and black dancing particles like gnats. All three subjects gained weight during the experiment, but their muscular strength diminished as they became more fatigued. The most noticeable effects were on mental performance: their memory became highly defective and they lost their ability to pay attention. One subject failed to memorise in 20 minutes material that he would normally have committed to memory in two minutes. In all cases, the symptoms disappeared after the experiment.

The first person to become internationally famous for self-imposed sleep deprivation was an American disc jockey called Peter Tripp, whose other claim to fame was inventing the Top 40. In 1959, Tripp managed the feat of staying awake under supervision for more than eight days and nights, a total of 201 hours. He did it to raise money for charity. Tripp even managed to broadcast live during his marathon, from a booth in Times Square, New York.

Peter Tripp suffered. As time went on, his friends and invigilators found it harder and harder to keep him awake. Constant vigilance was required to prevent him from lapsing into microsleeps. Three days into the experiment, Tripp became abusive and unpleasant. After the fifth day he progressively lost his grip on reality and started to experience visual and auditory hallucinations. His dreams broke through into his waking thoughts and he began seeing spiders in his shoes. He became paranoid and thought people were drugging his food. At one point he ran into the street and was nearly knocked down. These disturbing psychological symptoms were accompanied by physical changes, including a continuous decline in body temperature. By the last evening, Tripp's brain-wave patterns were virtually indistinguishable from those of a sleeping person, even though he was apparently still awake.

After 201 hours of continuous wakefulness Tripp had broken the record and halted the experiment. He immediately fell into a deep sleep that lasted 24 hours. When he finally did awake, his hallucinations had gone and he felt relatively normal. But something seemed to have changed within him. Those close to Peter Tripp felt his personality had altered permanently, and for the worse. His wife left him, Tripp lost his job and he became a drifter. Tripp's marathon of sleep deprivation certainly did him no good, but it was probably not the sole cause of his subsequent decline and fall. Tripp was taking large doses of Ritalin, an amphetamine-like stimulant drug, to keep himself awake during the marathon, and it is possible that the drug, combined with the sleep deprivation, helped to stimulate his paranoid delusions and hallucinations.

A few years later, Tripp's record was broken by a 17-year-old high-school student from San Diego called Randy Gardner. In 1965 Gardner stayed awake for 264 hours (or 11 days) in a successful attempt to break into *The Guinness Book of Records*. Scientists from Stanford University monitored most of his marathon. During the first two days Gardner's friends helped keep him awake and he did not use caffeine or other stimulants. By the end of the second day he was suffering from blurred vision, making it difficult for him to read or watch TV. By the third day he was irritable and wanted to be left on his own. His speech became slurred and his movements uncoordinated. On the fourth day he experienced memory lapses and mild hallucinations. After nine days without sleep he was unable to complete sentences and had lost the ability to concentrate. On the eleventh and final evening he had double vision.

Despite these temporary but unpleasant symptoms, Randy Gardner suffered remarkably few ill effects. Having stayed awake for 11 days and nights, he went to bed and slept for nearly 15 hours. When he awoke he felt fine. The following night he slept only slightly longer than usual, and within a few days his sleep had returned to normal. He did not go mad and, except for mild hallucinations, he never displayed any psychotic symptoms during the experiment. The experiences of Randy Gardner and others have demonstrated that going without sleep for several days does not generally result in mental illness or other long-term damage.

Some feats of self-imposed sleep deprivation have been endured for purely financial reasons, with not a single scientist in sight. In the depression-era USA of the 1920s and 1930s, a bizarre fad developed for dance marathons, in which people would compete for money. The rules were simple: keep dancing until you drop. This grisly social phenomenon was portrayed in the 1969 film *They Shoot Horses, Don't They?* The dance-marathon competitors were supposed not to sleep, although some competitions permitted one dancer to sleep provided their partner held them upright and both kept moving. They must have slept while dancing, since it was not uncommon for these nightmarish marathons to last for weeks. The world record was set by a couple in Chicago, who danced from 29 August 1930 until 1 April 1931 – a total of almost 215 days. The rules allowed them to close their eyes for no more than 15 seconds at a time, so they must have been adept at sleeping with their eyes open. Their reward for this outlandish spectacle of public torture was a paltry $2,000. These dance marathons were eventually made illegal. And talking of torture, let us turn to that uncomfortable subject.

Uses and abuses

It was always at night – the arrests invariably happened at night. The sudden jerk out of sleep, the rough hand shaking your shoulder, the lights glaring in your eyes, the ring of hard faces round the bed. In the vast majority of cases there was no trial, no report of the arrest. People simply disappeared, always during the night.

George Orwell, *Nineteen Eighty-Four* (1949)

Sleep deprivation is unpleasant and debilitating. Someone who has not slept for two or three days can feel as if they are losing their mind. That is why, throughout history, sleep deprivation has been exploited as a form of torture and coercion. Fatigue can bring people to their knees, both metaphorically and literally, without leaving a mark on them. It can even be fatal.

According to legend, King Perseus of Macedonia was put to death by being prevented from sleeping when held prisoner in Rome. Sleep deprivation is also said to have been a form of capital punishment in China in times past. The American writer and insomniac Bill Hayes cites a nineteenth-century account of a Chinese merchant who was sentenced to death for murdering his wife. Sleep deprivation was deliberately chosen as the method of execution, on the grounds that it would cause the maximum amount of suffering and would therefore serve as the greatest deterrent to other potential murderers. According to the account, which was written by an American physician, the prisoner eventually died on the nineteenth day, having suffered appalling torment.

Sleep deprivation has been employed for many centuries to soften up prisoners and make them talk. And it still is. Amnesty International found that more than half the torture victims they interviewed had been deprived of sleep for at least 24 hours. The secret police notoriously prefer to make their arrests in the small hours of the morning because that is when people are at their weakest and most confused.

When applied patiently and systematically, sleep deprivation is said to be the single most effective form of coercion. The victim is repeatedly woken at odd hours, allowing them little or no sleep. The pattern of awakenings is randomised, so the victim loses all control over when they sleep (an extreme version of what happens to parents of small babies). This unpredictability makes it impossible for the body's internal clocks to readjust. The circadian rhythms become disrupted, leaving the victim fatigued and in a state akin to severe jet lag. All sense of time and place depart. Even the strongest person can be reduced in this way to a state of helpless and tearful disorientation.

Exhausted people are very poor at making sound judgments based on complex information, including information derived from their own knowledge, beliefs and experience. Sleep-deprived prisoners are therefore much more susceptible to persuasion that their actions or beliefs are wrong. Herein lies the secret of 'brainwashing'.

In the Korean War of 1950–53, the Communists tortured American and Allied prisoners of war by systematically depriving them of sleep. A constant succession of guards would interrogate a prisoner at random times throughout the day and night, subjecting him to a constant barrage of questions and arguments. The effects on prisoners' behaviour and beliefs were often profound. Sixty per cent of the US airmen who were captured in the Korean War either confessed to imaginary crimes, such as using biological weapons, or collaborated with the enemy in condemning the USA. When the prisoners were eventually returned to the USA, the US government appointed a panel of experts to discover what had happened. The assumption at the time was that the Communists must have subjected the Allied prisoners to some mysterious and sophisticated form of mind control. Either that, or the men must be traitors and cowards. But the evidence uncovered by the review panel ruled out drugs, hypnosis or other forms of novel trickery. The truth was more prosaic. Just one device had been used to confuse and torment the prisoners until they were ready to confess to anything. That device was prolonged sleep deprivation. A combination of fatigue, confusion, fear and loss of control had produced profound changes in the men.

The practice of sleep-depriving military prisoners continues to this day. In April 2001 a US military surveillance aircraft collided in mid-air with a Chinese fighter. The American plane was forced to make an emergency landing in China. Meanwhile, the Chinese plane crashed, killing the pilot. The Chinese authorities were not happy: they impounded the American plane along with its crew of 24, sparking a diplomatic row between the two nations. The crew were eventually released after being detained and questioned for 11 days. They later revealed that the Chinese had used sleep deprivation as part of the interrogation process. The American pilot reported that the Chinese had questioned him for several hours on the first night, and thereafter had repeatedly woken him at various times of the night and day, forcing him to snatch a little sleep whenever he could.

Prolonged sleep deprivation lay behind many of the psychiatric casualties of World War One. Thousands of men had to be withdrawn from the horrific conditions of the front line with what was referred to then as shell shock. (Nowadays we would call it post-traumatic stress disorder.) Continuous shelling was undoubtedly a cause of severe stress. For hours or days at a time, men crouched defenceless in muddy trenches, constantly exposed to the threat of instant death or injury but powerless to do anything about it. Even the hardiest of minds could crack. But the shelling broke men for another reason as well: it prevented them from sleeping. Doctors often found that when a man with disabling shell shock was granted respite from the front line, he would rapidly recover and be able to return to his unit within days. Getting a few nights' sleep in the hospital probably did more good than the psychotherapy that went with it.

When we starve ourselves of sleep in the name of work or play, we go partway down a path that leads eventually to something horrific.

4

The Golden Chain

Sleep is that golden chain that ties health and our bodies together.

Thomas Dekker, *The Guls Horne Book* (1609)

We have seen some of the bad things that sleep deprivation does to our minds. What does it do to our bodies? Sleep and physical health are intimately intertwined, which means that inadequate sleep can cause all sorts of physical problems. When sleep deprivation is taken to the extreme, death ensues. Some scientists have argued that many of us neglect or mismanage our sleep to the extent that it damages our health, and that large numbers of people around the world die prematurely every year because of undiagnosed and untreated sleep disorders.

What, then, is the scientific evidence that sleep impinges on our physical health? Some (admittedly crude) indicators point towards a link between poor sleep and poor physical health. For instance, people who are sleepy during the day are more likely to use healthcare. And older people who complain of poor sleep are at greater risk of a heart attack. One American study that tracked thousands of elderly people discovered that individuals with sleep problems were more likely to suffer a heart attack over the following three years.

Lack of sleep and the resulting daytime tiredness are associated with greater sickness absence from work. An investigation of absenteeism among French employees found that those who reported feeling very sleepy at least three days a week were more than twice as likely to take sick leave as their less sleepy colleagues. Sleep problems are also predictive of long-term work disability. Norwegian research found that adults who were experiencing mediocre or poor sleep were more than twice as likely to face long-term work disability a few

74

years later. People who sleep badly also tend to eat badly, which may contribute to their health problems.

Good sleep, on the other hand, fosters mental and physical health. Psychological wellbeing, physical health and longevity are all statistically associated with healthy lifestyle practices, one of which is good sleep. The lifestyle factors associated with a lower risk of dying prematurely include taking physical exercise, not smoking and getting seven or eight hours of sleep a night. For example, recent research that investigated longevity in Japanese people uncovered three important factors, each of which was independently linked with a reduced risk of dying. These factors were walking for at least one hour a day, *ikigai* (a sense that your life is meaningful), and sleeping for at least seven hours a night. There is even some tentative evidence that people who habitually go to bed early live longer. A study of people aged over 80 found that these long-lived individuals all reported having gone to bed early throughout their lives. However, retrospective evidence of this sort must always be taken with a large pinch of salt. (But not too much salt, because that would be unhealthy.)

At the other extreme, excessive sleep is also linked statistically with poor health – probably because sleeping for unusually long periods is often a sign of illness. Scientists discovered in the 1970s that people whose normal nightly sleep duration was either unusually short (less than four hours) or unusually long (more than nine or ten hours) had a higher than average risk of dying prematurely. Similarly, a study of elderly British people found that those who spent 12 or more hours a day in bed had a significantly higher mortality rate, while those who spent the proverbial eight hours a day in bed had the lowest mortality rate.

Excessively long sleep is often a consequence of heart disease or other medical conditions, so it would be a mistake to generalise this finding very far. There is no reason to suppose, for example, that sleep-deprived teenagers or exhausted adults who lie in bed at the weekends will die younger as a consequence of snatching a few extra hours of rest. Too much sleep can make you feel temporarily below par, however. Experiments have confirmed that healthy people who normally feel refreshed after eight hours of sleep tend to feel groggy and perform badly after they have slept (on request) for 10 or 11 hours. On the other hand, how many healthy adults routinely sleep for 10 or 11 hours at a time? Few of us have any reason to fret about the dangers of sleeping too much.

A waking death

We term sleep a death, and yet it is waking that kills us, and
destroys those spirits that are the house of life.

Sir Thomas Browne, *Religio Medici* (1642)

What happens when sleep deprivation is taken to the extreme? If a
slight insufficiency of sleep makes us feel unwell, would a prolonged
absence kill us? Setting aside purely anecdotal accounts, science has
unsurprisingly not investigated whether forcibly depriving humans
of sleep is fatal. Ethical committees would tend to frown upon applica-
tions from scientists proposing to test this experimentally. But the
evidence from other species is clear. Animals that have been experimentally deprived of sleep for long enough invariably die. There is
no reason to suppose that humans are fundamentally different.

Some of the earliest experiments on extreme sleep deprivation were
performed in the late nineteenth century by a Russian scientist called
Marie de Manacéïne. She deprived puppies of sleep by keeping them
constantly active. They all died within four or five days, despite every
effort to keep them alive. The younger the puppy, the more rapidly
it succumbed. Marie de Manacéïne also noticed a progressive decline
in the body temperature of the sleep-deprived animals, a phenomenon
that is now known to be a standard symptom of prolonged sleep
deprivation in humans and other species. She concluded that sleep
was even more crucial for survival than food:

As a rule, the puppy deprived of sleep for three or four days
presents a more pitiful appearance than one which has passed
ten or fifteen days without food. I can speak from observation,
as I was obliged to make experiments on the results of want of
food as well as of sleep, and I became firmly convinced that sleep
is more necessary to animals endowed with consciousness than
even food.

Italian scientists working at the end of the nineteenth century kept
adult dogs awake by making them walk. The sleep-deprived dogs all
died after 9–17 days, regardless of how much food they ate.

One objection to experiments such as these (apart from the obvious
ethical one) is that the scientists had to use increasingly stressful

methods to keep the animals awake, so perhaps it was the stress that killed them rather than the sleep deprivation itself. Prolonged stress can impair the immune system and make an animal more vulnerable to infection. However, more recent research has managed to sidestep this methodological problem.

In a long series of experiments, Alan Rechtschaffen and colleagues at the University of Chicago systematically investigated how prolonged sleep deprivation affects rats. They used an experimental procedure known as the disc-over-water method, which works like this. Two rats – the experimental subject and the 'yoked control' – are placed on a turntable mounted over a shallow bath of water. The brain-wave patterns of both animals are continuously monitored to detect the onset of sleep. When the experimental rat's brain waves indicate that it is falling asleep, the turntable automatically revolves slowly, waking the unfortunate rat and forcing it to walk in the opposite direction to avoid being pitched into the water. The control animal, which is on the other side of the turntable and separated by a partition, receives precisely the same treatment at precisely the same times. The crucial difference is that the turntable movements are unaffected by its sleep. The control animal is therefore able to get some sleep when the experimental animal is awake. This cunning technique has the advantage – from the human experimenter's point of view – of preventing the experimental subject from sleeping without having to subject it to other noxious stimuli.

Rats that are prevented in this way from sleeping invariably die after two or three weeks. The control animals, which experience the same stimuli but not the complete loss of sleep, survive and display relatively few symptoms. Before they die, the sleep-deprived rats all exhibit the same horrible syndrome. This is characterised by a debilitated appearance, skin lesions, increased food intake, weight loss, increased metabolic rate, increased levels of the hormone noradrenaline, and declining body temperature. Some of these changes are symptomatic of excessive heat loss from the body, which has led some scientists to suggest that sleep is crucial, among other things, for the regulation of body temperature.

A progressive rise in metabolic rate (the rate at which the body consumes energy) is an early symptom of sleep deprivation. Sleep-deprived rats eat more to compensate for their rising energy expenditure, but their weight and body temperature nonetheless continue to fall. Feeding them an easily digestible diet helps to slow this process somewhat, but it does not prevent them from dying. An increase in appetite is one of the less obvious effects of sleep deprivation in humans as well. Might it be that chronic sleep deprivation is one of the factors helping to fuel the epidemic of obesity that is currently sweeping the USA, UK and other industrialised nations?

Body and soul

> Health may be as much injured by interrupted and insufficient sleep as by luxurious indulgence.
> William Kitchiner, *The Art of Invigorating and Prolonging Life* (1822)

What of people? Research on humans has stopped short of the lethal sleep deprivation imposed on rats and puppies, but it has delved systematically into the consequences of a few days' sleep loss. The results consistently show that moderate sleep deprivation has pervasive effects on the human body as well as the human mind. Sleep loss impairs vision, for example, causing blurring and errors in judging distances. It also triggers the familiar decline in body temperature that Marie de Manacéïne observed in her puppies, together with a reduction in blood glucose levels and changes in various hormones.

Set against this, sleep loss has surprisingly little impact on our ability to keep moving around and doing physical work. Moderate sleep deprivation does not greatly diminish our capacity for labour. Physically fit young adults can withstand several days of sleep deprivation without a substantial deterioration in their muscle strength, muscle endurance or cardiovascular responses to exercise. In one experiment, for example, the exercise capacity of young women was assessed following 60 hours without sleep. The sleep deprivation had no significant effect on their aerobic capacity or their endurance for exhausting exercise. In another study, researchers monitored two men while they played a marathon tennis match lasting a week, during which time the players got very little sleep. Although their mental performance deteriorated during the match, the players were able to sustain a high level of physical work. Our muscles can mostly keep going even when our brains are flagging.

Sleep deprivation does disturb many aspects of physiological functioning, however. Breathing is one example. A single night of sleep loss impairs breathing in healthy people, provoking a small but significant reduction in the maximum amount of air that can be exhaled after maximum inhalation. Sleep loss also leads to a substantial blunting of the normal respiratory responses to reduced blood-oxygen levels. After 30 hours without sleep there are marked deteriorations in the strength and endurance of the muscles used for breathing – as revealed, for example, by a reduction in the time for which people can breathe in against a sustained pressure. Such changes could be important in patients with respiratory diseases, who often suffer from chronic sleep loss. Sleep deprivation also slows the rate of cardiovascular recovery from intense exercise. When someone has been deprived of sleep for 24 hours, their breathing rate and oxygen uptake after a burst of intense exercise remain higher for longer.

Sleep loss is accompanied by many changes in body chemistry. People who have been kept awake for more than three days have altered liver functions, marked by large increases in the levels of key liver enzymes, changes in various types of fat and a rise in the amount of phosphorus circulating in the blood. Thyroid hormone levels are affected and biochemical changes can be detected at the level of gene activity.

Glucose metabolism is particularly perturbed by sleep loss. Healthy young men whose sleep was experimentally restricted to four hours a night for six nights became less tolerant to glucose. They took 40 per cent longer than normal to regulate their blood-sugar levels after eating high-carbohydrate food, and their ability to produce insulin fell by nearly a third – a condition resembling the early signs of diabetes. These abnormalities vanished after the men had slept for 12 hours. Fatigue-induced physiological changes like these could contribute to the development of chronic conditions such as diabetes, obesity and high blood pressure, all of which are associated with a shortened lifespan.

Sleep, immunity and health

> Our foster-nurse of nature is repose.
> William Shakespeare, *King Lear* (1605–6)

Some of the most interesting, least well understood, and potentially important consequences of sleep deprivation are found within the immune system. In short, lack of sleep can impair the body's immune defences and thereby make us more susceptible to infection by bacteria, viruses and parasites.

The evidence comes mostly from research with other species. In one experiment, for example, mice that were immunised against the influenza virus were resistant to infection if they were exposed again to the virus a week later. But if the immunised mice were deprived of sleep for seven hours immediately after being exposed to the virus, they were no more resistant to infection than mice that had not been immunised at all. A mere seven hours of sleep deprivation disturbed their immune response enough to erase the benefits of immunisation.

Some scientists have suggested that one reason why prolonged sleep deprivation is ultimately fatal is that it breaks down the animal's immune defences, making it vulnerable to infection by any opportunistic bacteria and viruses that happen to be in the vicinity. Experiments with rats have shown that following severe sleep deprivation, the lymph nodes and other organs are invaded by potentially dangerous bacteria, which appear to have migrated there from the intestines. However, the role of infection in killing sleep-deprived animals remains a controversial issue.

Sleep loss impairs the human immune system as well. Even modest sleep deprivation evokes measurable changes. One night of sleep loss lowers the activity of natural killer cells and reduces the numbers of several different types of white blood cells circulating in the bloodstream. (Natural killer cells are a special type of lymphocyte, or white blood cell, that attack virus-infected cells and certain types of cancer cells.) Depriving healthy adults of sleep for seven hours on one night suppressed their natural killer-cell activity by 28 per cent. It bounced back to normal after a night of uninterrupted sleep. Moderate sleep loss will also reduce the body's production of interleukin-2, a chemical messenger substance that plays an important role in regulating immune responses. After two or three days of sleep deprivation there is a marked decline in the responsiveness of lymphocytes and an even bigger fall in the activity of natural killer cells.

Sleep loss might play a role in the well-established connection between severe depression and impaired immune function. Depressed people generally sleep badly and have poorer immune responses. The more disrupted their sleep, the bigger the decline in their immune function. One study, for example, found that people who were suffering from depression following bereavement had fewer natural killer cells. The bereaved subjects were troubled by intrusive thoughts that often woke them or kept them awake during the night. The extent of the reduction in their natural killer-cell numbers was correlated with the amount of time they spent awake during the night: the more troubled someone was by their loss, the more disrupted their sleep and the fewer natural killer cells circulating in their blood. Sleep deprivation could be one of the mechanisms by which depression makes people more vulnerable to illness.

The relationship between sleep and immunity works in both directions. Not only does sleep affect the immune system, but the immune system also affects sleep. The immune reactions triggered by infection and illness can elicit alterations in sleep patterns. That is why infections are often accompanied by lethargy, loss of appetite, depressed mood and general malaise. Animals infected with influenza virus display a large increase in sleep about 24 hours after exposure to the virus. These changes in wakefulness are part of the body's defence mechanisms and assist the recovery process. Human experiments, in which noble volunteers were injected with bacterial toxins, found that sleep is highly sensitive to the activation of the immune defences. Low-level infection tends to promote deep sleep. However, a full-blown infection accompanied by fever induces lethargy but typically disrupts sleep. You might have noticed that you sleep more deeply for a night or two when your body is fending off a potential infection, whereas when you are in the throes of a galloping illness you feel exhausted but lie for hours without sleeping.

The immune response to infection stimulates the release of chemical messenger substances that act on the brain to induce malaise, drowsiness, loss of appetite and sleep. During infection, a substance known as interleukin-1 stimulates the brain to induce deep sleep, while other interleukins trigger the fever that often accompanies infections. They do this by adjusting the brain's temperature control centres – in effect, putting the body's thermostat on a higher setting. That is why we feel hot and sleepy when we have a bad infection. The fever response is a defence mechanism found in all animals: the rise in body temperature makes life harder for the offending bacteria or viruses, and the lethargy forces the infected organism to curl up in a dark corner and sleep until it has recovered. It all makes good biological sense.

The brain and the immune system are interconnected through an elaborate network of chemical and neural communication channels. One important link between sleep, immune function and psychological stress is the steroid hormone cortisol. Sleep deprivation and prolonged stress both provoke an increase in the level of cortisol. After one night of sleep loss, your cortisol levels would typically be raised by about 45 per cent the next evening. It is not good to have elevated cortisol levels for too long, since cortisol has a powerful suppressive effect on the immune system. The functioning of the immune system is also intimately bound up with the 24-hour sleep–wake cycle and the circadian rhythms in hormone levels. Various aspects of immune function fluctuate in tune with the circadian cycle. Anything that disrupts the normal cycle of sleep and wakefulness therefore tends to disturb the immune system, with potential consequences for the body's ability to defend itself against infection and disease.

The intimate relationship between sleep and immune function takes on a potentially huge practical significance when you consider how widespread sleep deprivation has become in society. Tired people are more likely to become sick people.

The Battle of Stalingrad

> O, I have passed a miserable night,
> So full of fearful dreams, of ugly sights
> William Shakespeare, *Richard III* (1591)

Prolonged sleep deprivation, uncontrollable stress and starvation make a lethal cocktail, as Hitler's troops found to their cost during the Battle of Stalingrad in World War Two. In June 1941 German forces invaded the Soviet Union and were soon threatening Moscow. The capture of Stalingrad on the River Volga became a key strategic objective. Stalin decreed that the city must be defended to the bitter end. The titanic struggle that ensued cost the lives of at least 800,000 Axis soldiers and 1.1 million Soviet soldiers.

The fight for Stalingrad (now renamed Volgograd) began in earnest in the summer of 1942, as the Germans advanced rapidly towards its suburbs. There was fierce Soviet resistance and the fighting dragged on into the harsh Russian winter. By September 1942 the battle was being waged at close quarters among the buildings, cellars, sewers and bunkers of 'the Stalingrad Academy of street-fighting'.

To increase the pressure on their opponents, the Soviet commanders ordered continual raids to be carried out by night. They did this partly because the Germans lacked protection from their air force at night, but mainly to induce exhaustion among the enemy. To augment the night raids, the Soviets fired flares indicating that an attack was imminent even when it was not. Their air force also attacked German positions every night. The Soviets kept up the psychological pressure throughout the night, with loudspeakers blaring out propaganda broadcasts, surreal tango music, or the sound of a ticking clock. The strategy was highly effective. 'We lie exhausted in our holes waiting for them,' wrote one German soldier. The German commanders begged for air support, citing their men's exhaustion.

The German troops' health started to deteriorate badly even before the dreadful Russian winter had begun to bite. There was a sharp rise in deaths from infectious diseases including dysentery, typhus and paratyphus. The actual prevalence of these diseases was not much worse than it had been a year earlier, but the numbers of infected men who were dying from them increased fivefold. It was as though the German soldiers had lost their capacity to resist infection. The Russians noticed this phenomenon, which they referred to as 'the German sickness'.

In November 1942 the Russians launched a huge and ultimately successful counteroffensive that soon had the Germans encircled within the ruined city. But the Germans were under orders from Hitler not to surrender, and so they fought on through December while the Russians gradually tightened the noose. Conditions for the German troops became appalling as their supply lines were cut off and the Russian winter froze them. There was hardly any food and little or no medical care.

In mid-December 1942 the German military doctors in Stalingrad noticed a new phenomenon: more and more apparently healthy troops were suddenly dying for no obvious reason. The Germans were unsure whether the deaths were the result of starvation, exposure, exhaustion or an unidentified disease. A German army pathologist named Girgensohn, who was sent to Stalingrad to investigate the problem, became convinced that a combination of exhaustion, stress, cold and lack of food was responsible for the much higher death rate. The Russian night attacks and round-the-clock activity had caused severe sleep deprivation, and Girgensohn concluded that this had amplified the effects of the food shortage by 'upsetting the metabolism' of the exhausted Germans. We know now that one symptom of prolonged sleep deprivation is a marked increase in metabolic rate and hence the requirement for food. Whatever the precise explanation, the pressure was too much for the Germans. In February 1943 the Battle of Stalingrad finally ground to a halt, as the crushed and starving remnants of the German army surrendered.

Sleepless in hospital

I have the feeling that once I am at home again I shall need to sleep three weeks on end to get rested from the rest I have had.
Thomas Mann, *The Magic Mountain* (1924)

Sleep is good for you and lack of sleep is bad. It therefore seems odd that hospitals, which are supposed to promote recovery, are usually dreadful environments for sleeping. 'The hospital bed,' wrote one historian, 'is one in which normal sleep is forbidden.' A *Punch* cartoon of 1906 shows a patient being told to 'wake up and take your sleeping-draught'. Things have improved since 1906, but not much.

Sick people really do benefit from sleep. We saw earlier how the brain and the immune system respond naturally to infection by inducing sleep. This helps the body cope with disease in several ways. The production of growth hormone occurs mainly during sleep, and growth hormone aids physical recovery by promoting the healing of mucous membranes and in other ways. The hormone melatonin, which is also produced at night, boosts immune responses, inhibits the growth of tumours and enhances resistance to viral infections. Conversely, sleep deprivation impairs immunity and slows the healing process. Given the importance of sleep for recovery, it is ironic that hospital patients are routinely subjected to conditions that make normal sleep almost impossible.

Sick people start with big disadvantages, of course. Pain is a powerful disrupter of sleep. Patients suffering from chronic, severe pain often become exhausted. Disrupted sleep is a common complication of burn injuries, for example. Studies have found that between half and three quarters of burns patients experience significant sleep disturbances. Sleep problems are common among cancer patients too. Fatigue can become one of the most distressing aspects of having cancer, severely reducing the quality of life. While medical treatments for cancer have advanced apace, efforts to improve patients' quality of life by alleviating their fatigue have lagged behind.

To make matters worse, tired people are more sensitive to pain. Sleep deprivation lowers pain thresholds, generating a vicious cycle in which pain disrupts sleep, the resulting sleep loss makes the pain feel even worse, and so on. An investigation of patients with burns injuries uncovered a systematic link between the quality of their sleep and subsequent pain. Patients who slept poorly during the night experienced more intense pain the following day, because the fatigue intensified their perception of pain.

One of the worst places imaginable if you need a good night's sleep is an intensive care unit (ICU). The combination of serious illness, serious drugs, constant monitoring, bright lighting and the after-effects of surgery ensures that ICU patients are often subjected to severe sleep deprivation. And yet the ICU houses the sickest people in the hospital, with the greatest need for sleep. American researchers conducted an experiment to see if implementing regular 'quiet times' during the day would help ICU patients to get more sleep. Each day for two two-hour periods, lighting levels in the unit were reduced and the staff made a concerted effort to minimise noise. The 'quiet-time' regime worked: the patients were 60 per cent more likely to sleep during these quiet periods than at other times of the day. More flexibility over when patients are given their medication can also help them to get more sleep.

Sadly, the apparent neglect of sleep in hospitals is another reflection of the general disregard for sleep in medicine and society as a whole. Lack of sleep really does have very little to recommend it.

PART III

Mechanisms

5

The Shapes of Sleep

Sleep rock thy brain.

William Shakespeare, *Hamlet* (1601)

Now it is time to peer beneath the surface at the strange state of existence known as sleep – or, to be more precise, the *two* strange states of existence known as sleep.

All of human life is spent in one of three states. You are very familiar with one of them: it is called the waking state, or consciousness, and it forms the subject matter for almost everything that has ever been said, written, acted, painted or composed about humanity. When scientists analyse the mind, when novelists dissect the human condition and when biographers portray the lives of eminent individuals, it is the waking state they almost invariably describe. However, there are two other distinct states of existence that together account for at least a third of each life. They labour under the workaday names of Rapid Eye Movement (REM) sleep and Non-Rapid Eye Movement (NREM) sleep, and we are about to take a closer look at them.

A night's sleep is a complex and cyclic process, comprising several distinct patterns of brain activity and behaviour, with alternating episodes of NREM sleep and REM sleep. We will follow the sleep cycle from the beginning, starting with the transition from the waking state. But before we do that, a quick word about how scientists know what is going on when we are asleep.

Measuring sleep

Brains wave.

Owen Flanagan, *Dreaming Souls* (2000)

The sleeping brain reveals what is going on inside itself in various ways, both electrically and chemically. Since the middle of the twentieth century, the main tool for monitoring sleep has been the electroencephalograph. This machine exploits the fortunate fact that varying patterns of electrical activity within the brain manifest themselves as varying patterns of voltage changes on the surface of the scalp.

The brain comprises billions of nerve cells, or neurons, and although the electrical activity of an individual neuron is too faint to be detected outside the skull, it is possible to monitor the gross patterns generated collectively by large numbers of neurons. These show up as minute voltage changes, which can be detected by electrodes stuck onto the scalp. (The very first electrodes were small pins that were stuck *into* the scalps of stoical volunteers.) Thus, the brain emits electrical signals revealing information about its inner state. These tiny voltage patterns are amplified and displayed as the familiar 'brain waves' of the electroencephalogram, or EEG. (Confusingly, the machine is called an electroencephalo*graph*, while the graph it produces is called an electroencephalo*gram*, or EEG. To avoid nausea, I will use EEG to denote both the machine and its output.)

The EEG was invented in the 1920s by a psychiatrist named Hans Berger. It really came into its own in the 1950s when, as we shall see, it enabled the discovery of REM sleep. Before the invention of the EEG, scientists could only assess sleep by observing overt body movements, or the lack of them. Scientists still find it useful to record sleepers' body movements, especially in studies of sleep patterns under natural conditions where the use of EEG would be too intrusive or too expensive. Nowadays, body movements are usually logged automatically, using a miniature recorder worn on the wrist.

Sleep laboratories use an extension of the EEG called the polysomnograph – a sort of somnolent variation on the polygraph. A polysomnograph records the EEG brain waves, together with other informative measures of the sleeper's physiological state and behaviour. Electrodes placed near the corners of the eyes detect movements of the eyeballs, producing a trace known as the electro-oculogram, or EOG. Other electrodes placed on the chin and neck monitor the muscle tone (producing an electromyogram, or EMG) while electrodes on the chest record the heart rhythms (electrocardiogram, or ECG). Additional devices may record whole body movements, breathing, the flow of air through the nose and mouth, and the concentration of oxygen in the blood. In the early days of sleep science, these measurements were recorded as continuous pen traces on miles of rapidly unfurling paper, but nowadays the outputs are usually stored digitally.

In recent decades, brain scanning has become an increasingly important tool in sleep research. One of the main brain-scanning techniques is called positron emission tomography (PET). PET scans reveal the local patterns of blood flow and oxygen uptake within small areas of the brain by measuring how rapidly the tissue is using energy. Unlike some brain-scanning techniques, PET does not require the subject to sleep inside a large, claustrophobia-inducing scanning device. It therefore allows scientists to monitor sleep under conditions that are slightly closer to normality. Even so, the sleeping subject's head needs to be kept absolutely still, which is usually achieved by pinning the head down with a special mask (the stuff of some people's nightmares).

Most measurements of sleep are made in specialised sleep laboratories rather than people's own homes. The underlying assumption is that the sleep patterns observed in the laboratory closely resemble the real thing. Fortunately, this turns out to be a broadly valid assumption. Comparisons have confirmed that for most people there is a reasonably good concordance between their sleep patterns at home and in the sleep laboratory. But there are some systematic differences. In particular, people tend to sleep for a slightly shorter period under laboratory conditions and to wake up slightly earlier than they would normally. They also have less bizarre dreams and fewer wet dreams. (Wouldn't you?)

Falling asleep again, what am I to do?

Warm beds: warm full blooded life.

James Joyce, *Ulysses* (1922)

Falling asleep is not an abrupt process, like turning off a light, although it can seem like that because you usually forget about it. Recordings of brain-wave activity and other physiological variables show that falling asleep is in fact a continuous process, which starts from a state of relaxed drowsiness and ends in the first or second stages of unequivocal sleep.

During that process of falling asleep you may find yourself temporarily suspended for several minutes between the worlds of waking consciousness and sleep. This transition phase is often accompanied by strange thoughts, dreamlike images and occasional hallucinations. In one of his short stories, Washington Irving described how the mind can roam far and wide while it is in this pre-sleep state:

> My uncle lay with his eyes half closed, and his nightcap drawn almost down to his nose. His fancy was already wandering, and began to mingle up the present scene with the crater of Vesuvius, the French Opera, the Coliseum at Rome, Dolly's Chop house in London, and all the farrago of noted places with which the brain of a traveller is crammed – in a word, he was just falling asleep.

These dreamlike experiences occur when we are in what is known as the hypnagogic state – a twilight zone partway between wakefulness and sleep. They are referred to as hypnagogic (or sleep-onset) dreams and they are distinct from ordinary dreams, which do not occur until much later in the sleep cycle. Hypnagogic dreams can contain all the basic elements of ordinary dreams, including bizarre plots, visual images and sounds, but there are fewer of these features in any one dream, suggesting that hypnagogic dreaming is a reduced version of normal dreaming. Similar dreamlike experiences can also occur at the other end of a night's sleep, during the transition from sleep to wakefulness, when they are known as hypnopompic dreams.

The hypnagogic and hypnopompic states are strange and fascinating. In comparison with true sleep and ordinary dreams, they are also poorly researched and poorly understood. Indeed, the English language does not even have a decent name for them – unlike Italian, which has a single word for both (*dormiveglia*, or 'sleep-waking'). In English, hypnagogic dreams are colloquially referred to by a variety of vague terms such as 'faces in the dark' or 'visions of half-asleep'. As we shall see in a later chapter, many famous creative flashes and inspired thoughts have come to people while in the hypnagogic state.

During hypnagogic dreams we may see strange sights, hear strange sounds and think strange thoughts. As our wakefulness fluctuates, we may wake up again and consciously remember the strange things we have briefly been dreaming. This hypnagogic nonsense sometimes includes bizarre, invented words. The sleep researcher Ian Oswald recalled waking from one hypnagogic dream with the phrase 'or squawns of medication allow me to ungather' running through his mind. On another occasion he found himself musing on the hypnagogic thought that 'it's rather indoctrinecal'. A British magazine once printed a collection of hypnagogic ramblings sent in by readers. These included the immortal verse 'Only God and Henry Ford have no umbilical cord.'

Hypnagogic thoughts and images can be more coherent, however. Charles Dickens often fell into a half-sleeping state while on one of his long nocturnal walks, and he could compose poetry while in this reverie. Dickens wrote of how, one night, he got out of bed at two in the morning and walked thirty miles into the countryside:

> I fell asleep to the monotonous sound of my own feet, doing their regular four miles an hour. Mile after mile I walked, without the slightest sense of exertion, dozing heavily and dreaming constantly ... It is a curiosity of broken sleep that I made immense quantities of verses on that pedestrian occasion (of course I never make any when I am in my right senses), and that I spoke a certain language once pretty familiar to me, but which I have nearly forgotten from disuse, with fluency. Of both these phenomena I have such frequent experience in the state between sleeping and waking, that I sometimes argue with myself that I know I cannot be awake, for, if I were, I should not be half so ready.

People who play lots of computer games sometimes experience 'screen dreams' as they fall asleep, in which they see vivid images of the game they have been playing. These screen dreams are also products of the hypnagogic state. The computer game Tetris, which requires the player to fit together coloured shapes as they cascade down the screen, is well known for provoking hypnagogic dreams. Scientists at Harvard Medical School investigated screen dreams by getting volunteers to play Tetris for several hours. Many of them experienced vivid dreams about Tetris as they fell asleep. Among the subjects in this experiment were five amnesiac patients who had extensive brain damage in their temporal medial lobes – brain regions crucial for conscious memory. Three of the five amnesiacs experienced hypnagogic dreams of Tetris even though they had no conscious memory of playing the game. This implies that the brain can generate hypnagogic dreams without input from conscious memory.

The length of time it takes you to fall asleep, once you have lain down and shut your eyes, is known as your sleep latency. It varies according to lots of factors. As we saw earlier, very short sleep latencies usually indicate sleep deprivation, whereas very long sleep latencies may signify other problems. A study of people living in rural Oxfordshire found that those with the longest sleep latencies typically described themselves as bored or mildly ill. You can make yourself fall asleep faster if you are minded to do so. Researchers proved this by giving volunteers a financial incentive to fall asleep quickly at various times during the day. The paid volunteers fell asleep faster than subjects who had no financial incentive.

Your body temperature has a big influence on how fast you fall asleep. A night's sleep is normally preceded by a drop in core body temperature, and scientists have established that this drop in temperature actively facilitates the onset of sleep. Under normal conditions, the maximum rate of decrease in body temperature occurs about one hour before the onset of sleep. If the onset of sleep is artificially delayed, the drop in body temperature is attenuated – further evidence that the two are closely linked. The polymath Benjamin Franklin realised the importance of a falling body temperature in triggering sleep. He set out this practical advice in a 1786 essay called 'The Art of Procuring Pleasant Dreams':

> Get out of bed, beat up and turn your pillow, shake the bedclothes well with at least twenty shakes, then throw the bed open and leave it to cool; in the meanwhile, continuing undressed, walk about your chamber. When you begin to feel the cold air unpleasant, then return to your bed, and you will soon fall asleep, and your sleep will be sweet and pleasant.

Benjamin Disraeli found that he was more comfortable when sleeping in hot weather if he used two beds, moving periodically from the hot, sweaty bed into the cooler one. Benjamin Franklin lit upon the same trick years earlier, but Franklin reckoned he needed *four* beds to be really cool. William Harvey, the seventeenth-century English physician who discovered the circulation of blood, similarly appreciated that cooling the body helps to induce sleep. According to his contemporary, the biographer John Aubrey, Harvey would tackle his insomnia by cooling himself down until he began to shiver:

> He was hot-headed, and his thoughts working would many times keep him from sleeping. He told me that then his way was to rise out of his Bed, and walk about his Chamber in his Shirt, till he was pretty cool, i.e. till he began to have a horror [began to shiver], and then return to bed, and sleep very comfortably.

Another scholar who stumbled across the sleep-inducing properties of cool air was Lord Monboddo, an eccentric eighteenth-century Scottish nobleman and pioneering anthropologist. When Samuel Johnson and James Boswell visited Monboddo, the great sage and his biographer were surprised by their host's behaviour. As Boswell recorded:

> Lord Monboddo told me he awaked every morning at four, and then for his health got up and walked in his room naked, with the window open, which he called taking *an air bath*; after which he went to bed again, and slept two hours more. Johnson, who was always ready to beat down any thing that seemed to be exhibited with disproportionate importance, thus observed: 'I suppose, Sir, there is no more in it than this, he awakes at four, and cannot sleep till he chills himself, and makes the warmth of the bed a grateful sensation.'

A less irksome way of achieving a similar effect is to take a hot bath an hour or two before bedtime. The bath will temporarily raise your body temperature. Over the following hours, your temperature will drop again and, all being well, this will help to trigger sleep. Experiments have confirmed that people do feel sleepier at bedtime after taking a hot bath. But the bath must not be too hot, too long or too close to bedtime, or it may have the reverse effect.

The fall in core body temperature that precedes sleep is accompanied by a small rise in the temperature of the hands, feet and other appendages. Blood vessels in your appendages dilate when you lie down to sleep at night, causing them to warm up. As they warm up so your body cools down, helping to send you off to sleep. Experiments have shown that warm feet assist the onset of sleep, bearing out another piece of folk wisdom. One of the best ways of predicting how quickly someone will fall asleep is to measure the temperature gradient across their body. The hands and feet are normally a degree or two cooler than core body temperature, but the temperature difference dwindles to nothing as sleep approaches.

A further demonstration of the linkage between warm appendages and the onset of sleep came from a study of people suffering from a disorder known as vasospastic syndrome. This condition is caused by faults in the physiological mechanisms controlling the peripheral blood vessels, which become less able to dilate. One of the main symptoms is cold hands and feet. As predicted, the cold-toed victims of vasospastic syndrome took longer than normal to fall asleep at night.

The importance of a declining body temperature means that artificial heat sources like electric blankets can disturb sleep. An electric blanket operating between the early hours of the morning and waking will typically increase your core body temperature by about 0.2 degrees Celsius. Even this small increase in body temperature is enough to disrupt sleep.

The sleep-inducing effect of a falling body temperature helps to explain why vigorous physical exercise, which raises body temperature, is not a good idea just before going to bed. It also reminds us why it is inadvisable to eat a large meal shortly before bedtime. The digestive processes that follow a large meal evoke a rise in metabolic rate, which in turn raises body temperature. In an ideal world, a large evening meal would be eaten at least three hours before bedtime. However, this helpful advice is of little use to the many people who work long hours and face long journeys to get home afterwards. They may have barely enough time to prepare and eat an evening meal before going to bed – another example of how lifestyles can conflict with good sleep.

One popular notion that reportedly fails to stand up to scientific scrutiny is that we fall asleep faster after orgasm. A group of enterprising researchers conducted an experiment in which they monitored the sleep of men and women under three different conditions: after the subjects had masturbated to orgasm, after they had masturbated without orgasm, and after they had simply read some nonsexual material. Recordings of their subsequent sleep yielded no evidence that masturbation, with or without orgasm, affected any aspects of sleep, implying that post-coital sleepiness has nothing to do with the attainment of orgasm. (You may find this hard to believe.) This is clearly an area crying out for more research.

The sleep cycle

> Sleep flooded over him like a dark water.
>
> Jorge Luis Borges, *Labyrinths* (1964)

Two broadly different states are conventionally bracketed together under the general heading of sleep: rapid eye movement (REM) sleep and non-rapid eye movement (NREM) sleep. NREM sleep is further subdivided into four different stages, based on their characteristic EEG patterns. Each sleep stage has its own distinctive pattern of brain activity. The various stages and types of sleep alternate cyclically throughout a night's sleep.

As you become sleepier, your EEG pattern changes. If you are tired enough, this can happen even when you are walking around and supposedly wide awake. The pre-sleep state of quiet restfulness is heralded by the appearance of brain waves of lower frequency and higher voltage, called alpha waves. If you are very sleepy, but still awake, your alpha waves will be accompanied by slow, rolling movements of your eyes. Nearly there. Then you are asleep.

The initial phase of sleep, which has the prosaic name of stage 1, typically lasts only a few minutes. Your muscles start to relax. If you are trying to sleep in a sitting position, the relaxation of your neck muscles will allow your head to slump forward, briefly waking you; your head straightens, you nod off again, and so on. That is why you 'nod off'. You can easily be roused into wakefulness from stage 1 sleep. If someone does wake you during stage 1 sleep you may be aware that you have been asleep, or you may be equally convinced that you have been awake the whole time. Stage 1 sleep is accompanied by a further slowing of the brain-wave patterns.

The next phase is known, predictably, as stage 2 sleep. This is signalled by the appearance on the EEG of two specific brain-wave patterns called K complexes and sleep spindles. The K complex is a single, strong wave that lasts less than a second. The sleep spindle is a brief burst of waves lasting less than a second. A sleep spindle on an EEG trace looks like a spindle moving along a loom, hence its name. During stage 2 sleep your eyes are still and your muscles are relaxed. You are less easily awoken by stimuli and you appear to an observer to be sound asleep. Altogether, stage 2 occupies about 45–50 per cent of a night's sleep.

About 10–15 minutes after the onset of stage 2 sleep, a new type of brain wave starts to appear on the EEG. These slow, rolling waves, known as delta waves, are of low frequency and high amplitude. They indicate the onset of stage 3 sleep, which lasts for only a few minutes. Stage 3 sleep is deep sleep. Your eyes are still and your muscles are relaxed.

Then comes stage 4 sleep, the deepest form of deep sleep. By now, your EEG consists largely of slow, rolling delta waves with a frequency of less than four cycles per second. Stage 4 sleep is accompanied by profound muscle relaxation. You are deeply asleep and least responsive to stimuli. This is when you are hardest to wake up. The first episode of stage 4 sleep usually lasts for 30–40 minutes. Stage 3 and stage 4 are both characterised by the distinctive low-frequency delta waves, and for that reason they are referred to collectively as slow-wave sleep. Together, they account for about 25 per cent of a night's sleep.

By now, a brain scan would reveal an overall decrease in activity in your brain. The reduction is by no means uniform, however: some areas have become much less active than others. The reductions in brain activity are most marked in certain areas of your cortex and upper brain stem. The regions of your cortex where activity is most muted are known to play major roles in the emotions and social behaviour. It is probably no coincidence that these faculties are especially impaired by sleep deprivation.

You have now been asleep for more than an hour. Then something extraordinary happens. The pattern of activity in your brain changes profoundly from the slow, rolling waves of deep sleep to what looks rather like the waking state. The delta waves disappear, to be replaced by a frenzy of high-frequency waves. From the EEG trace alone it might appear as though you have woken up. But you are still sound asleep. You have entered a new state known as rapid eye movement sleep, or REM ('rem') sleep. For obvious reasons, the state you have just left behind is called NREM ('en-rem') sleep.

The paradoxical world of REM

> That some have never dreamed is as improbable as that some have never laughed.
>
> Sir Thomas Browne, *On Dreams* (c. 1650)

REM sleep is sometimes called paradoxical sleep because it is so paradoxical. From the outside, REM sleep looks superficially like deep sleep: you still appear to be flat out and fast asleep. But your brain activity looks more like that of someone who is awake. REM sleep is characterised by a jumble of low-voltage electrical activity containing bursts of alpha and beta waves similar to wakefulness.

Your brain is intensely active and you are very probably dreaming, but your muscles are paralysed and your body is frozen. Your brain is being bombarded with stimuli, but these are not coming from the outside world: they are being generated within your brain. And, as the name suggests, your eyeballs are making rapid movements beneath your closed eyelids.

A brain scan now reveals a pattern of brain activity that is very different from when you were in NREM sleep. Many areas of your brain have become activated – indeed, some regions of your forebrain are now more active than when you were awake. The activity is particularly intense in areas that are involved in processing memories with an emotional content (the lateral hypothalamus and amygdaloid complex). The overall effect is one of greater brain activity, but the increase is not uniform: some parts of your brain, including your primary visual cortex, have become less active. Areas involved in processing visual associations when you are awake have formed a closed system. They are now operating in isolation from the sensory inputs and outputs that normally connect your brain with the outside world.

Your breathing is irregular and there are marked fluctuations in your heart rate and blood pressure. Your metabolic rate, pulse rate and blood pressure have all risen. The temperature of your brain has also risen. If you are male, your penis will be erect. If you are female, your vagina will be lubricating and your nipples will be erect. (The strange phenomenon of nocturnal erections will go under the micro-scope in chapter 7.) The rate at which your body is expending energy (your metabolic rate) is now much higher than it was during slow-wave sleep. REM sleep consumes a lot of energy, much of it in the brain. Your eyeballs are still moving rapidly under your closed eyelids, and the frequency with which they move depends partly on how tired you are. Your rapid eye movements will be somewhat less rapid if you are tired and catching up on your sleep.

If you are woken now and asked, there is a very high probability (80–90 per cent) that you will say you have been dreaming. REM sleep is closely associated with dreaming. Indeed, it is sometimes loosely referred to as 'dream sleep'. (There will be much more about dreaming in chapters 9 and 10.)

One theory is that during REM sleep we 'watch' our dreams, our eyes tracking the movements of the dream images. The rapid eye movements in REM sleep do sometimes relate to the content of a dream. For example, William Dement observed one dreamer in his Stanford sleep laboratory make 26 regular, side-to-side eye move-ments during an episode of REM sleep. When Dement woke the dreamer, the man reported that in his dream he had been watching two friends play table tennis. Unfortunately, the appealing idea of a direct correspondence between our dream images and our rapid eye movements does not seem to hold up more generally. More recent research with monkeys has found that their eyes move more or less independently of one another during REM sleep, and the two eyes are often aiming in different directions. Their lines of sight can be out of kilter by as much as 30 degrees, making it impossible for them to be 'watching' the same dream image. Moreover, the individual eye movements often occur out of synchrony. This casts doubt on the beguiling theory that we 'watch' our dreams.

You are still in REM sleep and your body is immobile, as though frozen. Your muscles are paralysed by special mechanisms that evolved to prevent us from moving around during REM sleep. This muscular paralysis is actively imposed by nerve signals sent out from your brainstem, which block sensory inputs and muscular action. The net effect has been likened to a car with its engine revving but the gears disengaged. Your eye muscles are excluded from this general paralysis because they are separately controlled by nerve fibres linked directly to your brainstem. That is why your eyes are able to move rapidly while the rest of your body remains rooted to the spot. Fortunately, the muscles responsible for your breathing are also free to move.

The fact that body movement is actively suppressed during REM sleep was first demonstrated by the French scientist Michel Jouvet, in a classic experiment with cats. Jouvet cut certain nerve fibres that transmit the inhibitory signals during REM sleep. This had the effect of removing the paralysis. When the sleeping cats entered REM sleep they would get up and move around, as though acting out a dream. Instead of lying still they would stalk, pounce on, kill and eat imaginary prey. EEG recordings confirmed that the cats were still asleep, and they would pay no attention to real stimuli such as food. They appeared as though they were sleepwalking, but in fact they were dreaming. Some unfortunate people who have brain damage in certain parts of their brainstem also lack the normal muscular paralysis in REM sleep. They consequently behave rather like Jouvet's cats. They get out of bed while still asleep and act out their dreams, sometimes doing wild and dangerous things. They may even need to be physically restrained at night to protect themselves and others from injury.

REM and NREM sleep are in many respects as distinct from one another as they both are from wakefulness. Apart from the dramatic differences in brain activity, for example, there is no muscular paralysis during NREM sleep. (For that reason, nocturnal activities such as sleepwalking, sleeptalking and tooth-grinding all occur during NREM sleep, not REM sleep.) The idea that emerges, then, is one of three different modes of existence: the waking state, REM sleep and NREM sleep. There is little new under the sun, of course. The modern concept of dream sleep as a third state of existence was foreshadowed in ancient Indian philosophy, which distinguished between the waking state, dream sleep and dreamless sleep.

REM sleep is a distinct state of being that occupies between one and a half and two hours of every day of our lives. And yet, astonishingly, its very existence remained unknown to science until the second half of the twentieth century. REM sleep was discovered in 1952 by a young graduate student called Eugene Aserinsky. He was working at the time in the world's first specialist sleep laboratory, at the University of Chicago. Aserinsky stumbled across REM sleep when he was monitoring the sleep EEG of his own eight-year-old son Armond. When he first saw the dramatic changes in the EEG that signify the onset of REM sleep, Aserinsky assumed his equipment must have malfunctioned. His research supervisor Nathaniel Kleitman (commonly recognised as the founder of modern sleep research) was even more sceptical to begin with, since the existence of REM sleep torpedoed his own theory that sleep was fundamentally a passive state. But evidence is evidence, and Kleitman was soon convinced that his student had discovered a profoundly important phenomenon. The following year, 1953, Aserinsky and Kleitman published their seminal paper, announcing the discovery of REM sleep, in the journal *Science*. 1953 was also the year that Francis Crick and James Watson unravelled the molecular structure of DNA, and almost half a century after Einstein published his special theory of relativity.

Why did it take science so long to stumble across a fundamental state of existence that accounts for several years of every person's life? The failure of scientists to notice REM sleep before 1952 is even more peculiar when you consider that its discovery did not even require the use of EEG or other electronic technology. Thanks to the bulging shape of the human eyeball, the rapid eye movements that are symptomatic of REM sleep are visible to an observer. The erect penis that normally accompanies REM sleep in males should have been even easier to spot. But no one was looking.

The reasons why REM sleep was overlooked for so long probably boil down to a mixture of lack of interest, unchallenged assumptions about the nature of sleep and old technology. The late Nathaniel Kleitman once said that REM sleep would have been discovered much earlier, but for the fact that sleep researchers always switched off their polysomnographs after their subject had been asleep for a while. They did this to save paper. Nowadays, polysomnographs can store the data digitally, but in the early days of sleep research the measurements were recorded as continuous pen traces on a roll of rapidly moving paper. A full night's EEG recording could easily generate more than half a mile of paper, so the desire for economy was understandable. Unfortunately, the first episode of REM sleep does not normally start until someone has been asleep for more than an hour, by which time the machine had been switched off and the researcher was probably tucked up in bed. From 1953 onwards, however, scientists started leaving their polysomnographs running all night.

The remarkable failure to notice REM sleep also reflected a common misconception that the brain became largely dormant during sleep. Scientists, including Aserinsky and Kleitman themselves, were simply not expecting to find anything like REM sleep and so they did not look for it. Soon after the discovery of REM sleep, further groundbreaking work by Nathaniel Kleitman and his students, including William Dement, uncovered the cyclical nature of sleep and the linkage between REM sleep and dreaming. But let us now return to your night's sleep. If you recall, you have just been through stages 1–4 of NREM sleep and your first episode of REM sleep. Congratulations – you have just completed your first sleep cycle.

The sleep cycle continued

> Man was not made to enjoy endless activity; Nature destined him for an interrupted existence.
> Jean-Anthelme Brillat-Savarin, *The Physiology of Taste* (1825)

A full night's sleep is made up of four or five sleep cycles. Each cycle starts with a descent into light sleep (NREM stages 1 and 2), then slow-wave sleep (NREM stages 3 and 4), then a brief semi-awakening to light sleep, then REM sleep. One complete cycle, from the start of stage 1 sleep to the end of REM sleep, usually lasts about 90 minutes in adults. (In babies it lasts about 60 minutes.) Eight hours of sleep, allowing for a few interruptions, should therefore give you enough time to complete five 90-minute cycles before you awake.

During successive sleep cycles the episodes of REM sleep become progressively longer, while the episodes of slow-wave sleep become shorter. Slow-wave sleep therefore predominates during the first part of the night and REM sleep predominates in the early morning. The final sleep cycle of the night often ends with an episode of REM sleep, which means you are more likely to remember dreaming if you wake spontaneously in the morning. If, on the other hand, you are rudely awoken at some arbitrary time by the clatter of an alarm clock, then it will be a matter of chance whether you surface from REM sleep or NREM sleep.

If you are a typical, healthy adult, your full night's sleep will comprise roughly 50 per cent shallow sleep (NREM stages 1 and 2), 25 per cent slow-wave sleep (NREM stages 3 and 4) and 25 per cent REM sleep. By the age of 75 you will therefore have spent about six years of your life in REM sleep – longer than some marriages.

The proportions of time occupied by the different types of sleep vary considerably according to how old you are. REM sleep progressively declines with age, both in absolute terms and as a fraction of total sleep. When you were a foetus you spent 60–80 per cent of your time in REM sleep. In the first few weeks after you were born, you probably slept for 12–16 hours a day, of which 50 per cent was REM sleep. By adulthood, however, REM sleep accounts for only 25 per cent of your sleep. If you are older, the amount of slow-wave sleep (stages 3 and 4) will have declined at the expense of shallow sleep (stages 1 and 2). In old age, slow-wave sleep diminishes to only 5–10 per cent of total sleep time.

The 90-minute sleep cycle has a less obvious echo during the waking state, in the form of rhythmic variations in your waking activity and behaviour patterns. Careful statistical analysis of your behaviour and mental state during the day would reveal a 90-minute cyclic variation in many activities, known as an ultradian rhythm. Similar ultradian rhythms in waking behaviour have been detected in other species as well. As in humans, they are roughly the same length as the sleep cycle. In smaller animals the ultradian rhythm and sleep cycle are usually shorter. In adult mice, for example, both cycles last about 20 minutes, whereas in baby mice both cycles last about 10 minutes.

Periodically throughout the night, you will partially awake, briefly bobbing up to shallow sleep or the edge of wakefulness before sinking down into deeper sleep again. These arousals are usually too brief and incomplete for you to remember, so you will rarely be consciously aware of them. Nonetheless, they enable your brain to scan your environment for signs of potential danger before submerging back into deeper sleep. They have been likened to a submarine periodically rising to periscope depth to check out the surface. During these partial awakenings you shift your body position. You will change your whole posture 30 or 40 times during the course of a night's sleep without ever noticing. The majority of people spend most time sleeping on their right side and least time sleeping on their front.

If your sleep is disrupted for some reason, your metabolic rate will rise. Measurements of energy expenditure in sleeping people show that the arousals and brief awakenings that punctuate a normal night's sleep are accompanied by brief increases in metabolic rate. If someone artificially disrupts your sleep by repeatedly waking you, your energy expenditure for the whole night will rise significantly, even if your total sleep duration is only slightly curtailed. However, this is not recommended as a technique for losing weight.

Other processes are also underway. During the course of your night's sleep the tear film over your eyes is undergoing complex changes in composition. There are progressive increases in the number of tissue cells and the amount of the protein albumin in your tears. These changes contribute to the telltale 'custard corners' that you may discover when you first open your bleary eyes in the morning.

While you have been sleeping, all your sensory systems – vision, sound, smell, taste and touch – have remained active. Sensory stimuli of all types are capable of influencing your sleep. But there have been major changes in how those sensory systems operate and how your brain processes the incoming sensory information. Your sleeping brain can recognise and respond to salient stimuli such as the sound of your own name.

More than 40 years ago, experiments with Oxford University students demonstrated that people could respond selectively to complex stimuli during sleep. Before they went to sleep, the students were instructed to respond, by clenching their hand, if they heard their own name or another, specified name during their sleep. While they then slept, the researchers played them recordings of long lists of names, read out at intervals of several seconds. The sleeping subjects often responded by clenching their fists after hearing their own name or the specified target name. And even if they did not make any physical response, their EEG patterns almost always changed when they heard the names, indicating that the brain had recognised them. The same sounds provoked no responses when they were played backwards. Occasionally, a subject would react to one of the other names in a list if it happened to be of personal relevance, such as the name of a partner or heart-throb.

More recent experiments have shown that the sound of a sleeper's own name elicits a highly specific response in the sleeping brain, very similar to the response it normally evokes during the waking state. More generally, different sounds evoke specific patterns of activity in the sleeping brain. A pure tone elicits activity in brain areas responsible for processing sound stimuli, while the sound of the sleeper's own name additionally elicits activity in areas that deal with language, memory and emotion. Sounds that have emotional significance for the sleeper trigger greater activity in certain areas of the brain (notably the left amygdala and left prefrontal cortex) than emotionally neutral stimuli.

People can also detect and respond to smells during sleep. In one experiment, volunteers were exposed to short bursts of peppermint odour while they slept. The distinctive smell provoked small but consistent changes in their EEG patterns, body movements, heart rate and breathing. They were responding to the smell even though they were sound asleep.

The basic features of the sleep cycle are broadly similar in other species, although the details vary. REM sleep is observed in mammals and birds, but the balance between REM and NREM differs considerably between species. Most species have more NREM and less REM than humans. Mice and other small mammals, for example, spend a large proportion of their time asleep, but most of that is slow-wave NREM sleep. The blind mole rat, which lives almost its entire life underground, spends on average 52 per cent of its life asleep, of which 15 per cent is REM sleep. The lizard *Ctenosaura pectinata* sleeps more than 16 hours a day, with REM sleep accounting for less than 1 per cent of this. The Amazonian manatee, a marine mammal, spends more than a quarter of its life asleep, but less than 4 per cent of its sleep is REM sleep.

Birds, especially those that roost or perch in high places, have little REM sleep. And the REM sleep they do have is not accompanied by the complete loss of muscle tone, which is just as well if you are trying to sleep on a cliff face or a slender branch fifty feet up a tree. Many birds spend several hours a day asleep, but the great majority of it is NREM sleep. Their REM sleep occurs in very brief bursts. In parakeets, for example, NREM sleep is punctuated by very short episodes of REM sleep that last on average only six seconds. Emperor penguins have a relatively large amount of REM sleep by bird standards. They spend 10–11 hours a day asleep, of which 12–14 per cent is REM sleep. As in other birds, their REM sleep is fragmented into many short episodes, each of which lasts only 8–10 seconds. Perhaps they dream of snow.

Waking up

> Golden slumbers kiss your eyes,
> Smiles awake you when you rise:
> > Thomas Dekker, *Patient Grissil* (1603)

The night is reaching an end and you are about to wake up. For some time beforehand your body will have been preparing you for the horrors of waking and the rigours of the day ahead. Your body temperature will be rising and various hormone levels will be changing to help get you in gear. Even so, waking may not be a pleasant experience. Some people claim to enjoy waking up every morning, though it is difficult to imagine how. Perhaps it is because they have time to lie there, dozing and contemplating their dreams, before prising themselves out of bed. This is how Charles Dickens described the limbo between sleep and wakefulness:

> There is a drowsy state, between sleeping and waking, when you dream more in five minutes with your eyes half open, and yourself half conscious of everything that is passing around you, than you would in five nights with your eyes fast closed, and your senses wrapt in perfect unconsciousness.

Most people are capable of waking themselves at a predetermined time without the aid of an alarm clock. This ability has been confirmed under laboratory conditions, although the accuracy is not great. In one experiment, seven people who claimed they could awake at a chosen time were tested in a sleep laboratory. They were asked to wake at 3:30 a.m. one morning and at 5:30 a.m. on another occasion. They mostly woke within 20 minutes of the target time. (It has to be said that asking anyone to wake up at 3:30 a.m. is a harsh test.)

Research has also revealed that people can exert some degree of conscious control over the normal daily cycle in their hormone levels. During the course of a night's sleep the pituitary gland produces increasing amounts of a hormone called ACTH, starting about an hour before the usual time of waking. ACTH controls the secretion of various other hormones and is an important component of the waking process. This early-morning rise in ACTH secretion prior to waking is part of the physiological alarm clock that prepares your sleeping body for the day ahead. Researchers instructed healthy volunteers to wake at different times and then measured their ACTH levels. When the volunteers were told to wake at 9 a.m. their ACTH levels duly began to rise at about 8 a.m., one hour in advance. And when the volunteers were told to wake at 6 a.m. their ACTH levels obligingly started rising in anticipation at 5 a.m. This demonstrated that conscious thought before going to sleep can influence physiological processes that occur during subsequent sleep.

Not all of us feel as sharp as a razor the instant we wake up. Indeed, some of us spend the first hour or two after crawling out of bed in a state of zombie-like befuddlement. Again, there are similarities with the effects of alcohol. In *Remembrance of Things Past*, Marcel Proust conveys the mental confusion we often experience when waking suddenly from sleep or a nap. Madame Cottard has fallen asleep in her chair after dinner. When she emerges blinking from the depths of sleep, it takes her a while to realise that she is awake and not dreaming:

> 'My bath is just right,' she murmured. 'But the feathers on the dictionary . . .' she cried, straightening up. 'Oh, Good Lord, how foolish I am. I was thinking about my hat, and I must have said something ridiculous. Another minute and I would have dozed off. It's the heat from the fire that does it.' Everybody began to laugh, for there was no fire.

(Actually, it is unlikely that Madame Cottard would have been waking from REM sleep after merely dozing in a chair. Entering REM sleep takes time and, for most people, a near-horizontal posture. Her dream would probably have been of the hypnagogic or hypnopompic variety.)

The temporary impairment in alertness and performance that we all experience immediately after waking up is known as sleep inertia. During the first hour or so after awakening the mists gradually clear, our alertness increases and our reactions become sharper. The improvement is most rapid during the first half-hour. Measurements have shown that our ability to make good decisions during the first few minutes after waking abruptly is about 50 per cent below par. Even half an hour after waking, mental performance may still be down by 20 per cent. Sleep deprivation amplifies the effects of sleep inertia, so if you have not been getting enough sleep your befuddlement on waking will be even worse. (Bear that one in mind if a sleep-deprived doctor is woken at four in the morning to care for you or a loved one.) Sleeping Beauty's ability to be sharp and lovely the instant she awoke from a hundred years' sleep was truly fantastic.

The final traces of sleep inertia can linger for up to three hours after awakening. Surprisingly, one experiment found that the rate at which sleep inertia dissipates is unaffected by whether you stay in bed, or immediately get up, shower and eat breakfast. (That will be my excuse from now on.)

The quality of sleep

The pleasant land of counterpane.
Robert Louis Stevenson, *A Child's Garden of Verses* (1885)

How can we assess the quality of sleep? Individuals have widely differing subjective experiences of sleep, but how do these relate to their actual sleep patterns? One basic measure of sleep quality is known as sleep efficiency. It is simply the percentage of time, between falling asleep and waking the next morning, that you actually spend asleep. Sleep efficiency is a measure of how continuous and uninterrupted your sleep has been. A normal night's sleep is punctuated by numerous brief arousals and occasional awakenings that can easily add up to 20 or 30 minutes over the course of eight hours. These arousals and partial awakenings, which are perfectly normal, are rarely remembered the next morning.

A bit of tiredness will help you to fall asleep and stay asleep, thereby improving your sleep efficiency. If, on the other hand, you spend too long lying in bed you will find it harder to sleep and your sleep efficiency will drop. A healthy sleeper would normally expect to have a sleep efficiency in the region of 90–95 per cent, whereas an insomniac might have a sleep efficiency of only 70 or 80 per cent. An unusually high sleep efficiency (say, more than 95 per cent) may indicate sleep deprivation.

Your sleep efficiency has a major bearing on your subjective sense of sleep quality. Researchers conducted an experiment in which volunteers slept according to irregular schedules in a sleep laboratory. The schedules were specially designed to provide normal amounts of time in bed but a variable quality of sleep. The results showed that the subjects' subjective assessment of their sleep quality was closely correlated with their actual sleep efficiency. Subjects rated their sleep as 'rather good' only if it had a sleep efficiency of at least 87 per cent. In another study, researchers compared polysomnograph recordings of individuals' sleep patterns against their subjective experiences, which they had recorded in sleep diaries. The main determinants of sleep quality were sleep efficiency and the depth of sleep, as measured primarily by the amount of slow-wave sleep. Two other factors that influence our subjective perceptions of sleep quality are total sleep duration and immobility. As we saw earlier, a normal night's sleep is made up of periods of immobility punctuated by shifts in body posture. Recordings have shown that the longer someone lies still, the better their sleep is perceived to be.

Meditating – or only sleeping?

> Close up his eyes and draw the curtain close;
> And let us all to meditation.
>
> William Shakespeare, *Henry VI Part Two* (1592)

Are there any other states of being besides waking consciousness, NREM sleep and REM sleep? For many centuries the Buddhist, Taoist, Vedic and other traditions have practised various forms of meditation that induce a state of profound relaxation and mental tranquillity. Might meditation constitute a fourth state? The second half of the twentieth century saw a surge of popular interest in a particular form of meditation known as transcendental meditation, or TM. This interest, which was partly fuelled by the enthusiasm of The Beatles and other cultural icons, stimulated scientific research into its biological basis.

Like sleep, meditation entails withdrawing the mind from the distractions of sensory stimuli and idle thoughts. In TM, for example, the meditator silently repeats a short word or phrase known as a mantra. This mental repetition of the mantra helps to expel the hurly-burly of conscious thought and allows the meditator to enter a deeper level of mental and physical relaxation, free from the normal cacophony of the waking mind. But is meditation a unique mental state, as some of its proponents claim, or might it be more akin to the preliminary phases of sleep?

Numerous scientific investigations of meditation have uncovered surprisingly little support for the belief that it invokes a unique state of consciousness. In fact, meditation (or at least, the forms practised by many Western adherents) looks more like relaxed wakefulness or the early stages of shallow sleep. More specifically, it resembles the hypnagogic state – that strange and poorly understood twilight zone between wakefulness and sleep. Recordings of EEG patterns during meditation often look like those of people on the verge of falling asleep. One study, for example, analysed the EEG patterns of 13 experienced practitioners of TM. Their EEG patterns were characteristic of a state between wakefulness and drowsiness. Meditation had at least one distinguishing feature, however, in that the experienced meditators displayed an average EEG frequency that was about one cycle per second slower than that of non-meditators. Other than this, meditation was not accompanied by a unique EEG pattern. More than one study of meditators has found that most of them were actually asleep for a considerable proportion of the time they were meditating.

The picture is confused, if not confusing. EEG patterns during meditation vary considerably, both between individuals and over time. Some people display patterns of brain activity that are virtually indistinguishable from the normal waking state. But some individuals who regularly practise TM exhibit subtle changes in their EEG traces, both during TM and when they are not meditating. This suggests that persistent meditation can induce long-term changes in brain activity. It would certainly be premature to conclude that meditation is no different from sleep.

Another common belief is that the meditative state produces a large reduction in metabolic rate. Again, the evidence is mixed. Meditation is accompanied by a physiological relaxation response, involving reductions in heart rate and breathing. However, the reductions in metabolic rate are fairly small and can probably be accounted for by muscle relaxation alone. Brain scanning supports the conclusion that meditation involves a profound relaxation response, producing a state of tranquil alertness. Brain scans have revealed that the level of activity actually increases in several regions of the brain during meditation; the more active regions are those involved in the control of attention and physiological arousal, as would be expected in a relaxation response. In line with this, researchers have also detected an increase in blood flow to the brain during meditation. Regardless of its physiological basis, many practitioners find that meditation is an effective way of reducing anxiety and stress.

Perhaps the final word should be left to Paul McCartney, an experienced practitioner of TM. In 1968 The Beatles travelled to India to imbibe the techniques of TM from the Maharishi Mahesh Yogi at the Maharishi's ashram in the foothills of the Himalayas. The four Beatles, and many others in Western countries at that time, saw meditation as a potential route to happiness, truth and inner knowledge. Some of them later became disillusioned. Paul McCartney nonetheless continued to practise meditation and regarded it as highly beneficial. Years later he described meditation as producing for a few minutes at a time the most pleasant and relaxed sensations he had ever experienced, and he believed that the world would be a happier place if more people took it up. But he also had a measured grasp of the reality. He gave his biographer this refreshingly balanced description of the meditation experience:

> You sit still, I suppose you regulate your breathing and, if nothing else, you rest your muscles for twenty minutes. It's like a lie-in . . . Are you dreaming or are you awake? There's a nice little state that they recognise halfway between it.

6

Morpheus Undressed

There ain't no way to find out why a snorer can't hear himself snore.

Mark Twain, *Tom Sawyer Abroad* (1891)

Once upon a time, sleep was thought to be a passive state, defined by an absence of waking consciousness rather than the presence of anything more interesting. According to this traditional but utterly wrong view of sleep, the brain simply lapsed into inactivity when it stopped receiving stimulation from the sense organs at night. Indeed, sleep was often likened to death. In ancient Greek mythology Nyx, the goddess of Night, had twin sons called Hypnos (Sleep) and Thanatos (Death). Morpheus, the bringer of dreams, was the son of Hypnos and hence the nephew of death.

We know better now. Sleep is not caused by an absence of stimulation and it is much more than merely the absence of wakefulness. The brain puts itself to sleep and remains busy while it is asleep; as we saw in the previous chapter, some parts of the brain are more active during sleep than when we are awake. The brain actively generates the state of sleep and then imposes variations on it. Our propensity to be awake or asleep varies according to a daily cycle, controlled within the brain. And individuals differ in their deep-rooted tendency to go to bed early or late, and to sleep a lot or a little. We shall now see why.

The rhythms of life

What hath night to do with sleep?

John Milton, *Comus* (1637)

Virtually all of our bodily functions, ranging from cell biochemistry to social behaviour, exhibit a 24-hour cycle governed by an internal clock. Scientists have known for at least two centuries, for example, that our body temperature fluctuates according to a regular daily cycle, or circadian rhythm. Assuming you do not routinely work night shifts, your temperature will peak in the afternoon or early evening and fall to a trough, about one degree Celsius lower, in the early hours of the morning.

The most obvious circadian rhythm is the daily cycle of sleep and activity displayed by all animals. But more or less any biological process that can be measured, including immune function, hormone levels, digestion and urine production, display some sort of daily rhythm. There is even a circadian rhythm in human body height. You shrink slightly when you are standing up and expand again when you are lying down in bed. Depriving adults of sleep blunts that rhythm and very slightly reduces their average height (another reason for getting enough sleep, albeit not the most compelling).

Circadian rhythms are a universal characteristic of the living world. They are found in insects and even bacteria. Plants too have circadian rhythms. Indeed, probably the first ever experiment to investigate biological rhythms was conducted on plants. In 1728 the French astronomer Jean de Mairan reported that the leaves of mimosa plants would open during the day and close at night, and that this daily cycle of opening and closing continued even when the plant was placed in a completely dark room, away from the alternating daylight and darkness. From this observation de Mairan correctly concluded that the plant had its own internal rhythm.

Humans and other animals are equipped with internal biological clocks that control these circadian rhythms. In mammals, the principal on-board clock is located in a pinhead-sized region of the brain called the suprachiasmatic nucleus, located deep in the centre of the brain above the pituitary gland. This internal clock controls our daily rhythms in sleep, activity, body temperature and hormone levels.

Someone once said that the 24-hour circadian rhythm *evolved* because the earth *re*volved. Obviously, it is no coincidence that the circadian rhythm coincides with the 24-hour cycle of darkness and light on Planet Earth. The biological clocks in humans and other species are built to run with a period of about 24 hours, but they are also responsive to environmental time cues, and they are reset each day by those cues to ensure that they remain in close synchrony with the physical world. Light is the main cue for resetting biological clocks, which is why, for example, nocturnal animals become deeply confused if they are exposed to bright light in the middle of the night.

There is a good biological reason why our internal clocks are readjusted every day with reference to environmental time cues. The duration of daylight varies with the seasons, and animals must therefore adjust their activity cycles so that they sleep and wake at the right times. If animals slept and woke according to a fixed and invariant internal clock, nocturnal animals would find themselves asleep in daylight during the summer, while species that are normally active by day would find themselves waking in darkness during the winter. The on-board clock therefore needs to resynchronise itself continually with the changing environment. A hypothetical animal that dispensed altogether with its on-board clock and relied solely on environmental cues, such as the amount of daylight, would also be at a disadvantage. The benefit of having an internal clock is that it enables physiological processes to adjust in anticipation of periods of sleep and activity. When we wake in the morning, our bodies have already been preparing us for wakefulness for some time. What is more, animals that depended solely on daylight to determine their daily periods of rest and activity would find themselves being frequently misled by unreliable cues – for example, during overcast or inclement weather (in other words, most days in the UK).

The biological benefits of circadian rhythms were highlighted by a serendipitous experiment, when scientists were observing ground squirrels living in a large desert enclosure. Half the squirrels had been treated surgically to destroy their suprachiasmatic nuclei, the site of the internal clock. Ground squirrels are normally active by day and sleep at night. But the treated squirrels, lacking their biological clock, were often awake and active at night. One night, a feral cat broke into the enclosure and killed many of the squirrels. When the scientists inspected the slaughter they found that the predator had killed twice as many of the clockless squirrels: 60 per cent of them had died, as against only 29 per cent of the intact animals. The intact squirrels had been sleeping safely in their burrows when the predator struck, whereas many of the treated squirrels had been up and about and therefore vulnerable.

The scientists went on to demonstrate that the circadian rhythm is important for survival in the natural environment. They monitored the activity patterns and mortality of chipmunks living in a forest in the Allegheny Mountains. A large number of chipmunks were fitted with miniature radio-tracking devices before being released into the area. Some of them had previously had their internal clock destroyed surgically. These clockless chipmunks were more active at night and they proved to be much more vulnerable to weasels and other predators, probably because they were much easier for nocturnal predators to find.

Back to humans. Scientists used to believe that when daylight and all other external time cues were experimentally removed, the human internal clock ran inherently slow, with a natural period of about 25 hours. This belief in a slow-running biological clock stemmed from German research in the 1960s, which found that when volunteers were isolated for weeks in a cave their daily cycle of sleeping and waking lengthened to about 25 hours. However, in the late 1990s Charles Czeisler at Harvard University established that this well-known fact is wrong. The subjects in the original experiments had been allowed to use artificial lighting that was of low intensity but still sufficiently bright and prolonged to affect their internal clocks. At that time scientists thought, incorrectly, that only the much higher light intensities of daylight could influence the human biological clock. The volunteers used to leave lights switched on until they felt sleepy, which had the unintended effect of lengthening their daily cycle by about an hour. It is now known that even the puny light intensity produced by electric light bulbs can eventually influence the setting of our internal clock.

More recent experiments, which were carefully designed to eliminate the distorting effects of artificial light cues, have proved that the human biological clock is actually rather accurate. It turns out to have an average free-running period of 24 hours 11 minutes, with relatively little variation between individuals.

The discovery that ordinary electric lighting can influence our internal clock has important practical implications. It means, for example, that our habit of exposing ourselves to artificial lighting until late at night will subtly affect our biological rhythms, just as it did for the cave-dwelling volunteers in the German experiment. Thanks to the misleading light cues, our internal clock will be trying to operate on a 25-hour day, while the real world will be reminding us that there are only 24 hours in the day. It has been suggested that the malign influence of artificial lighting might contribute to the early-morning grogginess that many of us experience, especially on Monday mornings after two or three late nights.

Light resets our biological clock every day, but light is not the same thing as vision. The parts of your brain that respond to the daily alternations between light and darkness are not the same as those that enable you to see. The eye conveys information to different regions of the brain for different purposes, and seeing objects is different from sensing whether it is night or day. Experiments with rats have shown that a region of the brain (called the superior colliculus-pretectum) that plays no role in vision is nonetheless crucial for mediating the influence of light on circadian rhythms.

Your eyes may not even be the only channel by which light reaches your brain and synchronises your biological clock. Scientists at Cornell University discovered that shining pulses of very bright light onto the backs of volunteers' knees provoked a shift in their circadian rhythms. One possible explanation for this curious phenomenon is that very bright light affects the blood in capillaries close to the surface of the skin, stimulating the release of a substance such as nitric oxide that could in turn influence the brain. The Cornell research remains controversial, however, so it would be premature to conclude that exposing the backs of your knees to bright light will cure jet lag.

In the seventeenth century the philosopher René Descartes postulated that the pineal gland performs the role of internal clock, coordinating the 24-hour cycle of waking and sleeping. Modern biology has confirmed that the pineal does indeed perform a clock-like function, even if it is not the location of the interface between the body and the soul, as Descartes believed. The pineal gland is a pea-sized clump of nerve cells lying deep within the brain, between the two cerebral hemispheres. It responds to changes in ambient light and day-length by altering the production of a hormone called melatonin, which plays a major role in regulating the sleep–wake cycle. Melatonin is released by the pineal gland during the hours of darkness. Peak levels of melatonin therefore increase during the winter, with its longer nights. Melatonin induces sleep, probably by inhibiting brain mechanisms that maintain wakefulness.

A study of melatonin rhythms in US submariners discovered that there are limits to how far our circadian rhythms can be shifted by alterations in the environment. For some reason, the crews of US submarines live on an 18-hour artificial day, comprising three six-hour shifts of duty or rest. This 18-hour 'day' is too short for their biological rhythms to adjust to, especially when living in a low and relatively constant level of artificial light. When researchers measured melatonin levels in crew members on a Trident nuclear submarine during a long undersea voyage, they found that melatonin continued to maintain a daily rhythm of just over 24 hours despite the 18-hour 'day' of submarine life. The lighting, social and other time cues in the submariners' environment were insufficient to shift their melatonin rhythms onto the artificial 18-hour cycle.

Growth hormone is another of the many hormones that are intimately connected with sleep. Growth hormone is secreted within the body during the first few hours of sleep, mostly in a large pulse during the first episode of slow-wave sleep. In men, more than two-thirds of their total daily production of growth hormone occurs during the early phase of sleep. (The pattern of secretion is more varied in women.) The longer you spend in slow-wave sleep, the more growth hormone your body produces. Between the ages of about 30 and 40, there is a two- to threefold decline in the amount of growth hormone produced each night, accompanied by a drop in the amount of slow-wave sleep. This parallel decline in slow-wave sleep and growth hormone seems to be part of the normal ageing process. Levels of the male sex hormone testosterone are also higher during sleep. In young men, testosterone levels start rising at the onset of sleep, reach a peak at around the first episode of REM sleep and remain high until awakening. Testosterone falls during the day and is restored by sleep. (Young men who sleep too little should ponder on that.)

The normal ageing process is accompanied by a deterioration in the biological clock, a fact that has some bearing on the deterioration in sleep that generally accompanies ageing. The peaks and troughs of the circadian rhythm in core body temperature are substantially smaller in old people than in young adults, and the low point occurs nearly two hours earlier in the day. Older people also tend to go to bed earlier and wake earlier, in line with this shift in their temperature rhythm. (We shall be looking at how sleep changes with age in chapter 11.)

Jet lag, another source of grief, is caused by the sudden creation of a large discrepancy between the internal clock and the external world.

Humans evolved in a world where the maximum distance anyone could possibly travel in a day would have been measured in miles rather than thousands of miles. Our biological rhythms take much longer to adjust than modern air travel allows. The hormonal and other physiological rhythms typically need about one day to adjust for each time zone crossed. We therefore take about five days to adjust fully to a transatlantic flight crossing five time zones. This is longer than most people realise or allow themselves. We also find it harder to adjust to a shorter day than a longer one, which explains why flying east creates worse jet lag than a westward flight.

Jet lag is not exclusively the result of a confused biological clock, however. Other ingredients add to the fuggy brew. Many people sleep badly the night before a flight, especially if they are not seasoned travellers. And they often have to present themselves at the airport at some sleep-unfriendly hour, perhaps after travelling a long distance from home. Long-haul passengers therefore tend to start their flight in a state of partial sleep deprivation. Once they are on the plane they will be unable to sleep deeply if they are jammed upright in a cramped steerage-class seat. Only the select few will have the luxury of going horizontal and sinking beyond the teasing foreplay of shallow sleep. Then there are the copious amounts of sleep-disrupting alcohol and caffeine that passengers are encouraged to consume during the flight. Combined with the dry, low-oxygen cabin air, the continuous loud noise, the in-flight entertainment and frequent offers of refreshments from the cabin crew, the overall effect is hardly conducive to slumber. Air travel and sleep are not natural companions.

Given the crucial importance of light cues in keeping our biological rhythms in synchrony with the outside world, it is not surprising that many blind people suffer from disrupted sleep. Total blindness deprives the body of information that normally couples the circadian rhythms to the 24-hour cycle of light and dark (knee pits notwithstanding). The circadian rhythms in body temperature, sleepiness and hormone levels in blind people consequently tend to become decoupled from their sleep–wake cycle, disrupting their sleep.

Nearly half of all blind people experience sleep disturbances – several times the incidence among normally sighted people. Sleep problems can affect partially blind people who retain the ability to perceive light, but they are most severe in the totally blind who have no light perception. The most common complaints are interrupted sleep, difficulty in falling asleep and short sleep duration. One clinical case illustrates the problem. A 63-year-old man who was totally blind had suffered from intermittent insomnia for 28 years. His body temperature and hormone levels fluctuated with a cycle of slightly more than 24 hours, but these rhythms were disconnected from his sleep–wake cycle. The man worked as a schoolteacher and adhered to a normal work schedule, which meant there was often a big mismatch between what his environment was telling him to do (get up, go to sleep) and what his internal clock was telling him to do (go to sleep, get up). On the other hand, many blind people do manage to maintain a conventional sleep schedule without suffering from insomnia.

Melatonin has been used with some success to treat the sleep disturbances associated with blindness, shift work and jet lag. Doses of melatonin taken at the right times can help to realign the body's biological rhythms with the outside world far more rapidly. Several studies have confirmed that melatonin can reduce or prevent jet lag on long-haul flights. To work, the hormone has to be taken at around the time of day that the traveller will want to go to sleep after reaching their destination. If melatonin is taken at the wrong times of day it can make jet lag worse. The benefits are greater for longer flights, crossing five or more time zones, and for flights travelling in an eastward direction. There is some evidence that taking melatonin may be harmful to people suffering from epilepsy and patients using the anticoagulant drug warfarin. Otherwise, it seems to be relatively free from side effects when used sparingly and in suitable doses. One potential hazard worth bearing in mind, however, is that the actual amount of synthetic melatonin contained in capsules that are sold commercially is variable and uncertain. (And in a few cases, the melatonin content has turned out to be zero.)

Like other hormones, melatonin has numerous different physiological actions within the body, many of them unrelated to sleep. Melatonin influences sexual and reproductive behaviour in some species, and in humans it plays a role in regulating the immune system, among other things. Synthetic melatonin has been sold (if not oversold) as a miracle drug that not only cures jet lag but also promotes sleep, protects against disease, rejuvenates the ageing body and stimulates the libido. It is licensed for sale in the USA on the somewhat dubious basis that it is a 'natural food supplement' and not a prescription drug. A number of eminent sleep scientists remain sceptical of claims that routinely taking melatonin will improve health, happiness and longevity.

So SAD

> In winter I get up at night
> And dress by yellow candle-light.
> In summer, quite the other way,
> I have to go to bed by day.
>> Robert Louis Stevenson, 'Bed in Summer' (1885)

The daily cycle of sleep and activity is the most obvious rhythm, but our biological functions also fluctuate over much longer time scales. Humans, like other animals, respond physiologically to seasonal changes as the days grow longer in summer and shorter in winter. People who are exposed experimentally to artificial 'days' of progressively changing lengths display corresponding shifts in their physiological cycles, such as the nightly secretion of melatonin.

In modern urban environments, the effects of artificial lighting and other stimuli usually mask the biological response to seasonal variations in day-length. Most of us do not notice big differences in our sleep patterns between winter and summer. For some individuals, however, the shorter days of late autumn and winter bring with them a peculiar form of unpleasantness, characterised by depression and lethargy and accompanied by increases in sleeping, appetite and body weight. Seasonal Affective Disorder, or SAD, as it is known, affects between 2 per cent and 10 per cent of people in northern Europe. Seasonal changes in appetite, body weight and sleep patterns occur in everyone to some extent, but for most the changes are not problematic. SAD is a matter of quantity rather than quality.

The usual treatment for SAD entails exposure to a very bright artificial light, usually at around the same time early every morning. The light sources used for this treatment generate light intensities of more than 2,000 lux, which is several times brighter than normal electric lighting. For comparison, the light intensity inside a conventionally lit office is a few hundred lux, while daylight is more than 10,000 lux. Bright light therapy usually alleviates the depressed mood associated with SAD, but the antidepressant effect takes a few weeks to kick in. Light treatment is safe and helpful for many sufferers, but it is not always sufficient for the most severely affected. One study of American SAD patients found that after almost nine years fewer than one in seven had fully recovered.

There is more to SAD than just inadequate light, however. Research has found that SAD sufferers are exposed on average to just as much natural light as non-sufferers. This implies that they differ in their sensitivity to variations in light, rather than the amount of light itself. Experiments have also revealed that at least some of the perceived benefits of bright light therapy might arise from the psychological placebo effect, because they can also be elicited by sham treatments.

Larks and owls

> For a long time I used to go to bed early.
> Marcel Proust, *Swann's Way* (1913)

Each of us has our own preferred sleep habits, and these vary considerably from person to person. Some individuals consistently prefer to rise early, some late; some like to sleep a lot, some less. President Calvin Coolidge, for instance, was allegedly a great sleeper. The journalist H. L. Mencken once wrote of Coolidge:

> Here, indeed, was his one really notable talent. He slept more than any other President, whether by day or night . . . Nero fiddled, but Coolidge only snored.

Ronald Reagan was another US president famed for slumbering. He once made the strangely admirable remark that 'It's true hard work never killed anybody, but I figure, why take the chance?' On the morning of his inauguration as president, Reagan had to be woken at nine o'clock. On another occasion, he nodded off in the presence of the Pope. In the Reagan White House it was well understood that the President was to be woken during the night only if something truly cataclysmic had happened. While those around him fretted and toiled, Reagan snoozed.

By the age of 30, Marcel Proust had developed a deeply eccentric sleep pattern that entailed going to bed at dawn and rising in the late afternoon. This lifestyle did not endear him to his family. And then there was the fictional sleepyhead George in Jerome K. Jerome's *Three Men in a Boat*:

Harris said he didn't think George ought to do anything that would have a tendency to make him sleepier than he always was, as it might be dangerous. He said he didn't very well understand how George was going to sleep any more than he did now, seeing that there were only twenty-four hours in each day, summer and winter alike; but thought that if he *did* sleep any more he might just as well be dead, and so save his board and lodging.

One dimension on which people differ systematically is their preferred time of waking. Individuals can be classified broadly as larks (morning types) or owls (evening types) according to whether they prefer to rise early or late. Larks have been defined as those who consistently prefer to go to bed before 11 p.m. and to get up before 8 a.m., while owls are those who prefer to go to bed after 11 p.m. and get up after 8 a.m., although there is nothing sacrosanct about those precise timings.

Most people are neither extreme larks nor extreme owls, but between 5 per cent and 10 per cent of the population has one or other characteristic to a marked degree. The extreme lark will be the one nodding off at the dinner table, and the true owl will be the person who stumbles into work late most mornings, bleary-eyed and craving large doses of caffeine. The lark's alertness (and body temperature) will peak around mid-afternoon, while the owl's will peak hours later. These traits are difficult to change: owls find it hard to become larks and vice versa. Or, as the proverb put it, 'In vain they rise early that used to rise late.'

A reassuringly large number of eminent and creative people have been owls (writes an owl). Oscar Wilde was a true owl. When a friend once asked Wilde to call on him at 9 o'clock the next morning, Oscar replied: 'You are a remarkable man! I could never stay awake as long as that. I am always in bed by 5 o'clock.' James Boswell was another owl. He loathed getting up in the morning and usually awoke feeling 'heavy, confused and splenetic' or 'dreary as a dromedary'. Boswell wrote that he:

. . . wished there could be some medicine invented which would make one rise without pain, which I never did, unless after lying in bed a very long time. Perhaps there may be something in the stores of Nature which could do this. I have thought of a pulley to raise me gradually; but that would give me pain, as it would counteract my internal inclination ... We can heat the body, we can cool it; we can give it tension or relaxation; and surely it is possible to bring it into a state in which rising from bed will not be a pain.

The Russian pianist and composer Anton Rubinstein habitually slept late in the mornings. It is said that to force him out of bed, Rubinstein's wife would play an unresolved chord on his piano. Her husband, who could not bear the sound, would leap out of bed and rush to resolve it into a perfect triad. Meanwhile, Madame Rubinstein would sneak in and steal his bedclothes to prevent him going back to bed.

The key difference between larks and owls resides in the natural length of their circadian rhythms. Scientists have compared the free-running circadian rhythms of larks and owls under conditions where all external time cues have been removed. They found that the circadian rhythms in body temperature and hormone levels in owls are consistently longer than 24 hours, whereas larks usually have a cycle that is close to 24 hours in length. This means that owls need to have their biological clock reset every day by environmental cues such as daylight and routines.

Sleeping at the right times is just as important as spending enough time doing it. Owls tend to have more variable sleep patterns than larks and are therefore more vulnerable to disrupted sleep. Research has shown that individuals who maintain a regular sleep routine tend to feel more alert during the day than those whose sleeping habits are more haphazard. In one experiment, two groups of American students were placed on routines that enabled them to sleep for at least seven and a half hours a day. One group was instructed to maintain a regular sleep–wake cycle, while the others could sleep whenever they chose. Those on the regular schedule slept more efficiently and felt more alert during the day than those who slept when they felt like it.

Life can be hard for owls. As well as suffering the daily nausea of being forced out of their nice warm beds far too early so that they can get to school or work, owls are made to feel guilty and inferior by smug larks. 'Early to bed and early to rise makes a man healthy, wealthy and wise,' opined Benjamin Franklin, to the irritation of generations of owls. But more thoughtful views have also been expressed. For example, in *The Anatomy of Melancholy*, Robert Burton advised that 'Waking that hurts, by all means must be avoided.' There is some evidence that larks enjoy a higher subjective quality of sleep than owls. However, a large-scale British study found no support for Benjamin Franklin's annoying maxim about early rising and wealth. In fact, it found that owls had larger average incomes than larks and were more likely to own their own cars.

Further consolation for owls comes from the discovery that larks have higher levels of the stress hormone cortisol. As a general rule, your cortisol level will rise to a peak about half an hour after you wake in the morning, then decline gently over the rest of the day. However, larks have higher cortisol levels, both immediately after waking and for the rest of the day. Prolonged high cortisol levels are generally associated with chronic stress, impairments in the functioning of the immune system and heightened susceptibility to illness. This might explain why larks also report having more minor illnesses than owls. James Thurber might have been onto something when he wrote that 'early to rise and early to bed makes a male healthy and wealthy and dead'. When it comes to sleeping late, I for one am with Thomas Dekker, who set out these trenchant views in 1609:

By the opinions of all Philosophers and Physicians, it is not good to trust the air with our bodies till the Sun with his flame-coloured wings hath fanned away the misty smoke of the morning, and refined that thick tobacco-breath which the rheumatic night throws abroad of purpose to put out the eye of the Element: which work questionless cannot be perfectly finished till the Sun's cart horses stand prancing on the very top of highest noon. So then, (and not till then) is the most healthful hour to be stirring. Do you require examples to persuade you? At what time do Lords and Ladies use to rise but then? Your simpering Merchants' wives are the fairest lyers in the world, and is not eleven a clock their common hour? They find (no doubt) unspeakable sweetness in such lying, else they would not day by day put it so in practice. In a word, midday slumbers are golden, they make the body fat, the skin fair, the flesh plump, delicate and tender: they set a russet colour on the cheeks of young women, and make lusty courage to rise up in men.

Individual sleep patterns differ in other ways as well, and for other reasons. Your sex makes a difference to your sleep, statistically speaking. On average, women go to bed slightly earlier, fall asleep slightly earlier and sleep slightly longer than men. Women also report more awakenings, more time awake at night, and poorer sleep quality than men. These sex differences are most apparent over the age of 50. Then there are individual differences in the habitual duration of sleep, although these tend to be heavily influenced by external factors such as work schedules and social pressures. Nonetheless, some people do seem to be able to cope with less sleep than others.

You might assume that those hardy individuals who habitually sleep less than the average amount ('short sleepers') are built differently and genuinely need less sleep than the rest of us. But this may not be true. When researchers monitored the waking EEG patterns of 'short sleepers' – people who habitually slept for less than six hours a night – they found that the short sleepers displayed more EEG brain-wave patterns of the type normally associated with resting or the early stages of sleep, even though they were awake. Their EEG patterns indicated a degree of sleep deprivation. One interpretation is that short sleepers need just as much sleep as the rest of us, and they are sleepy during the day (as indicated by their waking EEG patterns) because they do not get enough sleep. Their distinguishing feature may be that they are simply more tolerant of being tired and have adapted to a way of life that involves never getting enough sleep. The 'short-sleeper' pattern may be as much a product of personal choice, lifestyle and ability to withstand fatigue as some inherent biological preference for less sleep.

Individual sleep patterns also vary according to a number of strange and probably inexplicable influences. For example, scientists have unearthed statistical associations between sleep patterns, handedness and short-sightedness. A study of college students found that individuals who were neither strongly left-handed nor strongly right-handed tended to sleep less than individuals who were clearly left-handed or clearly right-handed. Goodness knows what that means, because statistical associations have also been found to exist between left-handedness and all manner of other things, including allergies, asthma, immune disorders, dyslexia, autism, homosexuality, alcoholism, birth complications, accidents and exceptional musical ability.

A separate study discovered that, on average, short-sighted people sleep slightly more than those who are not short-sighted. This association does have at least one plausible explanation. It may be that habitually sleeping longer than average slightly increases the risk of short-sightedness, since there is evidence that prolonged closure of an eye can contribute to the development of short-sightedness in that eye. People who have a drooping upper eyelid (a condition known as ptosis) tend to be more short-sighted in the affected eye. But these are only statistical associations, and weak ones to boot, so even if you are strongly left-handed and immensely myopic you will not necessarily be forced to spend longer in bed each night.

Genes and sleep

> He sleeps by day
> More than the wildcat.
>
> William Shakespeare,
> *The Merchant of Venice* (1596–8)

The techniques of molecular genetics are enabling scientists to identify individual genes involved in the control of sleep, including genes that influence biological rhythms. Several genes have been identified in fruit flies and other animals that are crucial for their normal circadian rhythms. Remarkable similarities have been found between species as diverse as fruit flies, mice, hamsters and humans in the genes controlling circadian rhythms. This is not as surprising as it might sound. Circadian rhythms are a fundamental feature of all organisms, even bacteria, and once evolution has constructed a biological mechanism that works well, that mechanism is usually incorporated (with variations) into many species.

Certain genetic mutations make the biological clock run fast, or slow. Strains of rodents with these mutations have circadian rhythms that consistently run either faster or slower than the normal 24-hour cycle. Research has identified a specific genetic mutation that dramatically alters the biological clock in hamsters, shortening their circadian period from 24 hours to 20 hours. This mutation is located in a gene similar to genes in humans that are also known to be involved in controlling circadian rhythms.

Individual human biological clocks differ slightly in their natural pace, for reasons that may in some cases reflect genetic differences. When someone's internal clock runs slow, with a period of more than 24 hours, the result is a condition known as Delayed Sleep Phase Syndrome. The individual concerned wants to go to sleep later and later each night, and to get up later and later each morning. This condition is not uncommon in a mild form (think of any teenager you know). More severe forms can often be treated successfully using bright-light therapy early in the mornings to reset the biological clock. The reverse problem is known as Advanced Sleep Phase Syndrome. In this case the biological clock runs with a period of less than 24 hours, making the individual feel like going to bed progressively earlier each night. People with this condition tend to fall asleep in the early evening and wake up in the small hours of the morning. In a milder form, it is relatively common among elderly people, as we shall see in chapter 11.

Scientists have begun to unravel some of the genetic influences on Advanced Sleep Phase Syndrome. American researchers discovered a rare example of a family with an inherited version of the disorder, known as Familial Advanced Sleep Phase Syndrome (FASPS). Three generations of one family – grandmother, mother and daughter – were all extreme larks. They would typically go to sleep and wake up four hours earlier than everyone else, with accompanying shifts in their melatonin and body-temperature cycles. When scientists analysed the women's DNA they found that all of them had the same mutation in a particular gene known as hPer2. This mutation is similar to one found in fruit flies and mice that causes the animal's body clock to speed up. Scientists have also isolated a specific genetic difference in humans that influences the biological clock, affecting the individual's propensity for waking early ('larkness') or going to bed late ('owlness'). Among a random sample of several hundred normal adults, individuals were found to have one of two genetic variants in a particular region of a gene known as the CLOCK gene. Individuals with one genetic variant had a tendency to be owls; they preferred going to bed later than the larkish individuals with the alternative form of the gene. The genetic owls also reported that they worked better at night than the genetic larks.

What other influences do genetic differences exert on individual sleep patterns? Twins tend to resemble each other in some aspects of their sleep patterns, as they do in some aspects of their waking behaviour. One study of identical twins found certain similarities in their sleep latency and the amount of REM sleep, but no such correlations were found for non-identical twins who share only half their genes. Other twin studies have also uncovered hereditary influences on variables such as subjective sleep quality, sleep disturbances and the propensity for 'short' versus 'long' sleeping. But the strength of these genetic influences is generally not dramatic.

Although the evidence suggests that heritable factors do help to shape individual sleep patterns, there is certainly no such thing as a gene 'for' sleep, any more than there is a gene 'for' being awake, a gene 'for' intelligence, or a gene 'for' language. Genes make proteins, not behaviour patterns. Many, many different genes affect an individual's sleep. Conversely, defects in any one of many different genes can have the effect of altering or disrupting sleep. Genes and environmental factors interact in complex ways throughout an individual's development, from conception to death, and it can be misleading to try to ascribe a particular characteristic or ability to a single gene or a single form of experience.

The hazard of adopting a simplistic nature/nurture approach to the roles of genes and environment is illustrated by the sleep of conjoined ('Siamese') twins. Despite being genetically identical and sharing essentially the same environment, conjoined twins do not sleep and wake at the same times. In one case, researchers studied a pair of newly born conjoined twins who were joined in the face-to-face position at the chest. The twins were genetically identical and shared the same blood circulation. Nevertheless, their sleep patterns were quite separate and distinct. Any correspondence between their two sleep states was at chance levels, showing that as far as their sleep patterns were concerned the twins behaved like two independent individuals. In another case, a pair of conjoined twins who shared a common heart and circulatory system also had fully independent sleep–wake patterns.

Genes reveal their hand most dramatically when something goes wrong. Certain genetic mutations or genetic damage can have devastating consequences for sleep and other forms of behaviour. Smith-Magenis syndrome, for example, is a rare inherited disorder characterised by severe learning difficulties, behavioural disorders and abnormal sleep patterns. It is caused by the loss of genetic material from a specific area of chromosome number 17. (Human cells contain 23 pairs of chromosomes.) Children born with Smith-Magenis syndrome display a range of behaviour problems which include hyperactivity, temper tantrums, aggression, head-banging, self-injury by biting hands or fingers, and excessive nose-picking. Sufferers develop characteristic facial features. The archetypal Smith-Magenis face is broad and square, with heavy brows and extended eyebrows. The eyes are close together, deep set and slanting upwards. The nose is short, with a wide, sunken base, and in older sufferers it takes on the shape of a ski jump. The Smith-Magenis mouth is particularly distinctive: it is wide and full with fleshy lips, and the central part of the upper lip is turned outwards, giving a tented appearance. The lower jaw grows faster than the upper jaw, producing a protruding jaw.

Sleep disturbances are a prominent feature of Smith-Magenis syndrome. Sufferers experience an array of sleep problems including difficulty falling asleep, frequent and lengthy awakenings from sleep, excessive daytime sleepiness, shortened sleep cycles, snoring and bed-wetting. Underlying these sleep disturbances is a disruption of the circadian rhythm, which is thought to result from changes in one or more genes controlling the biological clock.

The activity patterns of young Smith-Magenis sufferers resemble a state of permanent jet lag. They nap frequently during the day but go to bed early and wake before dawn. By early evening they are very tired. All display a large shift in the timing of their circadian rhythm in melatonin. Normally, melatonin levels start to rise at night and peak during the small hours of the morning, but in Smith-Magenis sufferers the rhythm is inverted; their melatonin levels start to rise in the early morning and peak around midday. This reversal makes them feel sleepy during the day. It may be that their hyperactivity and other behaviour problems stem from their constant struggle against daytime sleepiness. Treating Smith-Magenis sufferers with doses of melatonin at night can alleviate their sleep disturbances, reinforcing the view that abnormalities in the melatonin rhythm lie at the heart of this syndrome.

A sleeplessness that kills

> When the sick person became used to his state of vigil, the recollection of his childhood began to be erased from his memory . . . until he sank into a kind of idiocy that had no past.
>
> Gabriel García Márquez, *One Hundred Years of Solitude* (1967)

In the 1960s Gabriel García Márquez wrote a novel in which he imagined a plague of insomnia descending upon a village in the South American jungle. To begin with, the villagers are stoical about their inability to sleep. A few still pine for their lost sleep, but only because they are nostalgic for dreams. Most of them welcome their sleeplessness. So much the better, they say, if we never sleep again, because that way we can get much more done. They cease worrying about 'the useless habit of sleeping'. But the imaginary insomnia plague has a far more fearsome aspect than the mere impossibility of sleep. Its victims progressively lose their memories. First, the sick person's recollections of childhood are erased, then their memory for the names and functions of everyday objects and the identities of people, until finally their self-awareness goes and they sink into oblivion.

Years after the novel was published, medical science discovered a real disease with striking similarities to the insomnia plague imagined by Gabriel García Márquez. It is called Fatal Familial Insomnia (FFI). In this case, however, the reality is even nastier than the fiction. The fantasy plague of insomnia was infectious but curable; a gypsy arrived in the village and cured everyone with a magic potion that banished their insomnia and restored their memories. The real disease, however, is invariably fatal. FFI is an inherited disorder that stems from a genetic mutation in a known location. It is inherited in a dominant form – in other words, you need inherit only one copy of the aberrant gene from one parent to get the disease.

The clinical symptoms of FFI start with deteriorations in attention and memory, accompanied by a progressive, dreamlike state of confusion. The sleep–wake cycle, including the circadian rhythms in hormone levels, degenerates until sleep becomes severely disrupted. The victim cannot sleep and their brain is incapable of generating sleep-like EEG patterns. Even powerful barbiturate sleeping drugs cannot put them to sleep. These changes are accompanied by degeneration in a region of the brain known as the thalamus, which normally plays a central role in regulating the sleep–wake cycle. Death invariably results.

FFI is caused by the accumulation within the brain of abnormal protein molecules known as prions. Similar prions are also responsible for other degenerative brain diseases such as Creutzfeldt-Jakob disease (CJD) and kuru in humans, and scrapie in sheep. FFI was first described in 1986, two decades after Gabriel García Márquez wrote about the fatal plague of insomnia. As yet, no magic potion has been found to cure it.

7

Strange Tales of Erections and Yawning

We were so poor that if we woke up on Christmas Day without
an erection, we had nothing to play with.

Frank McCourt, author of *Angela's Ashes* (1996)

Nocturnal erections

We all have erections during sleep – you, me (I admit it), the US president (try not to think about it) and the people next door. The nocturnal
erection is a mysterious and neglected feature of sleep. Penile erections
occur during sleep in healthy males of all ages, even foetuses in the
womb. And before you accuse me of making outrageous assumptions
about the sex of my readers, let me stress that females have nocturnal erections as well – or, at least, the nearest equivalent without a
penis. The clitoris becomes engorged, the vagina lubricates, the nipples
spring erect, and physiological measures of sexual arousal ascend to a
peak. Some women experience orgasms during sleep, though without
leaving the physical traces of the male orgasmic wet dream.

Nocturnal erections are robust and ubiquitous things. Unlike their
daytime counterparts, nocturnal penile erections do not decline much
in frequency or girth with advancing years. The erectile performance
of men in late middle age does not differ in this respect from men in
their early twenties. They have just as many nocturnal erections, which
are just as stiff. A middle-aged man will typically have three or four
full erections during the course of the night, each lasting for about
half an hour. Their duration does decline slightly with age, in line
with the overall decline in sleep. Otherwise, the nocturnally erect
penis remains remarkably unbowed by age.

132

Nocturnal erections are specifically linked to REM sleep, rather than to the hours of darkness or to sleep in general. (The comedian Peter Cook once remarked, in character, that he didn't know much about Rapid Eye Movement sleep but he was all too familiar with Rapid Hand Movement sleep.) The close association between erections and REM sleep develops in babies between two and four months of age. 'Nocturnal' erections also occur during REM sleep in daytime naps, so the name is perhaps misleading. Either way, if you are male and you wake with an erection, you can deduce that you have probably woken from REM sleep and were dreaming.

These strange and seemingly pointless erections persist for most or all of each episode of REM sleep. This means that even the most ascetic and puritanical of men will sport a fully erect penis for between one and two hours every day. (Remember that fact next time you see some holy zealot preaching purity.) Even brain-damaged patients in a vegetative state have erections. An Israeli study of nine vegetative patients discovered that all nine still displayed penile erections during REM sleep.

The nocturnal erection is not a uniquely human feature either. Other species have them as well. Using ingenious miniaturised technology, French scientists were able to monitor penile erections in free-moving rats. They found that the rats' nocturnal erections occurred only during REM sleep, as in humans. Erections were observed in more than a quarter of all REM episodes and lasted on average for 11 seconds. Dogs also have erections during sleep.

For obvious but boring reasons, nocturnal erections have not featured prominently in the scientific literature. This easily visible phenomenon seems to have gone largely unnoticed – or, at least, unrecorded – for a disgracefully long time. Probably the first published scientific account of nocturnal erections appeared in 1944, when German scientists reported that the penises of sleeping men periodically became erect at intervals of about 90 minutes. The paper described how the first nocturnal erection of the night normally blossomed more than an hour after the onset of sleep, which we now know is when the first episode of REM sleep is underway. It also reported that these erections lasted on average 25 minutes, which we now know is roughly the length of a typical episode of REM sleep. Thus, a scientific paper describing perhaps the most easily visible hallmark of REM sleep was published eight years before Eugene Aserinsky discovered REM sleep from his son's EEG traces. Another bizarre aspect of this story is the image of German scientists scrupulously cataloguing nocturnal erections during the second half of World War Two.

Prudery aside, a general awareness of nocturnal erections obviously extends beyond scientific research and goes much further back in history than World War Two. Artistic representations of nocturnal erections can be found dating from prehistory. Ancient Egyptian art features nocturnal erections. For example, a bas-relief on the walls of the Great Temple of Denderah, dating from the Third Dynasty, portrays a sleeping man with a very obvious erection.

Among the cave paintings at Lascaux in France, which are thought to have been created around 16,000 years ago, is a strange picture of a man with a pronounced erection who is either lying down or falling over. The man, who is the only human figure depicted in that part of the cave, has the head of a bird. His arms are flung wide and his four-fingered hands are held open. This bird-headed man is part of a narrative scene involving a bison, a rhinoceros and a bird on a stick. The bison is drawn realistically, with its entrails apparently hanging out, whereas the man is drawn in a schematic, almost childish, way. Scholars have bickered over the interpretation of this bizarre, dreamlike concoction. Most have chosen to see it as a depiction of hunting or some shamanistic ceremony, despite its odd features. But there is a much more interesting interpretation. The French scientist Michel Jouvet has suggested that it is in fact a dream scene. The stick man is not killing the bison: he is *dreaming* about killing the bison, and that is why the painting depicts him sporting a nocturnal erection, the true hallmark of dream sleep. Our early ancestors must have been aware of nocturnal erections, and it seems reasonable to suppose that they might have represented them in their art.

Do nocturnal erections have anything to do with sex? They certainly occur during REM sleep, which is when most dreams occur. Sigmund Freud's encounters with young middle-European women patients in the late nineteenth century helped to foster the myth that most of our dreams are sexual in nature, even if that sexuality is supposedly disguised by the unconscious mind. In Freud's mental universe, sticks, guns, pens, umbrellas, pistols, snakes and almost any object longer than its width became a penis, while cups, cupboards, boxes and other vaguely concave objects became vaginas, and riding, shooting, ploughing, climbing stairs and almost all rhythmic activities known to nineteenth-century society signified copulation. The scientific discovery (or rediscovery) of nocturnal erections in the 1940s therefore delighted the disciples of Freud, because it seemed to confirm their faith that most dreams are fundamentally sexual in nature. But they were wrong.

Nocturnal erections have little or nothing to do with sex or erotic dreams. They accompany REM sleep irrespective of the dreamer's age or gender and irrespective of the contents of his or her dreams. Nocturnal erections are not linked to daytime sexual activity either. Scientists have tried, but failed, to uncover connections between people's sexual activity or fantasies during the day and the frequency or magnitude of their nocturnal erections during subsequent sleep. Some imaginative Greek scientists even conducted a laboratory experiment to see whether sexual intercourse had any dampening effect on nocturnal erections. The subjects were 12 male medical students who selflessly underwent sessions of overnight monitoring of their nocturnal erections. During some of the sessions they had sexual intercourse while in others they remained abstinent. Sexual activity made no significant difference to the tumescence or rigidity of their subsequent nocturnal erections. Strange, but reportedly true.

Further reasons for drawing a dividing line between waking erections and the nocturnal variety come from the effects of alcohol. In sufficient quantity, alcohol inhibits penile erections when we are awake. It is, after all, a central nervous system depressant. But alcohol has little or no such withering effect on nocturnal erections. Australian researchers found that giving volunteers a substantial amount of alcohol made no significant difference to the number, duration or size of their nocturnal erections. Smoking does impair nocturnal erections, however. A study of men with erectile dysfunction (impotence) uncovered a clear link between smoking and diminished nocturnal erections. The men who smoked the most cigarettes had nocturnal erections that were less rigid, shorter (in duration) and faster to subside. This suggests that men who have difficulty in maintaining erections may make their problem worse if they smoke, and not just because their breath smells.

You may be wondering how scientists have gone about probing and cataloguing nocturnal penile erections. The approved technique involves sophisticated electronic strain gauges. However, a charmingly low-technology way of detecting the presence or absence of nocturnal erections is known as the stamp method. This involves sticking a chain of postage stamps around the shaft of the flaccid penis at bedtime and checking them for breakages the following morning.

If during the night the penis has become erect and swollen in girth, the stamps will be torn in the morning. You could try it yourself, or with a friend.

The stamp method relies on the fact that the penis increases in circumference as it becomes erect. Girth is not everything, however. The stiffness or rigidity of the penis is another salient variable. A simple device used for assessing penile rigidity is known as the snap-gauge. This gadget consists of a series of plastic bands that are calibrated to snap at different forces. The snap-gauge is wrapped around the shaft of the flaccid penis at bedtime and, as the penis swells during the first nocturnal erection, the sections of the snap-gauge snap at progressively higher forces. The high water mark, as it were, represents the maximum force exerted by the erect penis during the course of the night's erections. One standard version of the snap-gauge has three bands that are calibrated to snap in quaint imperial units, at 10, 15 and 20 ounces of force. If all three bands are sundered, come the morning, then the sleeper has physical proof of a full erection.

An even less intrusive method of monitoring nocturnal penile erections is simply to watch or photograph the sleeper and gauge the magnitude of his tumescence by eye. Researchers at the University of Pittsburgh School of Medicine established that sleep laboratory technicians are reasonably accurate at judging the fullness of nocturnal erections, when compared against objective measurements made with electronic gauges. Male and female technicians differed consistently in their estimates, however. The women's visual estimates of erectile fullness were significantly higher and more consistent than those of their male colleagues. Draw your own conclusions.

Mostly, though, serious students of nocturnal erections use a portable electronic device with the beautifully descriptive name of the Rigiscan to measure stiffness and bulk. The measurement process has acquired its own acronym: NPTR, which stands for Nocturnal Penile Tumescence and Rigidity monitoring. The Rigiscan assesses penile activity by measuring the mechanical resistance to loops that wrap around the penis and gently squeeze it at regular intervals. The more rigid the penis, the less its shape is distorted by these rhythmic squeezes. (One could speculate whether all this gentle, rhythmic squeezing of the penis might subtly distort the measurements, or indeed the whole sleep pattern.) The mechanical properties of the nocturnally erect penis can be dissected in even more depth. For example, it is possible to measure the organ's longitudinal rigidity by waking the sleeper during an erection and pressing down along the length of the penis with a specially calibrated device. A few grams of pressure will cause a flaccid penis to bend, whereas a fully erect penis will easily resist a downward pressure of more than a kilogram before buckling. Do not try this at home.

The Rigiscan and other penile paraphernalia have been developed for good medical reasons rather than prurient interest. Nocturnal erections provide a valuable insight into the medical problems that under-lie erectile dysfunction, a distressing disorder that is estimated to affect more than 150 million men worldwide. Testing for the presence of nocturnal erections enables physicians to distinguish between cases of impotence that are largely psychological or emotional in origin, as opposed to purely mechanical. The fact that penile erections also occur during REM sleep in the daytime means that monitoring erections during long daytime naps (napturnal erections?) offers a convenient way of screening impotent men for erectile dysfunction.

If an impotent man has full erections during sleep, he does at least have living proof that there is nothing fundamentally wrong with the brute mechanics of his organ, implying that the problem lies else-where. That said, the dichotomy between mental and physical, or mind and body, is fundamentally misconceived and usually unhelp-ful, and it can be equally misleading when applied to impotence and nocturnal erections. Men suffering from major depression often have diminished nocturnal erections. Conversely, some men with mechan-ical problems still have erections while they are awake, but fail to produce nocturnal erections. Experiments found that men with mech-anical erectile problems, who had no erections while they slept, none-theless often responded to erotic videos with an erection when they were awake. There is more to erections than lewd thoughts and fluid pressure.

In some cultures, men find the thought of nocturnal erections and emissions upsetting. In India, for example, a culture-bound disorder known as Dhat syndrome has been described. The Dhat syndrome revolves around a fear of losing semen through nocturnal emissions. It reflects a belief that semen brings health and longevity and that losing it is therefore harmful. The sufferer becomes preoccupied with preventing nocturnal emissions, leading to severe anxiety and hypo-chondria. Preventing nocturnal erections and emissions is next to impossible, so the sufferer is doomed to frustration on at least two counts.

The biological purpose, if any, of nocturnal erections remains a mystery. A sleep scientist once said that he had no idea where dreams came from, but he thought the erect penis might serve as the antenna for receiving them. More seriously, it has been proposed that nocturnal erections might contribute to the maintenance of smooth muscle cells in the erectile tissues of the penis – a sort of automatic genital workout during the night. But nocturnal erections might be just a biologically irrelevant side effect of other physiological processes. Here at least is one new challenge for evolutionary biologists.

On the down side, nocturnal erections become a distinct nuisance following penile surgery. For the unfortunate man recovering from surgery on his organ, nocturnal erections are a painful and unwelcome intrusion. In extremis, nocturnal erections can be suppressed by injecting noradrenaline into the erectile tissues of the penis, a procedure that is presumably preferable to doing nothing.

The mystery of yawning

> Too tired to yawn, too tired to sleep:
> Poor tired Tim! It's sad for him.
>
> Walter de la Mare, *Tired Tim*

A simple thought experiment. Imagine yourself yawning. Feel your mouth stretch open as your nostrils flare, your ears pin back and you slowly inhale a massive lungful of air. Feel those jaw muscles really stretch as your eyes close and your head tilts back. After several luxuriant seconds agape, your jaws close and you emit a sigh of quasi-orgasmic satisfaction. The question is, can you keep this image in your mind or are you actually yawning? And if you are yawning in public, are others around you starting to yawn too? Yawning is truly a strange phenomenon.

Humans are not the only animals who yawn, as Charles Darwin observed. In one of his private notebooks, written long before *The Origin of Species* appeared, Darwin recorded the thought that, 'Seeing a dog and horse and man yawn, makes me feel how much all animals are built on one structure.' Mammals and birds yawn, as do many species of reptiles including tortoises and lizards. Even fish yawn. Yawning is infectious, as your thought experiment might already have demonstrated. It spreads through a group of people or animals like ripples on a pond. Science has paid scant attention to the phenomenon, which falls into a limbo between psychology and neurology. As we shall see, that is not its only similarity with nocturnal erections. But first, what is yawning for?

The conventional biological explanation for yawning is that it occurs in response to a drop in blood oxygen level, or a rise in blood carbon dioxide (which in practice amounts to more or less the same thing). According to this theory, the function of yawning is to reverse the sleepiness caused by shallow breathing, which in turn is caused by lack of sleep or boredom. Unfortunately, this theory does not stand up at all well to experimental scrutiny. The few scientific investigations of yawning have failed to find any connection between how often someone yawns and how much sleep they have had or how tired they are. About the closest any research has come to supporting the tiredness theory is to confirm that adults yawn more often on weekdays than at weekends, and that school children yawn more frequently in their first year at primary school than they do in kindergarten.

Another big dent in the tiredness theory is that yawning does not boost alertness or physiological arousal, as the theory would predict. When researchers measured the heart rate, muscle tension and skin conductance of people before, during and after yawning, they did detect some changes in skin conductance following yawning, indicating slight physiological arousal. However, similar changes occurred when the subjects were asked simply to open their mouths or to breathe deeply. Yawning did nothing special to their state of arousal. Experiments have also cast serious doubt on the belief that yawning is triggered by a drop in blood oxygen or a rise in blood carbon dioxide. Volunteers were told to think about yawning while they breathed either normal air, pure oxygen, or an air mixture with an above-normal level of carbon dioxide. If the theory was correct, breathing air with extra carbon dioxide should have triggered yawning, while breathing pure oxygen should have suppressed yawning. In fact, neither condition made any difference to the frequency of yawning, which remained constant at about 24 yawns per hour. Another experiment demonstrated that physical exercise, which was sufficiently vigorous to double the rate of breathing, had no effect on the frequency of yawning. Again, the implication is that yawning has little or nothing to do with oxygen.

A completely different theory holds that yawning assists in the physical development of the lungs early in life, but has no remaining biological function in adults. Ultrasound scans show that unborn foetuses yawn and hiccup in the womb; yawning is fully developed by the second half of pregnancy and has been observed as early as 11 weeks after conception. It has been suggested that yawning and hiccupping might serve to clear out the unborn child's airways. The lungs of a foetus secrete a liquid that mingles with its mother's amniotic fluid (along with its urine). Babies with congenital blockages that prevent this fluid from escaping from their lungs are sometimes born with deformed lungs. It might be that yawning helps to swill out the lungs by periodically lowering the pressure in them. According to this theory, yawning in adults is just a developmental fossil with no biological function. But, while accepting that not everything in life can be explained by Darwinian evolution, there are sound reasons for being sceptical of theories like this one, which frankly dodge the issue of what yawning does for adults. Yawning is distracting, consumes energy and takes time. It is almost certainly doing something significant in adults as well as in foetuses. What could it be?

The empirical evidence, such as it is, suggests an altogether different function for yawning – namely, that yawning prepares us for a gear-change in activity level. Support for this theory came from a study of yawning behaviour in everyday life. Volunteers wore wrist-mounted devices that automatically recorded their physical activity for up to two weeks; they also recorded their yawns by pressing a button on the device each time they yawned. The data showed a systematic tendency for yawning to occur about 15 minutes before a period of increased behavioural activity. Yawning bore no relationship to sleep patterns, however. This accords with anecdotal evidence that people often yawn in situations where they are neither tired nor bored, but are preparing for impending mental and physical activity. Incongruous yawning has often been observed: soldiers yawning before combat, musicians yawning before performing and athletes yawning before competing. Their yawning seems to have nothing to do with tiredness or boredom – quite the reverse – but it does precede a change of gear. Yawning even predicts increased activity in premature babies.

A similar link between yawning and impending activity has been noted in other species as well. Lions in zoos tend to yawn before feeding time – the most strenuous point in their otherwise languorous day. Similarly, rats that are used to being fed only once a day display a daily peak in yawning an hour or two before their normal feeding time. Siamese fighting fish often yawn before they fight. The gearing-up theory even fits with the observation that yawning is socially infectious. Mass yawning makes sense if everyone in the group is preparing for action. Indeed, the prime function of yawning may be to act as a social signal.

You might be wondering what yawning and erections are doing nestling together in the same chapter. There is a connection. In the 1980s, scientists noticed that rats of a strain that had been specially bred to yawn frequently (don't ask why) also developed erections whenever they yawned. The more the rats yawned, the more erections they had – a strange and potentially troublesome pairing.

The explanation for this bizarre linkage between yawning and erections lies in a particular region of the brain, called the paraventricular nucleus of the hypothalamus, which is involved in the control both of penile erections and yawning behaviour. A neurotransmitter substance called dopamine is one of the main chemical messengers in this system, and it mediates the central control of yawning and erections. When chemicals that mimic the action of dopamine are introduced into the relevant parts of rats' brains, they induce both penile erections and yawning. Conversely, substances that block the actions of dopamine inhibit both yawning and erections.

Comparable effects have been found in humans as well. In one experiment, men who were suffering from impotence were injected with a drug similar to a hormone called alpha-MSH. The drug evoked erections in eight out of ten men who had previously been unable to have an erection, and the drug-induced erections were accompanied by yawning. Fortunately, these potentially embarrassing side effects were judged to be a price worth paying.

8

Friends and Enemies of Sleep

It is said that the effect of eating of too much lettuce is 'soporific'.

I have never felt sleepy after eating lettuces; but then *I* am not a rabbit.

They certainly had a very soporific effect upon the Flopsy Bunnies!

Beatrix Potter, *The Tale of the Flopsy Bunnies* (1909)

You are what you eat. This trite phrase becomes a little less vacuous when applied to sleep. The majority of people on the planet daily consume a stimulant drug – caffeine – that suppresses sleep. Many also drink alcohol and smoke tobacco, and these too are sleep-disrupting substances. We all eat food, of course, and some people believe that diet affects their sleep. Then there is physical exercise, which is commonly thought to encourage slumber. In this chapter we shall consider how various things that we choose to eat, drink, and do to ourselves affect our sleep. We shall also look at some malign influences over which we may have little or no control, such as noise and shift work.

Brother caffeine

The insomnia caused by coffee is not distressing; one's perceptions are sharper and one has no wish to sleep; that is all.

Jean-Anthelme Brillat-Savarin, *The Physiology of Taste* (1825)

The human species is awash with a mind-altering drug. This drug is taken daily by billions of people and it is by far the most widely consumed psychoactive substance in the world. It helps us to stay alert and get through the day when we have not had enough sleep. It is, of course, caffeine. Around 75 per cent of the world's population, and 90 per cent of the American population, habitually take caffeine in one form or another. Between us we consume 120,000 tons of the stuff every year. Caffeine masks the symptoms of inadequate sleep and makes us feel more alive. And therein lies the universal appeal of this drug that even children consume.

Caffeine, otherwise known as 1,3,7-trimethylxanthine, is present in more than 60 species of plants, including coffee, tea, cocoa, cola and guarana (the source of a stimulant drink popular in South America). It is probably most familiar as the active ingredient of coffee, which accounts for just over half of all the caffeine consumed. Like many of the several hundred other chemical constituents that combine to give coffee its delicious taste, caffeine evolved originally as a toxin to protect the coffee plant from being eaten or parasitised. In other words, caffeine is a natural pesticide.

More than six hundred million kilograms of coffee beans are consumed every year, making coffee the second most valuable legally traded commodity on the planet. (The most valuable is oil.) Coffee has become one of the hallmarks of human culture, so ubiquitous that we take it for granted. Even the oldest and most famous image to appear on the Internet was that of a coffee pot. A live picture of the 'Trojan Room' coffee pot, located in the Cambridge University computer laboratory, first appeared on a computer network in 1991, before the World Wide Web was born. Two years later the camera was hooked into the embryonic Web. The coffee pot was subsequently viewed live over the Internet by hundreds of thousands of people around the world, making it the most famous coffee pot in history.

According to legend, coffee was discovered serendipitously by an Ethiopian goatherd, who noticed that his flock became strangely frisky whenever they browsed on coffee berries. From Ethiopia, coffee spread across the Red Sea to Arabia, where it was initially consumed as a medicine and an aid to prayer. Coffee was first mentioned in print in the tenth century by an Arab physician, but by then it had probably been under cultivation for hundreds of years. The first coffee houses opened in Mecca and spread from there throughout the Arab world and thence to Europe.

Coffee reached Europe in the seventeenth century. To begin with it was a rare and exotic substance, used mainly by the wealthy for medicinal purposes. But it rapidly infiltrated into everyday life. Coffee houses became popular venues for social and business meetings. One unexpected consequence of the craze for this new drug was a sharp drop in the consumption of alcohol in booze-sodden Europe. Many drinkers turned temporarily from alcohol to caffeine as their preferred means of altering their brain chemistry (a phenomenon that has some relevance to contemporary debates about legalising cannabis). Coffee did not reach England until after William Shakespeare died in 1616, which explains why there is no mention of it in his plays. One of the earliest descriptions of coffee in an English book appears in Robert Burton's *The Anatomy of Melancholy*, published in 1621:

> The Turks have a drink called coffa (for they use no wine), so named of a berry as black as soot, and as bitter . . . which they sip still of, and sup as warm as they can suffer; they spend much time in those coffa-houses, which are somewhat like our alehouses or taverns, and there they sit chatting and drinking to drive away the time, and to be merry together, because they find by experience that kind of drink, so used, helpeth digestion, and procureth alacrity. Some of them take opium to this purpose.

The earliest known account of someone actually drinking coffee in England dates from 1637, when a Cretan called Nathaniel Conopios prepared and drank coffee at Balliol College, Oxford. The first English coffee house was opened in Oxford in 1650, and by 1652 coffee was being advertised in London as a cure for headaches, coughs, consumption, dropsy, gout, scurvy and miscarriages. The only authentic claim made in the adverts of the time was that it would 'prevent Drowsiness and make one fit for business'.

The taste for coffee and its associated social rituals spread like wildfire, and by 1700 there were more than two thousand coffee houses in London alone. Lloyd's of London, the world's largest insurance market, began life in 1688 as a London coffee house catering mainly to merchants and mariners. Other London coffee houses spawned the London Stock Exchange and several newspapers. Surprisingly, the French did not discover coffee until somewhat later than the English. Coffee first went on sale in France in 1670 and the first café opened soon afterwards in Paris. The craze soon spread from England to North America, where the first coffee house opened in Boston in 1689. Nowadays, Americans drink so much coffee and caffeinated soft drinks that chemical traces of caffeine can be detected in the ocean off the US coast. Caffeine is a specific marker for human waste, because other species do not consume it, and American water engineers use caffeine traces to assess how well sewage disperses off the New England coast after it has been pumped into the sea.

The amount of caffeine in a cup of coffee varies enormously according to how the coffee is prepared, how strong it is and, of course, the size of the cup. As a very rough guideline, a typical 150 ml cup of instant coffee contains about 65 milligrams of caffeine. Real coffee contains about twice as much caffeine as instant, so a 150 ml cup will give you around 130 milligrams and may pack as much as 180. Bear in mind that most coffee addicts normally swill their brew from large mugs or capacious paper flagons rather than dainty china cups. The arithmetic is simple: double the cup size, double the caffeine. A 'medium' serving from an American coffee outlet may comprise more than 300 ml of coffee, containing anything from 150 to 300 milligrams of caffeine, while a bucket-sized *grande* of freshly brewed designer coffee can contain up to half a gram.

Tea also contains substantial amounts of caffeine. In fact, caffeine makes up about 4 per cent of the solid matter in tea-leaves. According to one estimate, 43 per cent of all the caffeine consumed in the world is taken in the form of tea. When brewed conventionally, tea tends to contain slightly less caffeine than a similar volume of instant coffee. A 150 ml cup of tea typically holds about 40–60 milligrams, which is still plenty enough to produce a noticeable boost in alertness. There may be some truth in the belief that tea is less likely to keep you awake at night than coffee. Researchers assessed the relationship between tea and coffee consumption, mental performance and sleep in healthy volunteers. They found that tea was as effective as coffee in maintaining alertness, but had a slightly less disruptive effect on sleep. Why the two drinks should differ in this way remains unclear; expectations and beliefs may play a role.

Coffee and tea are not the only sources of caffeine in our lives. Caffeine is added to most carbonated drinks and colas, and close chemical relatives of caffeine are present in chocolate. Caffeine is also added to many proprietary medicines, including pain relievers, cold and flu remedies and dieting aids. For those who are more single-minded about their caffeine, and want to cut out the middlemen, caffeine is also available in tablet form as an over-the-counter remedy for tiredness. Proprietary brands of caffeine tablets, as consumed by generations of students, clubbers and party animals, typically contain 50 or 100 milligrams of caffeine, making one or two tablets roughly equivalent to a cup of coffee. Altogether, someone who regularly drinks coffee, tea and soft drinks in moderate quantities can easily ingest more than 400 milligrams of caffeine a day.

Since the nineteenth century most carbonated soft drinks have included caffeine as a key ingredient. Americans drink more than 15 billion cans of soft drinks a year, of which about 70 per cent contain caffeine. Caffeine-free versions of Coca-Cola and Pepsi-Cola, still the two most popular soft drinks in the USA, account for only about 5 per cent of their total sales. Most cans of cola contain between 30 and 50 milligrams of caffeine – about half as much as a cup of coffee. Whereas coffee and tea are consumed mostly by adults, soft drinks are also aimed squarely at children. Caffeine must be one of the few psychoactive drugs that we actively encourage our children to consume.

Coca-Cola, the mother of all soft drinks, was invented in 1886 by John S. Pemberton, an American pharmacist in Atlanta. Pemberton wanted to produce a stimulating drink, but the chilling influence of the temperance movement convinced him to use something other than alcohol as his active ingredient. He therefore decided to base his drink around cocaine from the coca leaf and caffeine-rich extracts from the kola (or cola) nut – hence Coca-Cola. Kola nuts, by the way, come from tropical trees of the chocolate family; they contain caffeine and a related chemical called theobromine, which is the main stimulatory ingredient of chocolate. In parts of Africa, kola nuts are still chewed for their stimulant properties. When Coca-Cola was first marketed it was advertised, among other things, as a cure for 'slowness of thought'. Nowadays, of course, Coca-Cola and other soft drinks rely on caffeine instead of cocaine to deliver the kick. The cocaine was taken out in the early twentieth century. But the underlying concept is essentially the same. We like soft drinks because they give us a boost, not just because they taste nice.

Although caffeine now reigns supreme as humanity's stimulant of choice, other stimulants have enjoyed popularity in the past. Cocaine was hailed as a panacea in the nineteenth century. Sigmund Freud was infamously an evangelist for the drug, during what scholars later referred to as 'the cocaine episode'. One of the most popular beverages in nineteenth-century Europe was Vin Mariani, 'the world famous tonic for body and soul'. What made Vin Mariani so exceedingly popular were its key ingredients: a bracing combination of alcohol and cocaine. The Pope approved so much of Vin Mariani that he awarded its inventor a special medal. Amphetamines such as Benzedrine were widely consumed in the 1940s and 1950s. And people in east Africa have for centuries been chewing khat leaves, which contain amphetamine-like alkaloid stimulants. Cocaine and amphetamines had the attraction of combating sleepiness, like caffeine. But their popularity waned when it became obvious that they were also dangerous and addictive. (Even so, around four million people in the USA alone still regularly consume cocaine.) Fortunately, moderate consumption of caffeine seems to carry minimal health risks.

What exactly does caffeine do to us? In 1714, Alexander Pope noted that coffee 'makes the politician wise, and see through all things with his half-shut eyes'. Caffeine exerts subtle influences on learning, memory and performance. There is also some evidence that it improves the body's capacity for physical exercise by increasing the contractility of muscle tissue and enhancing muscular stamina. Its principal effect, though, and the reason why billions of us consume it every day, is that it makes us less sleepy.

Numerous experiments have confirmed the everyday experience that caffeine boosts alertness. It does this in a dose-dependent manner: the bigger the dose, the bigger the boost. The corollary is that caffeine disrupts sleep: it delays the onset of sleep, reduces total sleep duration and increases the number of awakenings during the night. A moderately large dose of caffeine (400 milligrams) will provoke a rise in metabolic rate during sleep, together with an increase in light (stage 1) sleep and a reduction in deep, slow-wave sleep. Caffeine shifts REM sleep forward to the earlier part of the night, at the expense of slow-wave sleep. It also changes our subjective perception of sleep, mimicking the effects of insomnia. Decaffeinated coffee does not do this. But although caffeine suppresses sleepiness, which is the main symptom of sleep deprivation, it does not eradicate the underlying problem. A sleep-deprived person is still a sleep-deprived person, whether or not they are wired on caffeine.

Despite claims to the contrary, no one is impervious to the stimulant effects of caffeine. Some people insist that they can drink strong coffee in the evening without affecting their sleep. But the reason why they can still fall asleep is not because their bodies are miraculously immune to the pharmacological actions of caffeine: it is because their sleepiness is sufficiently strong to overwhelm the stimulant effects.

Just as an exhausted person can fall asleep on a hard floor, so too can they fall asleep with high levels of caffeine in their bloodstream. That same exhausted person will probably also depend on caffeine to keep them awake the next day.

The great nineteenth-century gastronome Jean-Anthelme Brillat-Savarin noticed this insidious aspect of habitual coffee drinking. He observed that people who were not prevented by coffee from sleeping at night needed coffee to keep them awake by day, and never failed to fall asleep during the evening if they had not drunk coffee after dinner. There were many people, he noted, who slept during the day if they did not drink their morning cup of coffee. Although Brillat-Savarin was himself a connoisseur of coffee, he maintained that it was a far more powerful beverage than was commonly believed:

> A man of sound constitution can drink two bottles of wine a day, and live to a great age; the same man could not stand a like quantity of coffee ... It is the duty of all papas and mammas to forbid their children to drink coffee, unless they wish to have little dried-up machines, stunted and old at the age of twenty.

Caffeine acts on the central nervous system through biochemical mechanisms that are reasonably well understood. It boosts metabolism throughout the brain, while also reducing blood flow in the brain. In large doses, it increases the amount of the hormone adrenaline circulating in the bloodstream. But the main biochemical mechanism through which caffeine boosts alertness is by blocking brain receptor sites for a substance called adenosine. Normally, adenosine dampens the activity of other neurotransmitters. By blocking its dampening action, caffeine stimulates brain activity. To put it another way, caffeine sabotages the natural brake that would, left to its own devices, make us feel sleepy.

Virtually all the caffeine you ingest, in whatever form, is absorbed and spreads throughout your body. Caffeine acts quickly, starting to affect your brain within 15–30 minutes. It then lingers in your body for several hours. The half-life of caffeine – that is, the time taken for the concentration in your bloodstream to drop by 50 per cent – is between three and seven hours, with an average of about four hours. Therefore, if you drink several cups of strong coffee during your harassed working day, your blood will still contain substantial amounts of caffeine when you go to bed several hours later.

While no one is immune to caffeine, some individuals are unusually sensitive to it, and find that even modest amounts will disrupt their sleep. An individual's sensitivity to caffeine is partly determined by the rate at which their body metabolises it. Those with a history of being kept awake by caffeine tend to break it down it more slowly. In a true 'caffeine insomniac' the half-life of caffeine can be as long as seven hours, which means that the concentration in their blood will still be high when they are trying to get to sleep hours after drinking their last cup of coffee. There is no evidence that children are particularly more sensitive to caffeine than adults, which is just as well considering how much of it they push down their throats.

Manufacturers of soft drinks generally contend that they add caffeine to their products to improve flavour, and not because of the drug's stimulant properties. They would say that, though, wouldn't they? The scientific evidence suggests otherwise. An unbiased (double-blind) experiment conducted by scientists at Johns Hopkins University in the USA established that only 8 per cent of adults who regularly drank cola drinks could tell the difference between the caffeinated and decaffeinated versions. By comparison, they could all distinguish between a cola containing sugar and the artificially sweetened diet variety. Caffeine is not detectable by taste unless it is present in quite high concentrations, higher than those approved by the US Food and Drug Administration. Whatever the producers may claim, there seems little doubt that we prefer to have caffeine in our soft drinks primarily because of its stimulant effects, not its flavour. The same is true of coffee. In many countries, at many times, most coffee has been a foul-tasting brew. And yet people have continued to drink it every day. Taste cannot be the sole explanation. If your favourite coffee outlet secretly replaced its normal brews with decaffeinated, its loyal addicts would rapidly switch their custom elsewhere for their daily fix.

We sometimes joke about being 'addicted' to caffeine, but is it actually addictive? It is difficult to be wholly objective about this contentious issue, because if caffeine is addictive then the vast majority of the scientists investigating it, together with virtually all their experimental subjects, are themselves addicts.

At face value, caffeine does have the basic properties of an addictive drug. It acts on the brain, it produces pleasurable stimulatory effects on mood and behaviour, regular consumers become more tolerant of its pharmacological effects, and they suffer withdrawal symptoms if they stop taking it. Moreover, some individuals find they cannot break their habit, despite a strong desire to stop consuming caffeine. By these criteria, at least, most humans are caffeine addicts.

Caffeine resembles addictive drugs such as nicotine in its ability to exert a powerful, unconscious influence on our behaviour and choices. In one experiment at Johns Hopkins University, volunteers were given either red or blue capsules, containing either a dose of caffeine or an inert substance, on two successive days. Everyone received caffeine on day one. On day two, everyone received the inert placebo in a capsule of the opposite colour. Hence, if someone received caffeine in a blue capsule, they got the placebo in a red capsule, and vice versa. The subjects were all regular consumers of caffeine but they had no idea they were being given caffeine in the experiment. They were told simply that the experiment was to investigate the effects of chemicals found in common foodstuffs (which was true, after a fashion). On the following day each subject was asked to choose either a red or a blue capsule. A large majority (80 per cent) chose the capsule containing caffeine, regardless of its colour. They had unconsciously learned to associate the rewarding effects of the caffeine with the corresponding colour of the capsule. When given the choice, even an unconscious choice, they sought caffeine.

Another hallmark of addictive drugs is that they cause withdrawal symptoms, and caffeine is no exception. Habitual caffeine-dopers who suddenly stop taking their daily doses usually experience transient withdrawal symptoms, including headaches and impaired performance. In one experiment, researchers switched unsuspecting coffee drinkers to decaffeinated coffee without telling them. The duped subjects reported an array of unpleasant symptoms such as headaches, sleepiness, increased heart rate and a sense of reduced well being. These withdrawal symptoms subsided after a few days.

Withdrawal symptoms occur even in people who normally consume as little as 100 milligrams of caffeine a day, equivalent to one or two cups of instant coffee. In another experiment, scientists withdrew all caffeine from volunteers whose average daily intake had been 380 milligrams (which is regarded as moderate). After one night without caffeine the volunteers were assessed in a double-blind trial: some of them unknowingly received caffeine and some unknowingly received an inert placebo. Those who received the placebo ended up being deprived of caffeine for a stretch of 19 hours, though they were unaware of this. The caffeine-deprived subjects reported feeling unwell, worn out, less attentive and less helpful towards others. Some had headaches and even upset stomachs. The withdrawal symptoms were bad enough to make the placebo seem positively aversive to those who had received it, even though they did not know it was a placebo. When given the choice, they chose to forfeit money in preference to receiving the placebo again. Once again, caffeine was shown to exert a powerful unconscious tug on human behaviour. One of the reasons why consuming caffeine becomes a habit, if not an addiction, is that we unconsciously learn to avoid these unpleasant withdrawal symptoms. Some scientists have estimated that between 9 and 30 per cent of people who regularly consume caffeine satisfy the formal psychiatric diagnostic criteria for a substance-dependence syndrome.

Caffeine is clearly not in the same league of addictiveness as Class A drugs or nicotine. The caffeine highs of pleasure and the cold turkey lows of withdrawal are piffling in comparison to those of, say, cocaine or heroin. Experiments with rats have established that caffeine does activate the particular brain region that is thought to play a key role in addiction (the nucleus accumbens). But caffeine does this only at high doses, far beyond the amounts consumed in normal life. In contrast, cocaine, amphetamines, morphine, nicotine and other powerfully addictive drugs activate the nucleus accumbens at lower doses. The implication is that caffeine has a smaller potential for addiction.

Nonetheless, it is difficult to escape the conclusion that most of us are hooked on this friendly, civilised, apparently safe drug. And we are hooked mainly because caffeine provides a daily chemical crutch to keep us alert and upright in the face of inadequate sleep. If this is true, then how did people in Europe cope before coffee arrived in the seventeenth century? They coped without caffeine because they mostly rose with the sun and went to bed when it got dark. They got plenty of sleep and had no need of a chemical stimulant to keep them going.

Sister alcohol

> 'Drink, sir, is a great provoker of three things.'
> 'What three things does drink especially provoke?'
> 'Marry, sir, nose-painting, sleep, and urine.'
>> William Shakespeare, *Macbeth* (1606)

We consume caffeine to help us stay awake, and really for no other reason. We drink alcohol for all sorts of reasons, and one of them is to help us go to sleep. Alcohol is widely used as an aid to sleep and has been for many centuries. According to the seventeenth-century biographer John Aubrey, the great English philosopher and politician Sir Francis Bacon 'would often drink a good draught of strong Beer to bedwards, to lay his working Fancy asleep, which otherwise would keep him from sleeping a great part of the night'. In the fifth edition of his bestseller *The Old Man's Guide to Health and Longer Life*, published in 1764, Doctor Hill gave this timeless advice on how to slip the knot of insomnia:

> If he do not rest well any night, let him still rise at the same time the following day: and the next evening he will sleep better. If not let him next day go into a warm bath; and indulge himself with a glass more wine, a little before bed time. This will take off his watchfulness; and he will sink into the most pleasing slumber.

The tradition of the sleep-inducing nightcap continues to this day. For example, a survey in the 1990s found that 13 per cent of adults in the USA had used alcohol within the previous year specifically to help them sleep. The figure is probably higher in more bibulous European cultures.

Alcohol is a sedative, so it is no surprise that it helps us fall asleep. When you drink alcohol it is absorbed rapidly into your bloodstream, especially on an empty stomach, and it gets to work quickly. Even relatively small doses will facilitate sleep, although regular drinkers develop tolerance and therefore have to drink progressively more to achieve the same sedative effect. When consumed at lunchtime, alcohol deepens an afternoon nap by increasing the amount of slow-wave sleep.

Unfortunately, alcohol's influences on sleep are less uniformly benign than most of us would like to believe. Alcohol will help you to fall asleep faster, but it will also disrupt the sleep that ensues. In particular, alcohol reduces the amount of REM sleep during the second half of the night. Larger doses of alcohol cause more disruption to sleep, and doses greater than about 0.3 grams per kilogram of body weight (which is not much) usually do more harm than good as far as sleep is concerned.

Even a moderate amount of alcohol consumed several hours before bedtime will provoke measurable changes in sleep. In one experiment, healthy middle-aged men were given a middling dose of alcohol (0.55 grams per kilogram) six hours before bedtime. By the time they went to sleep there was no trace of alcohol on their breath. Nevertheless, EEG recordings revealed significant changes in their sleep patterns, compared with measurements taken on another day on which they had drunk only mineral water. Alcohol reduced their sleep efficiency, total time asleep and REM sleep. The men also felt subjectively that they had not slept as deeply. Studies of normal people in everyday life have discovered that, on average, the more alcohol someone drinks the less sleep they get. Alcohol can even disrupt the sleep of the unborn foetus. Researchers gave heavily pregnant women two glasses of wine each and monitored the sleep of their unborn babies. Even this small amount of alcohol had a detectable influence on foetal sleep patterns, especially REM sleep.

There is much folklore about the abilities of different sorts of alcoholic drinks to cause drunkenness and hangovers. To see whether whisky and vodka differ in their ability to disrupt sleep, researchers monitored healthy volunteers in a sleep laboratory after they had drunk either Scotch or vodka on three consecutive nights. The Scotch and vodka contained the same dose of alcohol and generated the same concentration of alcohol in the blood. The results showed that the two tipples were also equally effective at disrupting sleep and reducing blood oxygen levels. After drinking either Scotch or vodka (as compared to orange juice) the subjects spent less time in bed, less time asleep and less time in REM sleep; their blood-oxygen levels were lower and there were more episodes in which their blood oxygen fell below 90 per cent of normal. When it comes to disrupting sleep, the main culprit appears to be the alcohol content of the drink, not the vehicle in which it arrives – although if alcohol is consumed in the form of large volumes of beer, the drinker's sleep may be further disrupted by the urgent need to urinate during the night.

The disruption of REM sleep by alcohol might contribute to the self-loathing, depression, mood disturbance and general ghastliness that epitomise a really bad hangover. In chapter 12 we shall see that one of the main biological functions of REM sleep is to consolidate memories and regulate mood. Anything that disrupts this process is going to leave sticky fingerprints the next day.

When alcohol consumption tips over into alcohol abuse, the sleep-disruptive effects become hard to overlook. Insomnia is a notorious side effect of alcoholism. One study, for example, found that 61 per cent of alcoholics entering a treatment programme had been suffering from insomnia during the preceding six months. The intimate relationship between alcohol, sleep and insomnia can create a vicious cycle: alcohol makes it easier to go to sleep, which is attractive to insomniacs, but it also aggravates their insomnia by disturbing their sleep during the night. Tired alcoholics reach for the bottle as a way of treating their sleep problem and end up making both their insomnia and their alcoholism worse. Alcoholics receiving treatment for their dependency are twice as likely to relapse if they have insomnia. A study of alcoholics undergoing treatment found that those who relapsed took longer to fall asleep at night and had less slow-wave sleep than those who remained abstinent. Most alcoholics display long-lasting abnormalities in their sleep patterns even after they stop drinking. For the first few weeks after giving up, their sleep continues to be short, shallow and fragmented. It does gradually improve, but some aspects of sleep can remain abnormal even after two years of abstinence.

Another unwelcome aspect of drinking alcohol at bedtime is that it exacerbates a sleep-related breathing disorder known as sleep apnoea (which we shall be examining more closely in chapter 15). Alcohol does this by relaxing the muscles in the upper airway and impeding the flow of air. Even moderate amounts of alcohol will increase the resistance to airflow in the nasal passages, making it harder to breathe through the nose. The net result is disrupted breathing during sleep. Researchers measured the effects of alcohol on men who snored but had no other symptoms of sleep apnoea. After drinking only a moderate amount of alcohol there was a significant rise in the resistance to airflow in their upper airways, and a doubling in the frequency with which their breathing was interrupted. In people who have been diagnosed with sleep apnoea, alcohol makes the problem worse. Even a modest amount of alcohol consumed before going to sleep significantly increases the severity of sleep apnoea.

In view of all this off-putting evidence, many sleep experts recommend that anyone with a sleep disorder, especially sleep apnoea, should abstain from drinking alcohol. An alternative (but deeply reprehensible) conclusion might be that the determined drinker who still wants to get a good night's sleep should drink at lunchtime rather than in the evening.

Tobacco

A custom loathsome to the eye, hateful to the nose, harmful to the brain, dangerous to the lungs, and in the black, stinking fume thereof, nearest resembling the horrible Stygian smoke of the pit that is bottomless.

King James I, *A Counterblast to Tobacco* (1604)

Smoking tobacco is the riskiest thing that many people will ever do in their safe twenty-first-century lives. Smoking 20 cigarettes a day is more dangerous in the long run than working in most supposedly hazardous professions, let alone the occasional bungee jump or ski trip. Tobacco kills vast numbers of people around the world every year through coronary heart disease, cancer, emphysema and other illnesses. We all know that smokers are more likely to die of lung cancer and heart disease. We have all heard the dreadful statistics. But one of the lesser-known consequences of smoking is disrupted sleep.

Nicotine is a central nervous system stimulant, and it interferes with sleep. It increases the frequency of awakenings during the night, resulting in lighter, less refreshing sleep. The evidence shows that smokers experience more sleep problems than non-smokers. On average, they find it harder to get to sleep and harder to stay asleep. They also feel sleepier during the daytime. A study that assessed the sleep patterns of people living in rural Oxfordshire found that the smokers among them got significantly less sleep on average than the non-smokers. The direct effects of tobacco are compounded by the fact that smokers also tend to consume more caffeine than non-smokers.

One tiny consolation for smokers comes if they are forced to stay up all night. Scientists compared the mental performance of smokers and non-smokers during a night of total sleep deprivation. Every two hours throughout the night, volunteers were subjected to a battery of tests designed to measure their concentration, alertness, reactions, short-term memory and hand-eye coordination. As the night wore on, the smokers reported feeling almost as tired as the non-smokers, but their actual performance deteriorated less. The reactions of the non-smokers slowed down as they became more fatigued, whereas the smokers' performance remained relatively unaffected until the early hours of the morning. Like caffeine, nicotine provides a chemical boost when fatigue looms.

Human health is replete with irony. Smoking disrupts sleep, but so does giving up smoking. People who have recently quit smoking suffer nicotine withdrawal symptoms that interfere with their sleep. For the first few nights their sleep is more fragmented. Fortunately, these withdrawal symptoms do not last long and they can be eased by nicotine-replacement aids. In the long run, smokers' sleep improves when they kick the habit.

Food for sleep

> Any meat that in the daytime we eat against our stomachs, begets a dismal dream.
>
> Thomas Nashe, *The Terrors of the Night* (1594)

Eating a large meal is said to induce drowsiness. Is this true? Aristotle, who was otherwise far ahead of his time on the subject, argued that sleep was triggered by the intake of food, which made vapours rise from the stomach to the head and thereby cooled the heart. 'That is why,' he asserted, 'spells of sleep follow especially upon the intake of food: it is because the moist and solid matter are rising in a dense mass.' Sir Thomas Browne, writing in the seventeenth century, thought that different types of food had distinctive effects on dreaming:

> Physicians will tell us that some food makes turbulent, some gives quiet dreams. Cato who doted upon cabbage might find the crude effects thereof in his sleep; wherein the Egyptians might find some advantage by their superstitious abstinence from onions. Pythagoras might have more calmer sleeps if he totally abstained from beans. Even Daniel, that great interpreter of dreams, in his leguminous diet seems to have chosen no advantageous food for quiet sleeps.

According to an ancient but still prevalent belief, certain types of food engender nightmares. Cheese and spicy foods are often singled out for blame. When Ebenezer Scrooge first encounters Marley's Ghost in *A Christmas Carol*, Charles Dickens's old curmudgeon attributes the apparition to a food-induced nightmare:

> 'You don't believe in me,' observed the Ghost.
> 'I don't,' said Scrooge . . .
> 'Why do you doubt your senses?'
> 'Because,' said Scrooge, 'a little thing affects them. A slight disorder of the stomach makes them cheats. You may be an undigested bit of beef, a blot of mustard, a crumb of cheese, a fragment of underdone potato. There's more of gravy than of grave about you, whatever you are!'

Jean-Anthelme Brillat-Savarin was adamant that diet has a profound effect on sleep. He wrote that 'theory and experience are united in proving that the quality and quantity of food consumed exerts a powerful influence on work, rest, sleep, and dreams'. In Brillat-Savarin's view, certain foods were good at gently inducing sleep: he singled out lettuce and foods rich in milk. Other foods of a 'stimulating' nature caused dreams. Among the alleged dream-inducers, Brillat-Savarin listed red meat, pigeon, duck, hare, asparagus, celery, truffles, vanilla and spices. Nevertheless, the great gastronome believed it would be a mistake to banish these foods from the diet, since the dreams they engendered were mostly pleasurable and helped to 'prolong our existence'.

What light can science cast here? There is some empirical evidence that eating does induce daytime sleepiness, over and above the normal circadian variations in alertness. In one experiment volunteers ate meals at different times of day, so that the effects of eating and the circadian fluctuations in wakefulness could be teased apart. The subjects were significantly sleepier about 90 minutes after eating. Their sleepiness could not be attributed solely to the circadian rhythm, suggesting that eating had an independent influence. To distinguish between the effects of eating (as in chewing) and digestion (as in what happens after you swallow) researchers compared the effects of eating a solid meal versus sham feeding. On one day volunteers ate an evening meal, while on another day they chewed the food but spat it out before swallowing. They were then asked to nap. The subjects fell asleep more quickly after eating the real meal than after the sham feeding. In another experiment, volunteers fell asleep sooner after eating a solid meal than after drinking an equivalent volume of water. Both findings suggest that post-prandial sleepiness stems from digestion rather than ingestion.

One reason why we feel slightly sleepier after eating a large meal may be that digestion evokes a temporary rise in body temperature, followed by a decline. As we saw earlier, one of the physiological cues that helps to trigger the onset of sleep is a gently falling body temperature.

What you eat, as opposed to how much you eat, seems to have little systematic influence on the quality of sleep. Scientists have searched for connections between the composition of diets and the quality of sleep, but have found few. Fat and carbohydrates do not differ greatly in their effects on sleepiness. The consequences of eating high-fat, low-carbohydrate meals are broadly indistinguishable from those of low-fat, high-carbohydrate meals containing the same amount of energy. Very spicy foods can disturb sleep, however. Australian scientists conducted an experiment in which volunteers ate an evening meal laced with tabasco sauce and mustard. Their subsequent sleep was distinctly disturbed: they spent more time awake during the night and had less slow-wave sleep; their body temperature was also elevated during the first part of the sleep cycle. The rise in body temperature was probably caused by the chemical capsaicin, which is the main spicy component of red peppers. The raised temperature was probably responsible for the sleep disruption. (After a meal like that, anyone sharing a bed with the sleeper is likely to have a disturbed night too.)

The belief that certain foods stimulate nightmares has little solid evidence to support it, but this may be because no one has looked very hard. There are certainly plausible mechanisms to explain how it might happen. One possibility is that foods rich in certain chemicals could alter the balance of neurotransmitters in the brain. Many supposedly dream-inducing things, such as cheese, bacon, chocolate and wine, contain lots of tyramine – a chemical known to raise the level of the hormone noradrenaline in the central nervous system. A more prosaic explanation involves indigestion. High-fat foods such as cheese take longer to digest and can therefore disrupt sleep. Very spicy foods can also disrupt sleep, as we saw. If your sleep is disrupted, for whatever reason, you are more likely to remember your dreams, and therefore more likely to remember your *bad* dreams.

Exercise is bunk, isn't it?

> I have never taken any exercise, except sleeping and resting, and
> I never intend to take any. Exercise is loathsome.
>
> Mark Twain (1835–1910)

We live in an era when physical fitness is, quite reasonably, held in
high esteem and exercise is widely regarded as a panacea. There have
always been dissenting voices, however. Those who shun the flapping
shorts and sweaty trainers can always find succour in the Bible:

> But refuse profane and old wives' fables, and exercise thyself
> rather unto godliness. For bodily exercise profiteth little: but god-
> liness is profitable unto all things.

Far older than the belief that exercise promotes health is the belief
that it promotes sleep. But does it? Some research has found that
physical exercise elicits a modest increase in slow-wave sleep, with a
concomitant reduction in REM sleep. One study, for example, found
that long-distance runners spent more time in slow-wave sleep than
non-runners. Similarly, a group of athletes who had run a long-
distance road race displayed significant increases in slow-wave sleep,
while volunteers who had ridden a bike for four and a half hours had
less REM sleep. (Perhaps less encouragingly, their testosterone levels
were also lower.)

The evidence is far from consistent, however. Another study monitored the sleep of very fit men after they had completed a gentle
'training run' of 15–20 kilometres, and again after a race of twice that
distance. After the long race there was an increase in light (stage 2)
sleep, but the predicted increase in slow-wave sleep failed to appear.
The training run of 15–20 kilometres – which would be enough exercise to send some people to sleep for ever – had no discernible effect
on their sleep. It may just be that these men were already so fantastic-
ally fit that there was no scope left for exercise to improve their sleep.
More modest feats of athleticism might well have a bigger impact on
the sleep of ordinary, unfit people in their everyday lives.

Despite its apparently modest influence on objective measures
of sleep, many people find that regular exercise helps their sleep.
This experience is echoed by research which found that people who

exercised at least once a week were less likely to suffer from sleep disorders. Regular, gentle exercise can be especially beneficial for elderly people, who are generally more prone to insomnia. Work at the Netherlands Institute for Brain Research found that fitness training helped elderly people to sleep by reducing the disturbances in their circadian rhythms. Similarly, researchers found that men in their seventies who were more active during the day tended to sleep more efficiently than less active individuals.

Although regular exercise seems to improve sleep, it is often said that exercising vigorously just before going to bed will disrupt sleep and should therefore be avoided. But even this belief may be questionable. A laboratory experiment in which fit cyclists exercised for three hours before going to bed found that it made no difference to their sleep patterns. Once again, though, it may be that vigorous bedtime exercise would have a bigger impact on the sleep of someone who is not used to it.

In so far as exercise does modify sleep patterns, the mechanism probably involves body temperature. Artificially warming the body elicits a small increase in slow-wave sleep, similar to that seen following exercise. The importance of temperature was borne out by an experiment in which physically fit volunteers exercised by running. They ran both in hot conditions, which raised their body temperature by more than two degrees Celsius, and in cool conditions, where their temperature rose by only one degree. After exercising in the heat, the predicted rise in slow-wave sleep occurred when the runners slept that night. But there was no such rise in slow-wave sleep after exercising under cool conditions. It seems that exercise must make you hot if it is to affect your sleep.

The many benefits that exercise brings to physical and mental health have been widely publicised, with the result that most people now know that physical exercise is good for them (even if some cannot force themselves to do anything about it). A growing number are reaping the rewards of regularly using their bodies in the way that nature intended their bodies to be used. If only that were true for sleep. We might live longer and happier lives if we took our beds as seriously as we take our running shoes.

Things that go bump in the night

> Take thou of me sweet pillows, sweetest bed,
> A chamber deaf to noise and blind to light;
>> Sir Philip Sidney, *Astrophel and Stella* (1591)

We live in an increasingly crowded and noisy world. More people than ever before live within earshot of a busy road, an airport, a railway line, a factory, a bar, a late-night shop or some other source of sleep-disrupting sound. The traffic in our skies and on our roads grows relentlessly, while the 24-hour society means that many other sources of noise are active for larger portions of each day. In some urban environments it is never quiet, not even at four in the morning. Noise is probably the single biggest source of environmental pollution that most governments have done little to tackle.

We all know that noise can disrupt our sleep, just like we all know it is bad to smoke and good to take regular exercise. But we usually act as though noise made no difference. This may be a rational response, since noise is one of those sleep-eroding curses we can seldom control. Short of burying yourself in splendid rural isolation (if you can find any), there may be little you can do to block out the sounds that interfere with your sleep. Some people suffer considerable stress as a result. A British anti-noise lobby group has estimated that five people in the UK commit suicide every year because of persistent noise.

The disruption of sleep by noise is not a recent concern. It forms the basis of a pre-Biblical version of the Flood story that was recorded on Babylonian tablets dating from the seventeenth century BC. According to Babylonian mythology, a god called Enlil ruled over the earth while minor gods worked the fields. The minor gods eventually decided that they had had enough of working for Enlil, so they made humans out of clay, saliva and divine blood to act as their servants. Unfortunately, the noise of all the humans reproducing disturbed Enlil's sleep, as a consequence of which the angry god decided to destroy them with a flood. Fortunately, another god gave seven days' warning of the flood to a human called Atrahasis, who built a boat and loaded it with animals and birds. Atrahasis survived the flood, but the rest of humanity was destroyed. Enlil's apocalyptic reaction might seem a little harsh by modern standards, but if you have ever been woken in the small hours by persistent noise you might have longed for the option.

What effect does noise have on sleep? Clearly, we do not always wake up at the slightest sound, especially if the noise is familiar. When the same noises are repeated every night, as often happens in normal life, most sleepers habituate to some extent and become less likely to wake after repeated exposure. So, for example, people who have lived near an airport for a long time tend not to wake in response to routine aircraft noises. Only unusual noises will wake them. This does not mean, however, that sleepers become completely oblivious to a noise just because it is familiar. Even after weeks of repetition, a noise that fails to wake someone can still evoke measurable changes in their sleep EEG patterns. Noise can disrupt your sleep without waking you. As far as you are aware, you slept through the night, but you wake feeling less refreshed because your sleep was of lower quality. Noise tends to trigger a shift from deep sleep to shallow sleep, and it is particularly effective at curtailing REM sleep. Roads, railway lines, airports and other sources of noise still have a significant impact on the quality of sleep of people living nearby, even if many of them are not consciously aware of it.

The relationship between peak noise level and sleep disturbance is 'dose-dependent', which means the louder the noise, the worse the sleep. The threshold above which noise starts to affect sleep is about 40 decibels, which is not terribly loud. Noise levels of 60 decibels or more will often wake people and affect their mood the next day. Worse noise is not uncommon: a recent survey of noise levels in the UK found that people living under the flight path to an airport, or close to a motorway, were subjected to average noise levels greater than 65 decibels and often reaching 75 decibels. The quality of the noise also matters. Scientists have found a strong correlation between the extent to which people find a noise subjectively annoying when they are awake and its power to disrupt their sleep.

In a laudable attempt to tackle the problem, British inventors patented a soundproofed sleeping environment, consisting of a four-poster bed inside a triple-glazed Perspex capsule padded with sound-insulating foam. Ventilation is provided through sound-damping baffles and the capsule door opens automatically on a time switch. There are less drastic remedies, however. In *The Anatomy of Melancholy*, Robert Burton pointed out that soothing natural background noises are an aid to sleep, especially if they are just loud enough to mask the more irritating noises. He advised the would-be sleeper to lie near the pleasant murmur of gently trickling water – 'some flood-gates, arches, falls of water, like London Bridge, or some continuate noise which may benumb the senses'. On a similar note, James Boswell recorded that the laird of Dunvegan had chosen his bedroom specifically because there was behind it 'a considerable cascade, the sound of which disposed him to sleep'. Or you could just wear earplugs.

Shift work

For some must watch while some must sleep,
Thus runs the world away.

William Shakespeare, *Hamlet* (1601)

Humans are not built to work at night and sleep by day, but many do. Shift work disrupts sleep and makes people tired. It is bad for their health and makes them more likely to have accidents. None of this should come as any surprise. People working nonstandard hours, especially at night, are ignoring their biological clocks and constantly battling against their own physiology. They are forced to be active when their circadian rhythms say it is time for sleep, and to sleep when their bodies are geared up for activity. They consequently feel tired and perform worse when they are at work. And when they do go home to bed, they stay awake longer and sleep less.

Those who work night shifts or irregular schedules often complain of sleepiness. The research shows that they are not imagining it. Shift workers spend less time asleep on average than day workers and feel less alert when they are awake. One study, for example, found that workers doing on-call shift work got an average of 36 minutes less sleep each day than their colleagues. Even that small difference amounts to three hours' less sleep by the end of each working week. Many shift workers take naps to help compensate for their loss of sleep.

Shift work is not good for you. There is firm evidence that it heightens the risk of depression and cardiovascular disease. In one long-term study, researchers monitored the incidence of heart disease over a 15-year period, comparing the health of shift workers and daytime workers. The results showed that individuals who had worked shifts for 11 years or longer were more than twice as likely as day workers to develop heart disease. The longer someone had been working shifts, the greater was their risk of heart disease, regardless of age and smoking habits. Those who had worked shifts for more than 15 years faced almost three times the risk. The data also revealed that, having risen over time, the relative risk of heart disease then fell sharply after 20 years of shift work. The likely explanation for this fall was less than reassuring: it probably arose because many of the vulnerable shift workers had already developed heart disease and died by that late stage in their careers. Shift work had subjected them to a harsh form of selection in which only the more robust had survived.

Habitually working at night also increases the risk of certain cancers. Two recent scientific studies in the USA have established that women who often work night shifts face a substantially greater risk of developing breast cancer. Women who had routinely worked night shifts for several years were found to be up to 60 per cent more likely to develop breast cancer. The heightened risk appears to result from disturbances to the circadian rhythm in melatonin level. The hormone melatonin is normally produced at night, but its production is suppressed by exposure to bright light.

Some shift patterns are more disruptive than others. Overall, the evidence shows that evening shifts are less detrimental to sleep than night shifts or morning shifts. Where 24-hour operations require some people to work at night, the least disruptive way of achieving this seems to be by working slowly rotating schedules, with well-spaced alternations between morning, evening and night shifts. Slowly rotating shifts have less impact on sleep than permanent night working. Rapid changes in shifts are generally bad news: they affect sleep and alertness in ways that resemble jet lag, and for similar reasons. Workers on rapidly changing shifts exhibit more signs of stress and more emotional problems than those on fixed schedules. They also drink more alcohol. They can experience severe sleep disruption and are more likely to fall asleep at work. A study of hospital nurses found that those working rotating shifts were twice as likely to report having an accident or nodding off while driving than nurses who worked only days or only nights. One glimmer of hope comes in the form of melatonin, which has been used successfully to treat the sleep disruption caused by shift work.

At present, industrialised societies pay an indeterminate but undoubtedly heavy price for making people work according to artificial schedules that are at odds with their biological clocks and their environments. The individuals who work these shifts are generally less healthy, less happy, less productive and more accident-prone than those who work by day. These important considerations ought to form part of the balance sheet when society judges the costs and benefits of different work patterns.

Poppy, mandragora and drowsy syrups

> If sleep be wanting a small dose of syrup of diacodium every night will give it safely.
> Doctor Hill, *The Old Man's Guide to Health and Longer Life* (1764)

What of the many traditional remedies that have been used to assist sleep throughout the centuries? More and more people are turning back to herbal medicines for relief from insomnia. By the end of the 1990s one in eight American adults were using herbal preparations in some form.

Drugs that induce, prolong or improve the quality of sleep are known as hypnotics. There are hundreds of them, some ancient, some modern. William Shakespeare was familiar with several:

> Not poppy nor mandragora,
> Nor all the drowsy syrups of the world,
> Shall ever medicine thee to that sweet sleep
> Which thou owed'st yesterday.

The poppy seed is the source of opium, from which morphine, codeine and heroin are derived. Opium was long used as a sedative, even with its inherent drawbacks of inducing hallucinatory dreams and relaxing the smooth muscles of the gut. A self-help medical book from the early nineteenth century issued this pragmatic warning:

> We caution Bad Sleepers to beware how they indulge in the habit of exciting sleep by taking any of the preparations of Opium – they are all injurious to the Stomach, and often inconvenient in their effect upon the Bowels.

Plants of the genus *Mandragora*, better known as mandrake, were used in ancient times both as a narcotic and an aphrodisiac (which must occasionally have been frustrating both for the consumer and the object of their affections). The root of the mandrake was thought to resemble a small human and would, according to folklore, shriek when the plant was plucked from the soil.

Most other traditional hypnotics are also derived from plants. They include valerian (the dried root and rootstock of the garden heliotrope), skullcap, passionflower, wild lettuce, catnip, hops, lavender, and various herbal teas made from camomile, lemon balm or lemon verbena. Carpet moss (of the aptly named genus *Hypnum*) was believed to induce sleep and was sometimes used as the stuffing for mattresses. Other ancient remedies for insomnia have included anointing the soles of the feet with the fat of a dormouse and (my favourite) rubbing your teeth at bedtime with the earwax of a dog. Recommend that one to any gullible insomniac you secretly dislike.

Some of these remedies (but not, alas, the canine earwax) have been proved objectively to work, in a modest way. Several experiments have shown that an extract of valerian root can slightly improve subjective sleep quality. For example, a two-week programme of doses of valerian herbal extract enhanced sleep efficiency. The control subjects in this study, who unknowingly received an inert placebo instead of valerian, also displayed improvements in their sleep – yet again demonstrating the awesome power of the placebo effect to make the world a better place. However, the valerian produced some modest improvements over and above the placebo effect. Nutmeg extract has also been shown to evoke an increase in the duration of sleep – at least, in young chickens.

Another 'natural' and increasingly popular hypnotic is tryptophan, an amino acid that occurs naturally in our diet. Dairy products are rich in tryptophan, which might help to explain why hot milky drinks are traditionally recommended as a bedtime boon to slumber. The much-vaunted hormone melatonin is another supposedly 'natural' sleep-inducing product (natural in so far as any hormone is natural). Melatonin is present in substantial quantities in the first urine of the morning. This might be why some people drink their own urine every morning, believing it to bring physical and mental benefits. Urine-drinking is a traditional yogic practice that is still surprisingly widespread. But if in doubt, it may be best to stick to tea.

Smells can help too. A small-scale study found that the sweet aroma of lavender oil assisted elderly insomniacs to fall asleep and stay asleep. In fact, sniffing lavender was as effective as taking sleeping pills. How smell might influence sleep remains unclear. The placebo effect probably plays a role, since the consumer must presumably have some faith in the power of aromatherapy to help them sleep. It has been suggested that the chemical constituents of some aromas might exert a direct influence on the brain, but this remains conjecture.

Now to more conventional pharmacology. Non-prescription sleeping drugs are daily consumed in vast quantities. The biggest consumers are the elderly, who are more prone to insomnia. A Canadian investigation found that more than one in four elderly people had used non-prescription hypnotics over the previous year, including antihistamines, paracetamol, alcohol and herbal remedies. Antihistamines, of which there are many varieties, are normally used for treating congestion and allergies, but many also have a sedative effect. Antihistamines such as diphenhydramine, promethazine and pyrilamine reduce the time taken to fall asleep and increase sleep duration, making them reasonably effective treatments for mild to moderate insomnia. Diphenhydramine is the active ingredient in many non-prescription sleep medicines.

Specialised sleeping pills have somewhat fallen out of favour in recent years, largely because of their dubious past. Until the 1970s it was common for physicians to prescribe barbiturates for sleep problems. Unfortunately, these drugs were addictive and potentially dangerous. Users developed tolerance and required increasing doses, but overdoses could be fatal. The barbiturates were superseded by benzodiazepine drugs such as Librium and Valium, which were originally developed to treat anxiety rather than insomnia. They are less hazardous than barbiturates, but still far from ideal. Like alcohol, they assist the onset of asleep but alter the structure of the sleep that ensues. If you take benzodiazepine sleeping pills you are likely to get more shallow sleep (stages 1 and 2) at the expense of deep, slow-wave sleep (stages 3 and 4).

The latest generations of hypnotics have more to recommend them, however. In comparison with their chemical ancestors they are fairly harmless. Modern hypnotics such as zolpidem and zaleplon, which came into use in the 1990s, are short-acting, have little potential for addiction and produce few side effects. They offer an effective and relatively low-risk remedy for insomnia. Some experts have argued that they should be more widely prescribed for short-term relief. There is another school of thought that says you should try tackling the causes of insomnia before reaching for drugs to suppress its symptoms.

Hypnotic exotica

> If there's any illness for which people offer many remedies, you
> may be sure that particular illness is incurable.
>
> Anton Chekhov, *The Cherry Orchard* (1904)

Now for a brief foray into the wilder realms of curing and causing sleep problems. Tickling the soles of the feet was once practised in Russia as a means of encouraging sleep. Empress Anna of Russia, who reigned from 1730 to 1740, employed an old woman whose duty it was to tickle the soles of the Empress's feet as she was dropping off to sleep each night. In Nikolai Gogol's novel *Dead Souls*, a weary traveller is offered this service by his elderly hostess:

> 'Well, here's your bed all ready for you, sir,' said the old lady.
> 'Good night, sir, sleep well. Are you sure you don't want anything
> else? Perhaps you're used to having your heels tickled for the
> night. My late husband could not get to sleep without it.'

I have yet to discover whether heel-tickling really does make it easier to fall asleep, but it would not be a dangerous experiment to try.

In more recent times, various technical devices have been invented that purport to boost sleep by using electricity or electromagnetic radiation in some form. One type of device transmits small electric currents through the earlobes, using a technique that has been around since the 1960s called cranial electrical stimulation. This involves delivering pulses of high-frequency electric current through electrodes attached to the skin. Cranial electrical stimulation has a variety of measurable effects on human physiology, acting through mechanisms that are not well understood. These effects include boosting the action of some anaesthetics, altering the levels of certain neurotransmitters in the central nervous system, relieving pain, enhancing attention and, yes, relieving insomnia.

Other, more common, electronic devices may affect our sleep even though that is not their purpose. Frequent users of mobile phones should be aware of research indicating that the electromagnetic radiation emitted by their indispensable friend may subtly alter their brain activity during the early stages of sleep. Researchers exposed healthy young volunteers to the electromagnetic field generated by a mobile phone for half an hour, shortly before they went to sleep. The electromagnetic field mimicked the 900 MHz signal emitted by a typical GSM digital cellular phone and, as in real life, only one side of the head was exposed to the radiation. The outcome was a significant change in EEG patterns during sleep, principally an increase in NREM activity during the first hour of sleep. The changes in brain activity affected both sides of the brain equally, even though only one side of the head had been exposed to radiation. The fact that brain activity is altered for up to an hour after exposure to the electromagnetic field suggests that it affects the brain's chemistry in some way.

This discovery ties in with other work showing that mobile phone radiation affects the waking brain. Researchers found that exposure to a simulated GSM cellular phone signal significantly improved the reaction times of volunteers in choice tests. Whether the underlying changes in their brains were good, bad or irrelevant remains unclear. But in the absence of evidence to the contrary, it would probably be wise to presume they were not good. Evolution is unlikely to have equipped us with brains that improve on exposure to high-frequency electromagnetic radiation.

Low-frequency electromagnetic fields, of the sort generated by electricity power lines, also have a dubious reputation. For decades there have been claims that people living near overhead power lines are more prone to various health problems. The evidence suggests that there might be some basis to these concerns. Researchers monitored the sleep patterns of healthy volunteers who slept both with and without exposure to a low-frequency (50 Hz) electromagnetic field. There were some clear effects. When exposed to the electromagnetic field, the subjects spent less time asleep, got less slow-wave sleep and slept less efficiently. One might conclude from this that it is better not to sleep under an electricity pylon. Surely, though, the wisest of all words on how to nurture a good night's sleep must be this advice written in 1602 by one William Vaughan:

Let your night cappe have a hole in the top through which the vapour may goe out.

PART IV

Dreams

9

The Children of an Idle Brain?

I talk of dreams;
Which are the children of an idle brain,
Begot of nothing but vain fantasy.
William Shakespeare, *Romeo and Juliet* (1595)

To sleep, perchance . . .

Every night of our lives we spend two hours dreaming, possibly more.
Babies probably dream for seven or eight hours a day and it is possible
that foetuses do little else. By the time we die most of us will have
spent a quarter of a century asleep, of which six years or more will
have been spent dreaming. If you are, say, 30 years old, you will
already have notched up more than 30 months of dreaming, almost
all of which you will have forgotten. What is it all about?

Questions about the origins and meaning of dreams have exercised
the human mind throughout history. Aristotle set out the key issues
more than 23 centuries ago when he wrote:

We must inquire what dreams are, and from what cause sleepers
sometimes dream, and sometimes do not; or whether the truth is
that sleepers always dream but do not always remember; and if
this occurs, what its explanation is.

Thanks to scientific developments since the middle of the twentieth
century, we now know that dreaming occurs mainly, but not exclusively, during REM sleep and that our brains function in a fundamentally different way when it is happening.

If you are woken from REM sleep and asked, there is a probability of 80–90 per cent that you will recall dreaming. Your dream memories start fading very quickly as soon as you wake up, so you are more likely to remember dreaming if you are woken suddenly and asked about it straight away. Dreams extend throughout most of REM sleep, which is why there is a close relationship between the length of someone's REM sleep and the length of their dream reports. The experimental evidence also suggests that we dream in more or less real time. Dreams can be brief, but the old notion that they all occur in a flash at the moment of waking up has been discredited.

Our dreams are hugely variable and unpredictable, but several common strands can be discerned. Most dreams occur in the present and nearly all involve the dreamer. As Evelyn Waugh put it, 'One can write, think and pray exclusively of others; dreams are all egocentric.' Of course, dreams also involve other people, some familiar, some not. When Harvard researchers analysed hundreds of dream reports they found that about half the characters in dreams were identifiable individuals, known personally to the dreamer, while about a third were generic characters defined by their relationship to the dreamer or by their social role (for instance, an unspecified friend or a policeman). Fewer than one in six dream characters were completely novel.

Analyses of dream reports have also thrown up many curious observations, some of which have no obvious explanation. For instance, people with high verbal creativity have shorter dreams on average, while those with high blood pressure have dreams containing more hostility. Dreams that occur in sleep laboratories – the sort that have been most extensively studied – tend to be less rich and less surreal than dreams dreamt in the familiar surroundings of home. Nightmares and orgasmic wet dreams, which are relatively common in normal life, seldom occur in the sleep laboratory. Do males and females dream differently? Dylan Thomas apparently thought so. *Under Milk Wood* famously begins with this description of the small Welsh town at night, its inhabitants tucked up and dreaming:

> Young girls lie bedded soft or glide in their dreams, with rings and trousseaux, bridesmaided by glow-worms down the aisles of the organplaying wood. The boys are dreaming wicked or of the bucking ranches of the night and the jolly, rodgered sea.

The evidence on this point is sparse. However, scientists have established that, for reasons unknown, males dream of males more often than females dream of males. This peculiar sexual asymmetry in dreaming is universal; it has emerged from at least 29 different comparisons of male and female dreams and holds true for children, adolescents and adults in all parts of the world. Finally, research in the USA has discovered that unattached people who want to be in a personal relationship recall more dreams and dream more vividly than people who are either in a secure relationship or do not care. This fits with the idea, which we will explore later, that dreaming has something to do with processing memories and emotions.

Dreams certainly have a large emotional content, and the emotions are just as likely to be negative as pleasurable. Numerous studies have concluded that unpleasant emotions, such as anxiety or fear, occur in the majority of dreams. One of the earliest scientific investigations of dreams was conducted by two American women called Sarah Weed and Florence Hallam at the end of the nineteenth century. They analysed hundreds of dream reports written by women students, and found that unpleasant dreams outnumbered pleasant dreams by two to one. The most common negative emotions, in descending order of prominence, were 'perplexity and hurry', discomfort and helplessness, fear, anger, disappointment and shame. But despite this predominance of negative emotions, most of the subjects said they still looked forward to their dreams. Subsequent research has uncovered a similar preponderance of darkness over light in the emotional contents of dreams. The dream reports of insomniacs are even more depressing and more negative in content than those of non-insomniacs. This may be partly because insomniacs wake more often during the night and therefore remember more of their dreams, most of which are gloomy.

The strong emotional flavour of dreams is borne out by brain scanning. Measurements of brain activity using PET brain scanning have revealed that the brain regions most closely involved in mediating conscious emotions (the pontine brainstem and the limbic and paralimbic regions of the cortex) are also very active during REM sleep. Conversely, the brain regions involved in rational thinking and decision making (principally the prefrontal cortex) are less active. Judging from these activity patterns, the brain is focusing on the emotional rather than the rational during REM sleep.

Dreams are predominantly visual. We mostly *see* dream images rather than touching or tasting them. Nevertheless, all of the different senses are capable of featuring in dreams, and sounds are common. Sound occurs in more than half of all dream reports and about 10 per cent of dreams include sensations of movement, often falling. Tastes or smells feature in about 1 per cent of dreams. Most people experience sensations of taste or smell in their dreams at some stage in their lives.

Individuals who have been blind since birth or early childhood have few, if any, visual images in their dreams. Instead, their dreams are composed largely of emotions, sounds and touch. They also display few, if any, rapid eye movements during dream sleep, which again hints at some link between the visual images in dreams and the accompanying eye movements. However, people who lose their sight later than about seven or eight years of age usually retain extensive visual imagery in their dreams. In that sense, at least, some blind people can still see. A German called Heermann was probably the first scientist to report the absence of visual imagery from the dreams of blind people, in a paper published in 1838. Heerman studied the dreams of 100 blind people. He found that visual sensations and ideas were completely absent from the dreams of all 14 of the individuals who had become blind before their fifth year of life, whereas all those who had lost their sight after the age of seven still had visual elements in their dreams, in some cases after more than 50 years of blindness. One could speculate that the dreams of foetuses would be mostly composed of sound and touch sensations, given the dearth of visual stimuli in the womb.

Our dreams are often wildly fantastical. But on closer inspection they are nearly all composed of elements derived from our own experiences. Their strangeness comes from the ways in which these mostly mundane ingredients are mixed together in madcap sequences that defy the conventions of time, causality, custom and common sense. That is presumably why adults are found to have dreams that are on average more bizarre than children's dreams; the adult's richer stock of experience and memories provides the dreaming brain with more raw ingredients from which to concoct its surreal nightly dishes.

A nineteenth-century American psychologist and pioneer of dream research, Mary Whiton Calkins of Wellesley College, observed this correspondence between dreams and reality when she analysed the contents of hundreds of dreams. She noted that there was almost always a connection between dreams and waking life; in only 11 per cent of cases was it impossible to detect any relationship between a dream and the dreamer's waking experiences. It was almost always the unimportant trivia of recent experience that formed the basis of their dreams, rather than anything of great import. Perspectives on the contents of dreams would change radically when Sigmund Freud came along a few years later.

Do dreams occur only during REM sleep, or might we dream all night long? The idea that dreaming is exclusively linked to REM sleep has been overturned. A lot depends, however, on what is meant by 'dream'. We tend to associate the word with the bizarre and visually rich narrative dreams of REM sleep, in contrast to the more mundane thought processes of waking consciousness. When people are woken from NREM sleep and asked if they were dreaming, they reply yes on about 7–8 per cent of occasions. However, if instead they are asked whether they were *thinking* about anything, they say yes far more often. In fact, people woken from NREM sleep report some form of mental activity on 40–60 per cent of occasions. Incidentally, dreaming or mental activity can be recalled after waking from the first NREM episode of the night, before any REM sleep has occurred, which rules out the possibility that NREM 'dreams' are merely memories of earlier REM dreams.

The 'dreams' that occur during NREM sleep are different. They are generally less vivid, less surreal, less unpleasant and less 'dream-like' than the classic dreams we associate with REM sleep and Salvador Dali paintings. They lack the bizarre story lines, the strong emotions and the intense imagery. In fact, they are more like conventional waking thoughts or fragments of ideas. NREM 'dreams' are also shorter and less complex than REM dreams. Research conducted at Harvard Medical School found that when people woke from REM sleep and described what was going on in their minds, their accounts were on average twice as long as those reported after waking from NREM sleep. More surprisingly, REM dream reports were also longer than reports of *daytime* mental activity. By this yardstick at least, REM dreams are richer in content than conscious, waking thoughts. Brain scanning and EEG recordings have shown that REM dreams and NREM dreams also differ substantially at the neurobiological level, involving quite different patterns of brain activity.

Dreaming, then, is definitely not synonymous with REM sleep. We also dream, albeit differently, during NREM sleep, and during the transitions between wakefulness and sleep (the hypnagogic and hypnopompic dreams we encountered in chapter 5). Even on a conservative definition, NREM sleep contains some dreams. If we broaden the criteria and allow for the fact that our memories fade rapidly on awakening, we might conclude that some form of dreaming occupies most of a night's sleep. If so, then you might have spent nine or ten years of your life dreaming by the time you hit the age of 30 and more than a quarter of a century by the time you die.

Do flies dream?

> We can also see dogs yapping and twitching in their dreams.
> Montaigne, *Essays* (1580–1592)

Do other species dream, or is it a uniquely human activity? Throughout history, scholars have pondered on whether animals dream, and nearly all have concluded that they do. One of the fundamental lessons to emerge from biology since Charles Darwin has been that other species have much more in common with humanity than humanity once preferred to believe. There are certainly no grounds for supposing that our ability to dream sets us apart from other species, and there is plenty of contrary evidence to suggest that other animals dream like us. Aristotle was firm in his belief that other animals dream, based mainly on the evidence of his senses:

> It appears, moreover, that it is not only human beings who dream, but also horses, dogs and cows, as well as sheep and goats, and the whole family of four-legged animals that give birth to live young. Dogs show it by whining in their sleep.

Aristotle's observation reinforced his view that dreams have biological rather than divine origins. 'Since some of the other animals dream,' he wrote, 'dreams could not be sent by God, nor do they occur for that purpose.' The body movements of sleeping animals led the sixteenth-century French essayist Montaigne to a similar conclusion:

> The greyhound imagines a hare in a dream: we can see it panting after it in its sleep as it stretches out its tail, twitches its thighs and exactly imitates its movements in the chase ... Guard dogs can be found growling in their sleep, then yapping and finally waking with a start as though they saw some stranger coming.

Charles Darwin's grandfather Erasmus Darwin was equally confident that other species dream. He wrote of how he had observed dogs sleeping by the fire who, by their barking and movements, 'appeared very intent on their prey'. Seventy years later, Charles Darwin recorded these observations about the likely existence of dreaming in other species:

> As dogs, cats, horses, and probably all the higher animals, even birds have vivid dreams, and this is shown by their movements and the sounds uttered, we must admit that they possess some power of imagination.

The weight of scientific evidence strongly supports the view of Darwin, and others before him, that dreaming is not a uniquely human capacity.

A dream within a dream?

> Is *all* that we see or seem
> But a dream within a dream?
>
> Edgar Allan Poe (1827)

It is both conventional and comforting to assume that dreaming and being awake are two profoundly dissimilar and easily distinguishable states. But it may be wrong. Are these two states really so very different? Many writers have questioned the conventional belief. For example, in 1630 or thereabouts the Spanish playwright Pedro Calderón de la Barca wrote a play called *Life is a Dream*, in which he explored the idea that distinguishing between waking life and dreams is not straightforward. In life, as in dreams, extraordinary things happen that do not seem so extraordinary at the time:

> Life's an illusion.
> Life's a shadow, a fiction.
> And the greatest good is worth nothing at all,
> For the whole of life is just a dream
> And dreams ... dreams are only dreams.

The uncertain boundary between dreaming and wakefulness is explored in Arthur Schnitzler's 1926 novella *Dream Story*, which inspired the 1999 Stanley Kubrick film *Eyes Wide Shut*. A married couple confess to each other their sexual fantasies and adventures, whose dreamlike qualities leave a lingering question mark over their reality. Do the faintly surreal events that so upset the couple really take place, or are they just dreams? At one point the husband, Fridolin, asks himself whether his vivid experiences have all been imagined:

> Wasn't he perhaps lying at home in bed this very moment – and hadn't everything he believed he had experienced been nothing more than his delirium? Fridolin opened his eyes as wide as he could, put his hand to his cheek and brow, and felt his pulse. Scarcely above normal. Everything was fine. He was fully awake.

The supposedly sharp boundary between dreaming and wakefulness is actually rather fuzzy. For a start, and as the philosopher Schopenhauer pointed out in the early years of the nineteenth century, the waking and dreaming states are both products of the same brain:

> For the function of the brain which, during sleep, conjures up a completely objective, perceptible, and even palpable world, must have just as large a share in the presentation of the objective world of our waking hours. For both worlds, although different in their matter, are nonetheless made from the same mould.

As the psychologist and philosopher Owen Flanagan has pointed out, waking thought is not the neat, controlled, logical process that we usually assume it to be; it is actually more dreamlike and less coherent than is usually acknowledged. Conversely, dreams can contain all the features of waking consciousness. Systematic analyses of dream reports have unearthed no cognitive or emotional features that are unique to the dream state. The differences between what goes on in our minds during wakefulness and dreaming seem to be more quantitative than qualitative.

Some bold thinkers have taken this idea a step further, by questioning whether there is any difference at all between waking and dreaming. Perhaps waking consciousness is just another form of dream? And how could we know anyway? The Chinese Taoist philosopher Chuang Tzu, writing more than two thousand years ago, put it like this:

> While men are dreaming they do not perceive that it is a dream. Some men will even have a dream within a dream. And so when the great awakening comes upon us, shall we know this life to be a great dream. Fools believe themselves to be awake now!

In *Alice Through the Looking-Glass* Lewis Carroll gloriously pushed the idea beyond reason, by imagining that each of us might be nothing more than a figment of someone else's dream:

> 'Come and look at him!' the brothers cried, and they each took one of Alice's hands, and led her up to where the King was sleeping . . .
> 'He's dreaming now,' said Tweedledee: 'and what do you think he's dreaming about?'
> Alice said, 'Nobody can guess that.'
> 'Why, about *you*!' Tweedledee exclaimed, clapping his hands triumphantly. 'And if he left off dreaming about you, where do you suppose you'd be?'
> 'Where I am now, of course,' said Alice.
> 'Not you!' Tweedledee retorted contemptuously. 'You'd be nowhere. Why, you're only a sort of thing in his dream!'
> 'If that there King was to wake,' added Tweedledum, 'you'd go out – bang! – just like a candle!'

Alice aside, the notion that life is just a dream is not entirely without rational foundation. Professors Llinás and Paré at New York University have argued that wakefulness and REM sleep are essentially similar brain states, differing only in the extent to which they are shaped by sensory stimuli from the outside world. According to this view, REM sleep is a modified form of the waking state in which the brain's attention is turned away from sensory inputs and towards memories. Or to put it the other way round, wakefulness is essentially a dreamlike state that is moulded by sensory inputs. Brain scanning suggests that the difference between dreaming and being awake goes somewhat deeper, however. As well as receiving different sorts of inputs during dreams, the brain also processes those inputs in different ways. This is reflected in quite different patterns of brain activity during REM sleep, NREM sleep and wakefulness. When we are awake, the most active regions of our brains include those involved in rational decision making, planning and consciously controlled movement. These same brain regions (notably the frontal cortex) become relatively inactive during REM sleep.

Another reason for questioning the sharp distinction between the waking and dreaming states is the extensive interplay between the two. Our waking thoughts and experiences shape our dreams. And, as we shall see later, our dreams can influence what we think and do when we are awake.

The fact that our daytime experiences impinge on our subsequent dreams has been observed throughout history. In the fifth century BC the Greek historian Herodotus wrote that 'dreams in general originate from those incidents which have most occupied the thoughts during the day'. Four centuries later the Roman politician and writer Cicero observed that 'it is often the case that our daily thoughts and conversations are reflected in our dreams'. This common experience is borne out by research, which has established that the contents and style of people's daytime thoughts and experiences are correlated with the contents and style of their dreams.

Major emotional upsets such as bereavement or divorce can have a big impact on dreaming. They are often followed by an increase in REM sleep and dreaming, and the contents of those dreams often relate to the specific trauma. For example, a study of elderly people whose spouses had just died found that they had more REM sleep than normal. Another study of women undergoing divorce found that their dreams were more often about marriage and divorce than the dreams of women in stable marriages. The read-across from daytime experience to sleep and dreams seems to apply to other species as well. Experiments with rats have shown that different waking experiences, such as gentle handling, stress, or eating lots of nice food, elicit different changes in the animals' subsequent sleep patterns and EEG traces.

Our experiences during sleep itself can also influence our dreams. A sound or some other sensation may sometimes be incorporated into the current dream, often in a modified form. The sound of your alarm clock going off, for example, may be transformed into some other sound and then woven into the fabric of your dream. Experiments at Stanford University in the 1950s demonstrated that a variety of sensory stimuli, including touch, could also be incorporated into dreams. For instance, when sleeping volunteers were sprayed with a delicate mist of water during REM sleep, more than 40 per cent of their subsequent dream reports contained some form of water, often involving a shower of rain.

The nineteenth-century Russian scientist Marie de Manacéïne recorded several anecdotal accounts of physical sensations shaping dreams. In one case, an American gentleman dreamed he was being tortured by Mexicans. When he refused to reveal his dream secret of how copper could be transformed into gold, the Mexicans held his feet over a fire until he screamed in agony. He woke to find the hot water bottle resting on his feet. Another story concerned a doctor who spent the night in the 'very odorous' house of a cheesemonger. In his dream that night the doctor was incarcerated inside a giant cheese in punishment for some unspecified political offence. You will doubtless have better stories of your own.

Dreaming as madness

> All night long and every night,
> When my mamma puts out the light,
> I see the people marching by,
> As plain as day, before my eye.
>
> Robert Louis Stevenson,
> 'Young Night Thought' (1885)

The deep strangeness of dreams has led some observers to compare dreaming with madness. The philosopher René Descartes, for example, wrote that he was accustomed in his dreams 'to imagine the same things that lunatics imagine when awake'. Charles Dickens, a frequent victim of insomnia, similarly explored the idea that dreams are a form of madness. On one of his marathon late-night walks through the streets of London, Dickens found himself outside Bethlehem Hospital, better known as Bedlam. Gazing at the notorious lunatic asylum, Dickens mused on the similarities between dreaming and madness:

> Are not the sane and the insane equal at night as the sane lie a dreaming? Are not all of us outside this hospital, who dream, more or less in the condition of those inside it, every night of our lives? Are we not nightly persuaded, as they daily are, that we associate preposterously with kings and queens, emperors and empresses, and notabilities of all sorts? Do we not nightly jumble events and personages and times and places, as these do daily?

Are we not sometimes troubled by our own sleeping inconsistencies, and do we not vexedly try to account for them or excuse them, just as these do sometimes in respect of their waking delusions? Said an afflicted man to me, when I was last in a hospital like this, 'Sir, I can frequently fly.' I was half ashamed to reflect that so could I – by night.

The nineteenth-century French writer Gérard de Nerval, who is said to have been the inspiration for the Surrealist movement, tragically crossed the boundary between dreams and madness. He may well have been suffering from schizophrenia. From 1841, when he suffered a mental breakdown, until his suicide 14 years later, de Nerval was plagued by mental illness and spent long periods in a sanatorium. He was celebrated for walking around Paris accompanied by a pet lobster on the end of a blue ribbon. (When asked to explain why he had chosen a lobster as his companion, de Nerval replied, 'Well, you see, he doesn't bark and he knows the secrets of the sea.') In *Aurélia*, his semi-autobiographical account of his descent into madness, de Nerval explored the fuzzy boundary between dreams and reality:

> Here began for me what I shall call the overflow of dream into real life . . . For me the only difference between waking and sleep was that in the former case everything was transfigured before my eyes; everybody who approached me seemed changed; material objects possessed a kind of penumbra that altered their shape; the distortions created by the play of light and refractions of colour kept me occupied with a constant series of interlinked impressions, whose plausibility was further borne out by my dreams, less dependent as they were on external elements.

The intriguing parallels between dreaming and madness have also been explored by modern science. The Harvard neuroscientist J. Allan Hobson has pointed out that in many respects dreaming resembles the delirium caused by certain brain diseases. Dreaming and delirium share the same hallmarks of hallucination, amnesia, disorientation and the mixing together of disparate scenarios.

Given the madness of many dreams, it is just as well that we can normally separate them from our waking memories. Our brains work in such a way that we start forgetting our dreams as soon as we awake. Many people remember few, if any, of their dreams. This automatic amnesia helps us to keep a grip on sanity because it prevents us from confusing our dream fantasies with our waking lives. Occasionally, however, even the sanest person will mistake an element of a dream for reality, and unknowingly incorporate it into their waking memory. You may recall doing this yourself. If so, you may be relieved to discover that it is not uncommon.

An experiment demonstrated how fragments of dreams sometimes become confused with reality. On the first day of the experiment, volunteers listed words that had occurred in their dreams. They then studied a list of other words provided by the researchers. On the second day they were misled into thinking that some of the words from their dreams had appeared on the provided list. On day three they were asked to recall words from the first day's lists. The subjects erroneously 'recognised' words from their dreams almost as often as they recognised words that had actually appeared on the list. They genuinely believed that the words they had originally recalled from their dreams had actually been presented to them in the waking state, and they distinctly 'remembered' reading them from the list. Dream memories had become reality as far as they were concerned.

Some psychiatric patients find it much harder to distinguish their dreams from reality and they experience serious problems as a consequence. Indeed, it has been suggested that schizophrenic delusions might arise because of a defect in the mechanism that normally prevents our dreams from being stored in conscious memory. Fragments of dreams would then be 'remembered', misleading the patient into believing that the dream events had actually happened. Of course, failing to distinguish between dreaming and reality is entirely different from consciously recalling dreams. There is no suggestion that remembering your dreams – in the full knowledge that they *were* dreams – is a recipe for madness.

The writer Robert Louis Stevenson understood that our waking consciousness is built upon our memories of past experiences, and that those memories are mostly erased when they originate from our dreams. He also understood that only a gauzy boundary separates the worlds of waking sanity and dreaming insanity:

The past is all of one texture – whether feigned or suffered – whether acted out in three dimensions, or only witnessed in that small theatre of the brain which we keep brightly lighted all night long, after the jets are down, and darkness and sleep reign undisturbed in the remainder of the body. There is no distinction on the face of our experiences; one is vivid indeed, and one dull, and one pleasant, and another agonising to remember; but which of them is what we call true, and which a dream, there is not one hair to prove. The past stands on a precarious footing ... And yet conceive us robbed of it, conceive that little thread of memory that we trail behind us broken at the pocket's edge; and in what naked nullity should we be left! for we only guide ourselves, and only know ourselves, by these air-painted pictures of the past.

Are dreams meaningful?

There are two Gates of Sleep, of which one is said to be of horn, through which the spirits of truth find an easy passage. The other is made of gleaming white ivory, through which the gods send up false dreams to the upper world.

Virgil (70–19 BC), *Aeneid*, VI

Every culture in human history has entertained theories about the origins and meaning of dreams. A conventional view in many cultures for many centuries was that dreams are direct messages from the gods and predict the future. Prophetic dreams have been written about for millennia. The ancient civilisations of Babylon, Assyria, Egypt and Greece all regarded dreaming as a practical tool for predicting the future.

The ancient Egyptians were fascinated by their dreams. They made a lot of effort to record and interpret them, and erected dream temples in which expert priests would interpret people's dreams for them. The best known of these was built at Memphis in about 3000 BC. The British Museum houses an Egyptian papyrus called the Dream Book, which dates from around 1275 BC during the reign of Ramses II. The papyrus describes a succession of different dreams, together with an interpretation of each one and an overall diagnosis of whether the dream was good or bad. It states, for example, that 'if a man sees himself in a dream with his bed catching fire – bad; it means driving away his wife', whereas 'if a man sees himself in a dream looking out of a window – good; it means the hearing of his cry'.

The second-century soothsayer Artemidorus Daldianus devoted his life to cataloguing other people's dreams and their meanings. He travelled widely collecting information about dreams, which he collated in his magnum opus the *Oneirocritica*, or *The Interpretation of Dreams*. This drew heavily on the writings of earlier authors, which he had uncovered in his extensive field studies in Italy, Greece and the Near East. Many centuries later, Sigmund Freud acknowledged his intellectual debt to Artemidorus by using the same title for his great work on dreams.

Before the time of Aristotle, the Greeks conceived of dreams as direct messages from the gods. In Homer's *Iliad*, for example, Zeus sends a dream to Agamemnon telling him to attack Troy. By the fifth century BC, however, Greek rationalists had begun to challenge the notion of dreams as divine messages. Aristotle took this scepticism a stage further. Writing in the fourth century BC, he adopted a recognisably modern approach to sleep and dreams. Aristotle regarded dreams as biological phenomena, arising within the physical body and mind. He rejected the notion of dreams as messages from the gods for several reasons: its general irrationality, the apparent occurrence of dreaming in other species and (less admirably) the fact that everyone dreams regardless of how lowly they are. Aristotle presumed, somewhat undemocratically, that dreams sent from the gods would be directed only towards the finest and most intelligent individuals like him, rather than sprayed indiscriminately among the rabble. Unlike Freud, Aristotle did not regard dreams as deeply symbolic.

Other prescientific societies found a quite different way to make sense of their dreams: by regarding them as the direct experiences of the wandering soul. According to this world-view, the sleeper's spirit or soul left the body during sleep and actually visited the places seen in the dream. In his book *The Golden Bough*, the pioneering Scottish anthropologist Sir James Frazer noted that when a native Indian of Brazil or Guiana awoke from a dream, he would be firmly convinced that his soul had actually been off fishing, hunting, or whatever he had experienced in his dream, while his physical body had stayed sleeping in his hammock. Frazer observed that a whole village could be thrown into panic because somebody had dreamt that enemies were approaching. Because the soul was believed to leave the body during sleep, it was considered highly dangerous to disturb or suddenly awaken a sleeper in case his soul did not have time to get back into the body. If it was vital to wake someone then this had to be done gradually.

In the 1970s the French anthropologist Philippe Descola lived among the native tribes of Amazonia and recorded the prominent role that dreams play in their lives. The Achuar people in Ecuador, for example, extensively discuss and interpret their dreams, and have an elaborate system for classifying them. In the Achuar language the same word (*kara*) is used interchangeably to signify both 'sleep' and 'dream'. (Achuar is not the only language to fuse sleep with dreams; Spanish also uses a single word, *sueño*, to signify both.) The Achuar believe strongly that dreams convey crucial information about how they should conduct their waking lives, especially where to hunt. No man would go hunting unless he or a spouse had received a suitable dream omen. To that extent, wrote Descola, a hunting expedition would commence long before the hunter plunged into the jungle, because it would start with his nocturnal experiences in the world of dreams. The Achuar sometimes refer to great warriors as *kanuraur*, which means 'those who know how to sleep'.

Descola believed, with good reason, that the Achuar have developed such an impressive ability to recall their dreams because their sleep is fragmented into a series of relatively short episodes, punctuated by awakenings. Each time the dreamer surfaces from one of these brief 'excursions of the soul' the dreams are recalled and perhaps discussed with others before slipping back into sleep. By the time morning comes, each individual has accumulated a rich stock of dream material to interpret and discuss.

The fantastical nature of dream experiences must surely have reinforced humanity's traditional belief in the separation between the physical world inhabited by the body and the spiritual world inhabited by the soul. Dreaming fits snugly with the universal trait of believing in spirits and gods, and it might well have contributed in some way to the development and spread of formalised religions. The texts of the world's main religions, including Christianity, Islam, Judaism, Buddhism and Hinduism, all refer to dreams, many of which were said to foretell the future. The Bible contains numerous stories of prophetic dreams, including those of Jacob, Pharaoh, Nebuchadnezzar, Joseph and the Magi. In the Book of Daniel, for example, King Nebuchadnezzar has a dream that accurately foretells his downfall:

> I Nebuchadnezzar was at rest in mine house, and flourishing in my palace: I saw a dream which made me afraid, and the thoughts upon my bed and the visions of my head troubled me. Therefore

made I a decree to bring in all the wise men of Babylon before me, that they might make known unto me the interpretation of the dream. Then came in the magicians, the astrologers, the Chaldeans, and the soothsayers: and I told the dream before them; but they did not make known unto me the interpretation thereof.

According to Buddhist legend, the Buddha was conceived after his mother saw a white elephant enter her side in a dream. Dream divination had such a pervasive influence on daily life in the pre-Islamic Arab world that Muhammad forbade it.

Equally, fiction throughout the ages has reflected humanity's belief in dreams as prophecies. Shakespeare referred to portentous dreams in several of his works. In *The Merchant of Venice*, for instance, the moneylender Shylock is uneasy because he has had an ominous dream:

> There is some ill a-brewing towards my rest,
> For I did dream of money bags tonight.

And in *Julius Caesar*, a soothsayer memorably warns Caesar to beware the Ides (fifteenth) of March. Caesar unwisely dismisses the soothsayer as a mere dreamer, and is duly slain on that very day. Geoffrey Chaucer, on the other hand, seems to have been sceptical about the supposedly predictive nature of dreams. In the Nun's Priest's Tale from *The Canterbury Tales*, Chanticleer the cock has a terrible portentous dream in which he is seized by a fox. But Chanticleer's wife scornfully dismisses his fears:

> Dreams are just empty nonsense, merest japes:
> Why, people dream all day of owls and apes,
> All sorts of trash that can't be understood,
> Things that have never happened and never could.

Dreams foretelling diseases have been recorded throughout history. In modern times there have been a few documented reports of individuals dreaming about cancer before being diagnosed with the disease, leading some to infer that a person's dreams might provide an early clue to the presence of a tumour in their body. This is an intriguing thought that should not be dismissed out of hand, although the evidence to date is not overwhelming. But when it comes to the more general proposition that dreams foretell the future, there are many good reasons to be deeply sceptical.

Sir Francis Bacon was scathing about the notion of prophetic dreams. He wrote that 'dreams, and predictions of astrology ought all to be despised, and ought to serve but for winter talk by the fireside'. Bacon and other thoughtful souls realised that there is a much simpler and more obvious explanation for supposedly predictive dreams – namely, brute statistics. The Roman writer and politician Cicero, writing in the first century BC, put it like this:

> For what person who aims at a mark all day will not hit it? We sleep every night, and there are very few on which we do not dream; can we wonder then that what we dream sometimes comes to pass?

So many billions of different dreams are dreamt every night that it would be statistically incredible if, once in a while, one of those dreams did not coincidentally resemble subsequent events. In the UK alone, there are some 60 million people, each dreaming several dreams every night. At a conservative estimate, that makes at least 100,000,000 potentially prescient dreams every night of the week in one nation. Occasionally, one of them is bound to hit the target and be remembered. Add to that fact the strong propensity of the human mind to reshape memories in the light of experience, let alone the likelihood of wishful thinking and outright fraud, and you can be certain that most noteworthy events will have been 'foreseen' in a dream by someone, somewhere.

No discussion of dreaming would be complete without some mention of Sigmund Freud. So here goes. One of Freud's undoubted contributions was putting dreaming back on the map, both in terms of scientific and popular interest. Above all others, Freud championed the idea that dreams provide a route into the human mind. His enormously influential theories helped to focus scientific attention on the fact that dreams are generated within the mind, not delivered to the dreamer from the outside world, although he was certainly not the first person to have this insight. Aristotle was convinced that dreams are generated by the sleeping mind, and the Roman author Petronius, writing in the first century, was equally clear that 'each man makes his own dreams'.

Freud viewed dreams as a unique source of knowledge about the human mind. *'The interpretation of dreams,'* he wrote (in italics), *'is the royal road to a knowledge of the unconscious activities of the mind.'* He believed that by analysing dreams, including his own, he was discovering universal truths about the human mind and at the same time collecting data to support his theories. Freud developed his views about dreaming in the closing years of the nineteenth century and set them out in his book *The Interpretation of Dreams,* published in 1900. It has been suggested that Freud's interest in dreams was kindled by his own use of cocaine.

One of Freud's central tenets was that dreams were disguised wishes that had been repressed and held within the unconscious mind. According to Freud, these wishes were mostly sexual in nature. The mind therefore censored and disguised these troubling urges, lest they disrupt the dreamer's sleep. By decrypting the symbols, the psychoanalyst could reveal the desires concealed within the dream. Freud's singular approach to the relationship between theory and evidence put him in a no-lose situation. No matter how unlikely the dream, he could always interpret it in a way that supported his theory of wish fulfilment. The theory was therefore ultimately incapable of being refuted by empirical evidence. For that reason, Freud's approach is now widely regarded as unscientific.

An inescapable feature of Freud's theories about dreaming was his seemingly boundless capacity to interpret dreams in sexual terms. Consider this example from *The Interpretation of Dreams,* which also illustrates his characteristically confident use of assertive terms such as 'unquestionably' in lieu of testable evidence:

A little time ago I heard that a psychologist whose views are somewhat different from ours had remarked to one of us that, when all was said and done, we did undoubtedly exaggerate the hidden sexual significance of dreams: his own commonest dream was of going upstairs, and surely there could not be anything sexual in *that*. We were put on the alert by this objection, and began to turn our attention to the appearance of steps, staircases and ladders in dreams, and were soon in a position to show that staircases (and analogous things) were unquestionably symbols of copulation. It is not hard to discover the basis of the comparison: we come to the top in a series of rhythmical movements and with increasing breathlessness and then, with a few rapid leaps, we can get to the bottom again. Thus the rhythmical pattern of copulation is reproduced in going upstairs.

Now you know. Popping upstairs will never seem the same again. Freud could see latent sex anywhere and everywhere. Sharp weapons and long, stiff objects such as tree-trunks and sticks became the male genitals, while cupboards, boxes, carriages and ovens all turned out to be the uterus. In a world composed of objects that are mostly either convex or concave, the erotic possibilities are endless. An early sceptic, writing in 1913, highlighted this genital leitmotif:

> There is absolutely nothing in the universe which may not readily be made into a sexual symbol . . . All natural and artificial objects can be turned into Freudian symbols. We may explain, by Freudian principles, why trees have their roots in the ground; why we write with pens; why we put a quart of wine in a bottle instead of hanging it on hooks like a ham; and so on.

To be fair, Freud's thinking reflected the middle-European society in which he lived. It was an era of sexual repression, with unspent passions simmering just below the surfaces of many minds. The pioneering English anthropologist and psychologist W. H. R. Rivers felt that the sexual obsession of Freud and his followers was partly a reaction against the prudery of the nineteenth-century medical profession, and partly a protest about sexual ignorance in his society. Nonetheless, wrote Rivers, 'the mistake which is now being made by many is to regard this excess as a necessary part of the Freudian scheme instead of an unfortunate excrescence'. The philosopher Ludwig Wittgenstein put his finger on another weak spot in Freud's dream world of repressed sexual desires:

> Freud very commonly gives what we might call a sexual interpretation. But it is interesting that among all the reports of dreams that he gives, there is not a single example of a straightforward sexual dream. Yet these are as common as rain.

Sex aside, Freud's theories about the mind and dreaming reflected an equally nineteenth-century view of how the brain works. Freud hoped that his theories would eventually be supported by discoveries in neurobiology. But they were not. Neurobiology has established that the human brain does not appear to work in the way that Freud or Jung supposed, and it offers only the most tenuous support for Freudian concepts such as ego, somatic drives, cathexes of psychical energy and wish fulfilment.

Freudian theories about dreaming also run into trouble when confronted with empirical evidence (or the facts, as some might say). They have difficulty explaining, for example, why the proportion of time that we spend dreaming declines with age, so that an infant spends several hours a day dreaming while a very elderly person may hardly dream at all. What suppressed sexual desires could a small infant be dreaming about? You might expect a randy adolescent or an octogenarian with a lifetime of salacious experiences under his or her belt to have far more to dream about than a baby, but the reverse is true. And if dreams are repressed wishes, then why is it that hungry people do not invariably dream of food and celibates do not invariably dream of sex?

Current scientific knowledge, which is admittedly very far from complete, suggests that dreaming is bound up with the processing and consolidation of memory, not the suppression of wishes. It also suggests that we should not expect to find deep meaning in the contents of every dream. Neurobiology is perhaps more in tune with the view of Wilhelm Wundt, a pioneer of scientific psychology, who described dreams as 'unmusical fingers wandering over the keys of a piano'. We shall look more closely at the biological functions of REM sleep and dreaming in chapter 12.

Like wine through water

> I've dreamt in my life dreams that have stayed with me ever after, and changed my ideas; they've gone through and through me, like wine through water, and altered the colour of my mind.
> Emily Brontë, *Wuthering Heights* (1847)

Sigmund Freud was wrong to suppose that all dreams are rich with symbolic meaning. Dreaming did not evolve as a mechanism to provide psychoanalysts with insights into the human mind. However, it would be equally wrong to suppose that our dreams are mere random nonsense, devoid of all meaning. Dreams can offer interesting glimpses into our thoughts and emotions. We do tend to dream more about the things that are preoccupying us and we are more likely to remember dreams that are personally significant.

The writer William Hazlitt believed that dreams can help to reveal our true thoughts and feelings. 'We are not hypocrites in our sleep,' he wrote. William Smellie, an eighteenth-century Scottish naturalist and antiquary who compiled the first edition of the *Encyclopaedia Britannica*, held similar views about the occasionally revealing nature of dreams. In some ways foreshadowing Freud, Smellie argued that imagination prevails over reason during sleep, and that an understanding of one's dreams is therefore a route to understanding oneself. In his treatise *The Philosophy of Natural History*, Smellie described these 'capital scenes' from his own dreams, which he had noted down thirty years earlier:

> The third night, I found myself in the midst of a brilliant company of ladies and gentlemen ... the music struck up; each took his fair partner by the hand, and a sprightly dance immediately commenced ... But, in the midst of this enchanting scene, while setting to a young lady, my breeches fell plump to my heels! I quickly attempted to lay hold of them; but in vain. The very power of reaching forth my hand was abstracted from me. I remained fixed as a statue, and the dance was interrupted. The blushes of the company discovered how sensibly they felt my misfortune; but none had the courage to assist me ... But here I must stop, lest I should discover more of my own character than would be consistent with prudence.

When you remember a dream and subsequently think or talk about it, your waking mind will interpret that dream in ways that can sometimes be illuminating. Your dreams can reveal something about you, even if their raw contents are little more than a hotchpotch of your thoughts and experiences. Dreams are neither a special mode of consciousness that exposes our hidden inner selves, as Freud supposed, nor random noise devoid of all meaning. The truth lies boringly somewhere in between.

What is more, our waking memories of our dreams can sometimes impinge upon our thoughts and behaviour. A particular dream may change the way we think or act. Certain dreams may even help to influence or bring about the very events they 'foresaw'. In Charles Dickens's *Christmas Carol*, the curmudgeonly Ebenezer Scrooge is transformed by a dream in which he is confronted with chilling visions of his past, present and future lives, in the forms of the Ghosts of Christmas Past, Christmas Present and Christmas Yet To Come. A changed Scrooge awakes on Christmas morning, drained of bile and full of the milk of human kindness. Like wine through water, some dreams stay with us and alter the colour of our minds.

Can dreams be sinful?

Such fantastic images give us great delight, and, since they are created by us, they undoubtedly have a symbolic relation to our lives and destinies.

Goethe, *Italian Journey* (1816–17)

Do we have free will when we are in the dream state? And if we do have free will while dreaming, are we then responsible for all the bizarre, embarrassing and occasionally illegal acts that we commit there?

Saint Augustine wrestled with the conundrum of whether we are morally accountable for our thoughts and actions when we dream. Is lust still a sin when it occurs in a dream? What about the dream act of adultery? Augustine fretted because he suspected that the same kind of mental activity was occurring in his dreams as in his conscious thoughts. Like any healthy human, he sometimes dreamt of fornication and enjoyed it while it was happening. But as a man of God, this bothered him. (Had he known about his nightly nocturnal erections he would probably have been even more upset.) In his *Confessions* he wrote:

There live yet in my memory the images of such things which haunt me, strengthless when I am awake: but in sleep, not only so as to give pleasure, but even to obtain assent, and what is very like reality ... Am I not then myself, O Lord my God? And yet there is so much difference betwixt myself and myself, within that moment wherein I pass from waking to sleeping, or return from sleeping to waking!

Saint Augustine managed to wriggle off this moral hook by convincing himself that the lecherous thoughts he entertained in his dreams were involuntary, and that a man could only be held responsible for his voluntary thoughts. His lubricious dreams of fornication were therefore not sinful. And with one mighty philosophical bound he was free.

If one was inclined to do so, one could beg to differ with Saint Augustine. The fact is that we can learn to influence our dreams to some extent. In the next chapter we shall look at the curious phenomenon of lucid dreams, in which the fortunate dreamer is both conscious of dreaming and able to exert some limited control over the course of the dream. Moreover, to the extent that a waking memory of a dream reveals something about the dreamer's mind, it does cast some light on his or her experiences, thoughts, attitudes and preoccupations. Dreams are not entirely value-free. Sir Thomas Browne, writing in the seventeenth century, was clear in his view that 'death alone, not sleep, is able to put an end to sin'. If we were held legally responsible for what we do in our dreams, most of us would be locked up.

10

A Second Life

Dream is a second life.

Gérard de Nerval, *Aurélia* (1851)

If you are too tired or too indifferent to remember your dreams then you really are missing out on something rather splendid. We all have within us the capacity to experience, every night of our lives, a form of virtual reality that the designers of computer games could only dream about. Dreaming is both a source of pleasure and a powerful engine of creativity. And it is available to every one of us at no cost and with no risk.

A creative state

You see things; and you say 'Why?'
But I dream things that never were; and I say 'Why not?'
George Bernard Shaw, *Back to Methuselah* (1922)

Dreaming is rather like playing. When we dream, and when we play, we create novel patterns through a form of safe simulation. Dreaming and playing both entail the experimental mixing together of conventional elements of behaviour or thoughts in new and fanciful combinations, and doing this in a context where the conventional consequences do not apply.

The play of children and young animals is a concoction of 'serious' behaviour patterns, but performed in a special context where they are insulated from their usual outcomes. When two infants play-fight they are playing and not trying to injure each other, and when they play at driving a car they are playing and not trying to make a real journey. Play can involve thoughts and ideas as well as physical actions, and much of the play that adults engage in is of this mental variety. One could regard humour as a form of mental play, relying as it does on the incongruous juxtaposition of mundane ideas.

The concept of play casts some novel light on dreams. The psychologist Jean Piaget touched upon something important when he likened dreaming to a game played inside the brain. In dreams, as in waking play, conventional ingredients are mixed together in bizarre combinations, shorn of their normal consequences. We can dream of doing things we could never do in reality. We can defy the laws of physics and the laws of lawyers, and wake up the next morning entirely unharmed. The writer Arthur Koestler maintained that humans are at their most creative when rational thought is suspended, which happens most commonly in dreams. Only then can the mind receive insights by discovering previously hidden connections. The dreaming mind, argued Koestler, forges links between disparate things that we would not conceive of linking when we are awake. Research in artificial intelligence has found that the ability of computers to solve problems creatively can be enhanced by software that mimics the process of daydreaming, where the mind mingles unrelated thoughts and conducts playful experiments with them.

Many of the most creative and original thinkers in human history have been playful. They have also been dreamers. One of the most successful songs ever written came to its composer in a dream. On a May morning in 1965, Paul McCartney woke up in his bedroom in Wimpole Street, London, with the melody of 'Yesterday' running through his mind. McCartney immediately recognised that it was a gorgeous tune, but he assumed he must have heard it somewhere else. He got out of bed and started to play the notes on an upright piano in his bedroom. Still he could not quite believe that this magic melody had simply come to him in his sleep. Over the following days, McCartney tried hard to discover whose song it was and where he might have heard it, until eventually he became convinced that it was truly original and truly his. To begin with the song had no words, so for a while 'Yesterday' was 'Scrambled Eggs' ('Scram-ble-d Eggs . . .'). 'Yesterday' went on to become the most played and the most recorded song of all time, as well as one of the Beatles' biggest hits. Their recording of 'Yesterday' was played more than six million times on US radio alone, and more than 2,200 other artists went on to record the song.

The muse visited Paul McCartney more than once in his dreams. He dreamt up the idea for 'Yellow Submarine' one night in the twilight zone as he drifted off to sleep. Another of his most enduring songs, 'Let It Be', was written after he had a dream featuring his mother Mary, who had been dead for ten years.

Samuel Taylor Coleridge famously dreamt the central elements of his great poem 'Kubla Khan', though probably not in a conventional REM dream. In his preface to the poem, Coleridge described how he came to write it. In poor health, he had retreated in the summer of 1797 to an isolated farmhouse on Exmoor, between Porlock and Linton. Owing to a 'slight indisposition' he took opium, which caused him to fall asleep in his chair (and, presumably, to have weird, hallucinatory thoughts). According to Coleridge's account he was reading a work by Samuel Purchas, a compiler of travel books, at the moment he drifted off. The sentence he was reading described how the Khan Kubla had commanded a palace to be built. Coleridge's dreaming mind picked up that thought and ran off with it. As Coleridge later put it:

> The Author continued for about three hours in a profound sleep, at least of the external senses, during which time he has the most vivid confidence, that he could not have composed less than from two or three hundred lines; if that indeed can be called composition in which all the images rose up before him as *things*, with a parallel production of the correspondent expressions, without any sensation or consciousness of effort. On awaking he appeared to himself to have a distinct recollection of the whole, and taking his pen, ink, and paper, instantly and eagerly wrote down the lines that are here preserved. At this moment he was unfortunately called out by a person on business from Porlock, and detained by him above an hour, and on his return to his room, found, to his no small surprise and mortification, that though he still retained some vague and dim recollection of the general purport of the vision, yet, with the exception of some eight or ten scattered lines and images, all the rest had passed away like the images on the surface of a stream into which a stone has been cast, but, alas! without the after restoration of the latter!

Coleridge's account became one of the most famous and most widely disputed descriptions of creative composition ever written. Some authorities have suggested that the unnamed 'person from Porlock' may have been a convenient fiction. Others have pointed out that Coleridge had already written many previous versions of the poem. A sleep scientist would add that no one has a single dream lasting three hours, especially while sitting upright in a chair. Coleridge's dream would have been of the hypnagogic variety, in which case it would have lasted minutes rather than hours. As we shall see, many famous instances of artistic and scientific creativity during 'dreaming' have almost certainly occurred in the hypnagogic state, during the transition from wakefulness to sleep. The opium probably had something to do with it as well.

A revealing feature of Coleridge's account is his reference to the thought that was already going through his mind as he fell asleep, and which formed the key ingredient of his dream inspiration. The thought came from Samuel Purchas's *Pilgrimage,* in which these lines appear: 'This Citie is three dayes journey Northeastward to the Citie Xandu, which the great Chan Cublay now raigning, built; erecting therein a marvellous and artificiall Palace of Marble.' Coleridge's eventual poem famously begins: 'In Xanadu did Kubla Khan / A stately pleasure-dome decree.' The point is that inspirational dreams need plenty of preparatory thinking and good raw material to work on, and even then they seldom produce complete solutions. The waking mind still has to do most of the hard work.

The seductive notion that visionary poems or prose could be composed effortlessly while asleep, or in an opium-induced reverie, became highly fashionable following the success of 'Kubla Khan'. Many writers, including Thomas de Quincey, Baudelaire and Jean Cocteau, imitated Coleridge's drug-assisted approach to creativity. Opium, incidentally, has been smoked for more than two thousand years as a means of inducing vivid dreams – hence the term 'pipe dreams'.

Many other writers have drawn inspiration from their dreams. Mary Shelley claimed that the central idea for *Frankenstein* came to her in a dream. It happened in 1816 while she was taking a summer holiday in Switzerland with her husband Percy Bysshe Shelley and Lord Byron. She was provoked into inventing the story when Byron proposed a ghost-story competition. Mary Shelley's contribution was the story of Frankenstein, which she subsequently developed and published in 1818.

In her introduction to a later edition of the book, Mary Shelley described how she came to think of 'so very hideous an idea'. The initial ingredient for her dream was a long conversation between Lord Byron and her husband Percy, to which Mary listened in near silence. They discussed the nature of life and whether perhaps a corpse could be brought to life again. This was, after all, the age of galvanism. Perhaps the component parts of a creature might be manufactured, assembled and then artificially imbued with life? The conversation went on into the late hours, until Mary finally went to bed. Here is her description of what happened next:

> When I placed my head on my pillow, I did not sleep, nor could I be said to think. My imagination, unbidden, possessed and guided me, gifting the successive images that arose in my mind with a vividness far beyond the usual bounds of reverie. I saw – with shut eyes, but acute mental vision – I saw the pale student of unhallowed arts kneeling beside the thing he had put together. I saw the hideous phantasm of a man stretched out, and then, on the working of some powerful engine, show signs of life, and stir with an uneasy, half-vital motion ... behold, the horrid thing stands at his bedside, opening his curtains and looking on him with yellow, watery, but speculative eyes ... Swift as light and as cheering was the idea that broke in upon me. 'I have found it! What terrified me will terrify others; and I need only describe the spectre which had haunted my midnight pillow.' On the morrow I announced that I had *thought of a story*. I began that day with the words. 'It was on a dreary night of November,' making only a transcript of the grim terrors of my waking dream.

Mary Shelley's dream gave birth to material that was by itself sufficient only for a few pages. Fortunately, her husband urged her to work on the idea and flesh it out. But for his encouragement, she later wrote, the story would never have taken the form in which she eventually presented it to the world. Shelley's account also makes it clear that even if her dream produced the seed of the idea, the final story required much additional effort:

> Invention, it must be humbly admitted, does not consist in creating out of void, but out of chaos; the materials must, in the first place, be afforded: it can give form to dark, shapeless substances, but cannot bring into being the substance itself.

Charlie Chaplin is said to have been another creative genius whose ideas sometimes came to him through dreams. Chaplin's unconventional working habits reportedly included composing music for his films while in bed, using a recording device at his bedside. During the night he would wake up, hum a few bars, and go back to sleep.

Creative writers and musicians are not the only ones to have claimed inspiration from their dreams. Scientists have also benefited. The nineteenth-century German chemist Friedrich Kekulé was convinced that he solved the puzzle of the molecular structure of benzene in a dream he had in 1865:

> I turned the chair to the fireplace and sank into a half sleep. The atoms flitted before my eyes . . . wriggling and turning like snakes. And see, what is that? One of the snakes seized its own tail and the image whirled scornfully before my eyes. As though from a flash of lightning I awoke. I occupied the rest of the night in working out the consequences of the hypothesis.

Kekulé had been grappling with the benzene problem for years. The image of the snake biting its tail gave him the crucial idea that the two ends of the benzene chain were joined together and lo, the moderately important scientific puzzle of benzene's molecular structure was duly solved. Kekulé later remarked: 'Let us learn to dream, gentlemen, and then we may perhaps find the truth.' If Kekulé's account is correct, his insight almost certainly came not in a conventional REM dream, but from a hypnagogic dream during the transition from wakefulness to light sleep. It would be unusual for someone to enter REM sleep while dozing upright in a chair in the way that Kekulé described.

In 1920 the German physiologist Otto Loewi saw in a dream how he could test his theory that nerve impulses are transmitted chemically. The original idea had occurred to him years before, but he had been unable to devise any way of testing his theory experimentally. The dream showed Loewi how such an experiment would work:

> The night before Easter Sunday of that year I awoke, turned on the light, and jotted down a few notes on a tiny slip of thin paper. Then I fell asleep again. It occurred to me at six o'clock in the morning that during the night I had written down something most important, but I was unable to decipher the scrawl. The next night, at three o'clock, the idea returned. It was the design of

an experiment to determine whether or not the hypothesis of chemical transmission that I had uttered seventeen years ago was correct.

Loewi got out of bed, went to his laboratory, and performed a simple experiment on a frog's heart according to the scheme he had imagined in his dream. Its results became the foundation of the theory of chemical transmission of nerve impulses, a major advance in understanding that eventually helped Loewi to win a Nobel Prize. His experience once again illustrates the principle that creative solutions occur in dreams only after the dreamer has already thought extensively about the problem while awake.

Albert Einstein, who was a great sleeper, maintained that one of his dreams had been instrumental in helping him formulate the theory of relativity. He recalled how he had dreamt of speeding down a mountainside on a sled, accelerating ever faster until the light from the stars refracted into strange colours as he neared the speed of light. Einstein always insisted on having a good ten hours of sleep a night, and eleven if he was planning to do some really serious thinking the next day.

The inspiration for the key design feature of the modern sewing machine came to the American inventor Elias Howe in a dream. The location of the eye of the needle had been a major obstacle in the development of Howe's machine. He had been unable to solve the problem and was running out of money. One night, Howe had a dream. He dreamt he was about to be executed for failing to produce a sewing machine for the king of a strange country. He was surrounded by guards carrying spears that were pierced near the tip. Howe realised, while still dreaming, that this was the solution to his design problem. He woke up, rushed to his workshop and completed the basic design of what became the first effective sewing machine.

The catalogue of dream-fuelled creativity continues. William Blake invented a new engraving technique when his dead brother demonstrated it to him in a dream. The philosopher Condorcet awoke from a dream with the solution to a mathematical equation that he had been wrestling with for days. René Descartes (another great sleeper) wrote of how on the night of 10 November 1619 he had a dream that was 'the most important affair' of his life; Descartes' dream formed the inspiration for his whole philosophical system. The Russian chemist Dimitri Mendeleyev saw how to arrange the elements into the Periodic

Table while dozing in his chair one afternoon (almost certainly therefore another hypnagogic dream). And the Italian violinist and composer Guiseppe Tartini is said to have dreamt his sonata 'The Devil's Trill' in a dream at the age of 21.

Remembering more of your dreams will not guarantee that you give birth to groundbreaking ideas (though you might). But neglecting your sleep, and therefore your dreams, will almost certainly do the reverse. As we saw in chapter 3, poor sleep severely impairs creativity and mental flexibility. If you do not get enough sleep you will not realise whatever creative capacity you possess, let alone expand it.

Stevenson's Brownies

> The day supplies us with truths, the night with fictions and falsehoods.
>
> Sir Thomas Browne (1605–82), 'On Dreams'

Robert Louis Stevenson revelled in his dreams and made good money out of them. In 1887 he wrote that 'there are some among us who claim to have lived longer and more richly than their neighbours; when they lay asleep they claim they were still active; and among the treasures of memory that all men review for their amusement, these count in no second place the harvests of their dreams.' Dreams were more than just a source of idle pleasure to Stevenson: they became an important tool of his profession as a writer.

As a child and young man, Stevenson was plagued by dreadful nightmares. He was, he wrote, 'an ardent and uncomfortable dreamer', who would 'struggle hard against the approaches of that slumber which was the beginning of sorrows'. However, his nightmares dwindled as he grew older and he began to dream in stories rather than fragmented images. And thus, as he put it, he started to live a double life – one of the day, one of the night.

When Stevenson became a professional storyteller his dreaming developed into a major asset. As he lay down to sleep each night he was no longer seeking mere amusement from his dreams: he expected his dreams to generate 'printable and profitable tales'. When Stevenson needed new stories he would spend his waking and sleeping hours literally dreaming them up.

Stevenson's acts of creation were assisted by characters in his dreams, whom he referred to as his Brownies or the Little People. Stevenson's Brownies must have visited him mostly in hypnagogic dreams rather than ordinary REM dreams, given the technique he used for encouraging them. He would lie in bed with his arms at right angles to the mattress. This enabled him to slip into the twilight zone populated by his Brownies, but if he sank into deeper sleep his arms would fall onto the mattress and wake him up. When Stevenson the professional writer was hard-pressed for money his Brownies would usually come up with the goods, delivering to the author 'better tales than he could fashion for himself'. God bless the Brownies, wrote Stevenson, 'who do one-half my work for me while I am fast asleep'.

Stevenson credited his hypnagogic Brownies with many of his greatest literary inventions, including the main plot line for *The Strange Case of Dr Jekyll and Mr Hyde*. Stevenson's account of its genesis reveals that the Brownies provided some of the main ideas and images, but his waking mind added a great deal more. As with Mary Shelley's *Frankenstein*, a story like *Dr Jekyll and Mr Hyde* did not spring fully formed into the author's dreaming mind; it had to be painstakingly constructed during the day from odd fragments left behind by his dream assistants:

> For two days I went about racking my brains for a plot of any sort; and on the second night I dreamed the scene at the window, and a scene afterward split in two, in which Hyde, pursued for some crime, took the powder and underwent the change in the presence of his pursuers. All the rest was made awake, and consciously, although I think I can trace in much of it the manner of my Brownies. The meaning of the tale is therefore mine, and had long pre-existed in my garden of Adonis ... Mine, too, is the setting, mine the characters. All that was given me was the matter of three scenes, and the central idea of a voluntary change becoming involuntary.

Dreams, then, can be a rich source of creative thinking and inspiration, providing some of the basic imaginative ingredients on which our waking minds can go to work. Creativity is a benefit of sleep that many of us foolishly ignore. Conversely, creativity and mental flexibility are among the first things to suffer when we get insufficient sleep. Great ideas do not come effortlessly and complete, in a flash of inspiration, but only after preparation, consideration and effort. Creativity and innovation therefore also depend upon motivation, persistence, attention and a suitable mood – and all of these faculties are impaired by lack of sleep as well. So dream when you can.

Lucid dreams

> Row, row, row the boat
> Gently down the stream
> Merrily, merrily, merrily, merrily
> Life is but a dream.
>
> Nursery rhyme

There are dreams, and then there are lucid dreams. If you have never played host to a lucid dream, you should try one. And having read this book, you might. But what, you may be asking, is a lucid dream?

A lucid dream is a special sort of dream in which the dreamer is fully aware at the time that he or she is dreaming. Indeed, the lucid dreamer is both aware of dreaming and aware of being aware of dreaming (a state of consciousness that psychologists refer to as meta-awareness). During ordinary dreams we are unable to reflect on our current state or imagine a different one; we are just dreaming. But in lucid dreams, as in waking life, we can do both. The existence of lucid dreams further erodes the conventional dichotomy between dreaming and waking consciousness.

Lucid dreams differ from the common-or-garden REM variety in other ways besides their defining feature of self-awareness. They are more vivid and more memorable than ordinary dreams. Lucid dreams can be astonishingly lifelike and realistic. Moreover, the lucid dreamer is often able to think and behave rationally during the dream, retaining an awareness of the true facts of their waking life. During a lucid dream you might notice, for example, that the person who is your dream lover is not your lover or partner in real life. You might also be able to exert some influence over the nature and course of your dream, although this facility is by no means complete.

By and large, lucid dreams are more enjoyable than ordinary dreams and have a more positive emotional tone. As we saw in the previous chapter, the majority of ordinary REM dreams are not particularly pleasant; dreams of falling, being pursued or being attacked are common, and the predominant emotions are anxiety, fear, anger and frustration. Scholars who catalogue REM dreams generally find that around two thirds of them are more unpleasant than pleasant. In contrast, lucid dreams seldom contain any unpleasant sensations or negative emotions. Pain is rare, while pleasurable sensations are common. Lucid dreamers find it very difficult, if not impossible, to injure or kill themselves in lucid dreams, even when they try hard to do so (as some brave seekers after knowledge have done). Lucid dreams tend to be more straightforward and less loaded with the dark emotional undercurrents that typify ordinary dreams.

In many respects, lucid dreams are more like acting out fantasies and desires. Their most common theme, you will be amazed to discover, is sex. Other regular subjects of lucid dreams include encounters with deceased loved ones, flying, daredevil acts such as running naked down a crowded street, experimenting with the dream itself (for example, by deliberately walking through brick walls) and meeting famous people. Lucid dreamers who have written about their nocturnal adventures have stated that certain lucid dreams have been the most enjoyable and satisfying experiences of their lives.

Sex is the most frequently reported feature of lucid dreams, even allowing for the prudish tendency to under-report it. Adept lucid dreamers claim that they can deliberately conjure up delicious erotic adventures, safe in the knowledge that they will not be held to account the next morning. True aficionados say they can choose when, how and with whom (or with what) to attain their multiple orgasms. Sexual lucid dreams can include ultrarealistic sensations of touch and smell, offering the fortunate dreamer a source of free, harmless virtual-reality pleasure. The sexual partner is often a clearly identified individual known to the dreamer. The diarist Samuel Pepys was evidently one of this happy band of erotic lucid dreamers. In his diary entry for 15 August 1665, Pepys described the sensual delights of one of his lucid dreams, featuring Lady Castlemayne, a famous beauty of the time and mistress to the King:

> Walked to Greenwich, where called at Capt. Cockes and to his chamber, he being in bed – where something put my last night's dream into my head, which I think is the best that ever was dreamed – which was, that I had my Lady Castlemayne in my arms and was admitted to use all the dalliance I desired with her, and then dreamed that this could not be awake but that it was only a dream. But that since it was a dream and that I took so

much real pleasure in it, what a happy thing it would be, if when we are in our graves (as Shakespeare resembles it), we could dream, and dream but such dreams as this – that then we should not need to be so fearful of death, as we are in this plague-time.

But hold fast. Before you become too flushed at the prospect of starring nightly in your own private pornographic films, there are limitations. Lucid dreams are relatively infrequent treats, even for those who are accustomed to having them. Moreover, lucid dreamers do not exercise complete control over their dreams, as though they were omnipotent directors of computer animations. Proficient lucid dreamers can exert some influence over the course of a dream, but their power is by no means absolute. Each lucid dream has a life of its own. The dreamer may try to shove their lucid dream in one direction, only to find that it charges off in a quite different direction. And deliberate efforts to insert sex into lucid dreams often fail. Indeed, some lucid dreamers find that too much erotic excitement disrupts their lucid dreams and either wakes them up or, more commonly, plunges them back into the ordinary REM dream state. Lucid dreams are good, but not that good.

Sober scientific research into lucid dreams has established that they occur during REM sleep and have many other features in common with ordinary REM dreams. The self-awareness and most other aspects of lucid dreams are occasionally present to a limited extent in non-lucid dreams, suggesting that lucid dreams and ordinary REM dreams occupy different regions of the same spectrum. How do scientists know that lucid dreams occur during REM sleep? Waking someone up to ask them would break the spell and could not be relied upon as objective evidence that they were actually conscious while dreaming.

The key to establishing the link between lucid dreaming and REM sleep was conceived independently by two scientists: Keith Hearne in England and Stephen LaBerge at Stanford University in California. In the 1970s both men had the inspired idea of arranging for a lucid dreamer to give a prearranged signal while still in the lucid dream state. The key to doing this was using the eye muscles – the only muscles, apart from those responsible for breathing, that are not paralysed during REM sleep. Hearne and LaBerge separately devised an experiment in which a lucid dreamer signalled to observers from the lucid state by making deliberate eye movements. Those eye movements were detected by a polysomnograph, which simultaneously confirmed that the dreamer was in REM sleep.

Keith Hearne claims to have been the first person ever to witness signals from a dreamer in the lucid state, in an experiment that took place on the morning of Saturday 12 April 1975 at Hull University in England. Hearne's subject, an experienced lucid dreamer, gave the prearranged signal – a series of left–right eye movements – while the EEG confirmed that he was unequivocally in REM sleep. Meanwhile, Stephen LaBerge had also started investigating lucid dreams. He became adept at lucid dreaming and could experience them while hooked up to a polysomnograph. LaBerge was similarly able to indicate that he had entered a lucid dream by making left–right eye movements, and EEG recordings again confirmed that he was in REM sleep at the time. In a later experiment, LaBerge and William Dement demonstrated that a lucid dreamer could deliberately alter his breathing rate during a lucid dream.

Other experiments employing variations on this theme have repeatedly confirmed that lucid dreams occur during REM sleep and are therefore true dreams rather than temporary lapses into waking consciousness. Under laboratory conditions, lucid dreams typically last between one and six minutes, with an average duration of about two and a half minutes. Under natural conditions they might well last longer. The characters who appear in lucid dreams sometimes behave as though they too possess a form of consciousness. In one experiment, experienced lucid dreamers were asked to instruct the characters they met in their lucid dreams to perform various tasks, set beforehand by the researchers. These tasks included composing verses, drawing, solving arithmetical problems and finding rhyming words. Some of the dream characters actually agreed to perform these tedious chores. And, like real people, they were poor at arithmetic.

If you have never knowingly experienced a lucid dream, your appetite might by now have been whetted. How can you cultivate lucid dreams if you have never had one before? Most people are capable of having the occasional lucid dream, given the desire and a little practice. Simply reading about them and becoming aware of their existence is often enough, so perhaps this chapter might bring a little extra fun into your life. Keith Hearne began researching lucid dreams without ever having experienced one, but he had his first lucid dream soon after. As it happens, I had my first lucid dream after researching this part of the book. But I will not bore you with an account of my dream, since I agree with William S. Burroughs's dictum that dreams are as boring and commonplace as the average dreamer.

The starting point for lucid dreaming is simply knowing that they exist – which you now do – and deciding that you want to experience one. Then you can gently encourage lucid dreaming by thinking about it during the day and as you lie in bed before falling asleep. Focus on your intention of having a lucid dream as you drift off. If you are really determined, there are other things you can do to facilitate lucid dreams. For a start, you can develop the habit of recalling your ordinary dreams when you wake in the mornings. Experiments have found that people are more likely to experience lucid dreaming if they frequently recall their ordinary dreams, and they can increase their chances of experiencing lucid dreams by paying more attention to their ordinary dreams. Train yourself to think about your dreams first thing every morning when you wake up, before your memories evaporate and waking thoughts replace them. (You should be perfectly capable of distinguishing between your dreams and your conscious, waking memories, so recalling your dreams is unlikely to drive you mad.)

Try to get plenty of sleep so that you wake naturally in the morning, since a full night's sleep usually ends with an episode of REM sleep. Lucid dreams, like ordinary dreams, occur mainly during the final couple of hours before waking, when REM sleep predominates. Some lucid dreamers aim to wake up some time before they have to get out of bed, ponder on the ordinary dreams they have been having, and then return to sleep with the firm expectation of experiencing a lucid dream. Since REM sleep occurs at the end of the 90-minute sleep cycle, it may help if you contrive to wake up at a multiple of 90 minutes after going to sleep – that is, after approximately seven and a half hours or nine hours of sleep. The timing of sleep cycles is pretty variable, so hitting the right point in your own cycle may require some trial and error. Taking a long nap during the day also works for some people.

Another technique for fostering lucid dreams involves learning to recognise when you are dreaming. Proficient lucid dreamers become highly attuned to 'dream signs', such as incongruous events or surreal images, that tell them they are dreaming. You can test whether you are in a lucid dream by trying to do something that would be physically impossible in real life, such as flying through the sky or pushing over a brick wall. The anthropologist and New Age guru Carlos Castaneda claimed to have learned a technique for inducing lucid dreams from a Yaqui Indian shaman called Don Juan Matus. This involved staring at his hands during a dream.

Various electronic devices have also been produced to assist the would-be lucid dreamer. These generally work by sensing the onset of REM sleep from changes in eye movements, breathing or muscle tone, and then signalling this information to the sleeper with gentle flashing lights or sound cues. The idea is to help the dreamer to recognise the dream state, as a precursor to learning how to influence it. Both Keith Hearne and Stephen LaBerge invented such devices. Hearne's 'dream machine' sensed the onset of REM sleep from changes in breathing rate, detected by a miniature sensor taped to the nose. (Breathing becomes more rapid and less regular in REM sleep.) The machine could even be set to 'nightmare' mode, in which an alarm would sound and wake the dreamer if their breathing became too rapid.

A simpler technique is to practise, when you are awake, asking yourself if you are dreaming. The answer should obviously be 'no', but the habit of asking the question and thinking about the differences between waking and dreaming seems to spill over into sleep. Some lucid dream aficionados ask themselves several times a day whether they are dreaming, in order to hone their nocturnal skills. But perhaps the simplest advice of all is to hang on to the very basic fact that when you are awake you should not need to ask yourself if you are dreaming. (The only exception might be if something deeply horrible is happening to you, as some people do enter a dreamlike state when they are exposed to severe trauma.) So, if you find that you are asking yourself whether you are dreaming, then you almost certainly *are* dreaming.

Stephen LaBerge and other enthusiasts have advocated lucid dreams as a freely available way of having fun and improving the quality of life. The adept lucid dreamer can enjoy delightful experiences in their dreams and carry the memory of these experiences across into the waking state, helping to bolster their mood during the day. When lucid dreaming works like this, it goes a little way towards realising Samuel Taylor Coleridge's fantasy of bringing back a memento from the dream world:

> If a man could pass through Paradise in a Dream and have a flower presented to him as a pledge that his Soul had really been

there, and found that flower in his hand when he awoke – Aye!
and what then?

Lucid dreaming has been used successfully as a technique for helping
people to deal with severe, recurrent nightmares. However, broader
claims that lucid dreams bring profound insights into the human mind
remain unproven. There is no compelling reason to suppose that they
are any more revealing than ordinary dreams. But lucid dreams are
a potential source of free and presumably harmless pleasure. With
practice and luck, a prolific lucid dreamer might regularly experience
the heights of pleasure that are sought by users of hallucinogenic
drugs, and without any of the expense or obvious risks. Lucid dreaming is not even illegal.

Keith Hearne and Stephen LaBerge might have been the first scientists to analyse lucid dreaming under laboratory conditions, but they
were certainly not the first people to investigate lucid dreams in
a systematic way. As we shall now see, at least one scholar before
them had applied an experimental approach to investigating the
phenomenon.

The great dreamer

Through the combined action of attention and will during dreams,
you can take the first steps in directing and modifying the course
of dreams as you wish.
Hervey de Saint-Denys, *Dreams and How to Guide Them* (1867)

The Marquis Hervey de Saint-Denys was a master of lucid dreaming.
By day a distinguished scholar, he lived by night a second life devoted
to the exploration of dreams. Saint-Denys acquired his life-long passion for dreaming in childhood, and eventually distilled more than
30 years of dreaming experience into his book *Les Rêves et les Moyens
de les Diriger* (Dreams and How to Guide Them), which he published
anonymously in 1867.

Hervey de Saint-Denys was born in Paris in 1822, the son of Baron
d'Hervey. On his mother's death in 1844 he became the Marquis de
Saint-Denys. He taught Chinese and Tartar-Manchu at the Collège de
France, where he became a professor in 1874. During a long and
distinguished career as an Oriental scholar he published numerous

books on the Far East. At the age of 13 he had the idea of drawing a picture based on his recollections of a strange dream that had made a deep impression on him. Henceforth, he kept a regular diary of his dreams. Spurred on by his enthusiasm for the project, he found his dreams easier and easier to remember. He captured them in an illustrated dream diary, which eventually grew to fill 22 notebooks.

From the moment his lifelong project started at the age of 13, Saint-Denys found that the more attention he paid to his dreams, and the more he practised recalling them, the more clearly he remembered them the next day. For over 20 years after he developed this extraordinary faculty, Saint-Denys never once woke without being able to recall his dreams in detail. This led him to conclude that we dream every night, and that our commonplace failure to remember dreaming reflects a failure of memory rather than an absence of dreams – a conclusion echoed by modern science.

Soon after he started recording his dreams, Saint-Denys encountered the state that we now know as lucid dreaming:

> In my fourteenth year . . . I found myself developing a faculty to which I owe most of the observations recorded herein: that of often being conscious of my true situation while sleeping, of retaining in my dreams a sense of my preoccupations from the preceding day, and of then having sufficient control over my ideas to guide their further development along whatever course that suited me.

The term 'lucid dream' was not coined until some years later. One Frederik Van Eeden seems to have been the first person to use it, in a paper he delivered to the Society for Psychical Research in London in 1913. Like Saint-Denys, Van Eeden kept a detailed dream diary, and more than two thirds of the 500 or so dreams he recorded in it were of the lucid variety.

Once Hervey de Saint-Denys had experienced his first lucid dream, the faculty developed rapidly. His second lucid dream came a week later; six months after that, they were occurring on average two nights out of five. After 15 months, Saint-Denys found he was dreaming lucidly almost every night – an unusually high rate that most lucid dreamers do not achieve. He proceeded to spend much of his life, when not performing his academic day job, investigating the phenomenon.

Saint-Denys believed that almost anyone could learn to experience lucid dreams, with a little effort. A good way to start, he found, was to think about lucid dreaming while awake and keep a dream diary. But to master the faculty, you had to be able to discern that you were dreaming while you were dreaming. One way of doing this, he discovered, was to imagine you were closing your eyes. If you were not dreaming, closing your eyes would simply produce lasting darkness; in the dream state, however, new images would soon start to appear out of the blackness.

Friends and colleagues were recruited to help Saint-Denys in his investigations of lucid dreaming. Their experiences confirmed his belief that most people could do it. After Saint-Denys spoke to 14 people about lucid dreaming, two of them started experiencing lucid dreams immediately, nine achieved a similar result after a little delay and two made no effort. That left only one person out of 14 who seemed genuinely unable to have lucid dreams.

Unlike Freud, Saint-Denys did not quarry for hidden meaning or symbolism in his dreams. On the contrary, he found that whenever he thought about something during a lucid dream, an image related to that thought would pop up in his dream. He concluded that dream images are representations in the mind's eye of the objects that occupy our thoughts when we are awake. On that basis, Saint-Denys advised that anyone wishing to make their dreams more pleasant should fill their thoughts with agreeable images from their daily life.

Hervey de Saint-Denys dissected dreaming like a true scientist, systematically experimenting with his own dreams to test his hypotheses, and judging those hypotheses against the empirical evidence. For example, to test his theory that dream images are all drawn from waking experience, he repeatedly tried to dream of experiences he had never had, such as committing suicide. He found he could not. He also arranged to be woken at various times of the night. From this he ascertained that whenever he was awoken he always had the feeling of being interrupted in the middle of a dream, implying that he dreamt throughout the night.

Another of Saint-Denys's dream experiments was designed to see whether a sensory stimulus that was associated with a particular waking experience would trigger dreams about that same experience. Saint-Denys was going to spend a fortnight with friends in the mountainous Ardèche countryside. The day before leaving, he bought some distinctive but unfamiliar scent and took it with him in a sealed bottle. Throughout his vacation Saint-Denys constantly exposed himself to the scent by soaking his handkerchief in it, enduring his friends' jokes in the cause of science. He sealed the scent away when he left the Ardèche and smelt it no more. Several months later, by prior arrangement, his servant sprinkled drops of the scent on his pillow early one morning while he slept. For the first time, Saint-Denys dreamt he was back in the Ardèche. Mountains dotted with large chestnut trees rose up before him; a basalt rock appeared so clearly that he could have drawn it in the finest detail. When he awoke he could still smell the scent that had been sprinkled on his pillow and which he had breathed in while asleep.

In another experiment Saint-Denys used sound stimuli to trigger what might have been erotic lucid dreams. He started the experiment by paying an orchestra leader to play one of two waltzes every time he danced with one of two women. Each woman therefore became associated in Saint-Denys's mind with a particular tune. After repeatedly exposing himself to these conditioning stimuli, rather like Pavlov's dogs learning to associate the sound of a bell with the arrival of food, Saint-Denys tested their effect on his lucid dreams. He had a special musical box made for him that played the two tunes. The music box was linked to a clock mechanism that triggered one or other tune early in the morning. As predicted, each tune infallibly triggered dreams about the corresponding woman. Saint-Denys's memoir is coy about the precise nature of these dreams, and it may be unfair to assume they involved sex. He does, however, note that the ladies concerned were 'two women about whom it was particularly agreeable for me to dream'.

Many other such experiments and observations left Saint-Denys convinced that dreams are influenced by bodily sensations that occur during the dream, and that the sleeping mind weaves these sensations into the dream. As we saw in the previous chapter, laboratory experiments conducted a century later at Stanford University reached the same conclusion.

As Saint-Denys acquired more experience he was increasingly able to control his lucid dreams, unlocking a huge hidden potential. He discovered that his dreaming mind could retrieve vast numbers of buried memories that his waking mind was incapable of recalling:

The hidden depths of our memory are like immense subterranean passages, in which the mind can never see more clearly than when there is no light shining from outside. It is no surprise, then, if we can see again in our dreams, with marvellous clarity, people who have been dead or absent for a very long time, and if we recollect in fine detail places we have visited, tunes we have heard, or even entire pages of books that we have read years beforehand ... In that isolation from the external world, which allows the mind to condense within itself, as it were, all the heat and power, all the liveliness of our emotions and thoughts; under the influence of that state called sleep, which shuts the body's eyes to new perceptions and opens those of the mind to the buried treasures of the memory; when night falls outside while all is light inside, cannot the mind achieve a far greater sensitivity than when it is in the waking state?

Saint-Denys described how the dreaming mind, having dredged up all manner of previously inaccessible memories, would then rearrange those memories in strange new combinations, of a sort that would never be experienced in reality. The imagination, he wrote, could *create* in a dream, by generating images that have never been seen before, like the chance combinations of a kaleidoscope. The analogy with play behaviour springs to mind once again.

Another basic principle championed by Saint-Denys was that dreams affect our waking thoughts and behaviour (like wine through water, if you recall). He wrote that the influence of dreams on conscious thinking and action was vastly greater and more frequent than most people believed. Pursuing this theme, Saint-Denys pondered on Saint Augustine's quandary about free will and sin in dreams. He highlighted the potentially awkward fact that we can exert some control over our dreams, and yet we sometimes commit reprehensible or evil acts in them. Saint-Denys concluded that dreams usually produce a temporary suspension of free will, of the sort that would persuade any jury to acquit someone even though they had undoubtedly committed the misdeed. The problem for the dreamer was that merely thinking about something in a dream would automatically trigger an image of that thought. In dreaming, wrote Saint-Denys, to think of an act is inevitably to carry it out; and while we can prevent ourselves from committing evil acts, we cannot prevent ourselves from *thinking* about them. Therefore, he concluded, we are deprived of free will in our dreams, in the sense that we often act without intention. This seems like an entirely sensible conclusion to an otherwise abstruse philosophical debate.

Last, but not least, Saint-Denys recognised the sheer pleasure that can be derived from dreams:

> I am sure many people would admit that they have sometimes experienced in their dreams such surges of affection and warmth, such indescribable states of joy and pleasure, as could never be found in real life. And I would not hesitate to say that if our aim is to experience the ultimate feelings of pleasure and beauty, then the state of dreaming would be the most conducive for them.

PART V

Origins

11

From Egg to Grave

Thy cries, O baby, set mine head on aching.
>Sir Philip Sidney (1554–86), 'Sleep, baby mine, Desire'

Sleep, in all its many aspects, changes and develops continuously throughout the course of each individual's lifetime, from conception through childhood to old age and death. The foundations of our individual sleep patterns are laid down very early in life, long before birth.

The foetus spends most of its time immersed in REM sleep, dreaming about who knows what. The mechanism for suppressing muscular activity during REM sleep has not yet developed at this early stage, so the dreaming foetus kicks and twitches in its mother's womb. Seventy years later, that same individual will have very different sleep patterns; he or she will spend much less time asleep, and REM sleep will occupy a much smaller proportion of that diminished time. In between, during the course of childhood, adolescence and adulthood, the patterns of sleep will continue to change.

Individuals differ considerably in their characteristic sleep patterns. Some people are quite happy with only seven hours' sleep, while others need nine to feel human; some sleep like logs, unperturbed by the loudest din, while others snap-to at the slightest rustle; some sleep serenely while others toss and snore like flatulent pigs. These sorts of idiosyncrasies are often apparent from an early age, and some of them remain as recognisable threads running through a person's life, despite all the day-to-day variations. The fictional eponymous narrator of Thomas Mann's *Confessions of Felix Krull* is an extreme case of consistency in sleep style. Since before his birth, Felix has always felt more at home in darkness and slumber, and as an adult he will happily lose himself in sleep and dreams for most of the day:

This reluctance to exchange the darkness of the womb for the light of day is connected with my extraordinary gift and passion for sleep, a characteristic of mine from infancy ... Even without being physically fatigued I have always been able to fall asleep with the greatest ease and pleasure, to lose myself in far and dreamless forgetfulness, and to awake after ten or twelve or even fourteen hours of oblivion even more refreshed and enlivened than by the successes and accomplishments of my waking hours.

How and why each of us develops our own distinctive sleep characteristics remains largely unknown. Science cannot yet offer solid explanations of why, for example, some of us need more sleep than others, or why some of us are owls. Research has barely begun to scratch the surface of understanding how the manifold raw ingredients of genes and experience combine to shape our individual sleep characteristics, and why mine are different from yours. However, a fair amount is known about how, in general, human sleep patterns change during the lifespan. Let us start close to the beginning.

Screaming babies

Sleep deprivation is a popular torture device used by the Indonesian secret police and small babies.
John O'Farrell, *The Best a Man Can Get* (2000)

Someone once said that the art of being a parent consists in sleeping when the baby isn't looking. New parents soon experience the brutal reality that babies' sleep patterns are not at all like adults'. Young babies sleep for an average of 14–16 hours a day (with large individual variations) of which about half is REM sleep. The problem for parents is that their baby's sleep is fragmented into several short bouts, scattered across the day and night. Many animals sleep that way too. Adult humans, on the other hand, have grown accustomed to taking all their sleep in a single, uninterrupted block of seven or eight hours during the night. These radically different sleep patterns clash badly when parents go out to work and cannot easily carry their sleeping baby around with them all day.

Living with a small baby is a tiring business. Most new parents lose at least two hours' sleep a night for the first four or five months, and around one hour a night thereafter, amounting to several hundred hours of lost sleep over the first year of their baby's life. What sleep they do get is repeatedly interrupted, amplifying the effects of deprivation. Only a fraction of the sleep loss can be recovered through napping during the day. The net effect can be draining. This description by novelist John O'Farrell of the mind-numbing sleep deprivation of early parenthood may ring bells with some new parents:

> I would give anything just to have eight hours' solid, uninterrupted sleep. Not this violent bungee jumping in and out of half-consciousness, but real, deep, deep, proper sleep. That's the only drug I need: sleep.

The continual thwarting of the overwhelming urge to sleep has driven a tiny minority of parents and caregivers to violence. Anton Chekhov explored this dark theme in his story called 'Let Me Sleep', published in 1888. Varka, a 13-year-old nursemaid, has sole responsibility for her master's baby and must also perform all manner of domestic tasks. She has not slept in a long time and is exhausted:

> The baby is crying. For a long while he has been hoarse and exhausted with crying; but he still goes on screaming, and there is no knowing when he will stop. And Varka is sleepy. Her eyes are glued together, her head droops, her neck aches. She cannot move her eyelids or her lips, and she feels as though her face is dried and wooden, as if her head has become as small as the head of a pin.

Varka hums a lullaby to the screaming baby while her master snores in the next room. She knows he would beat her if she were to fall asleep. Morning comes and there is work to do, but at least the need for sleep does not feel quite so pressing when she is running errands. Another night comes and her master goes to bed, once more leaving Varka to tend the crying baby. She lapses back into her state of waking dreaming, with visions dancing before her eyes. In this exhausted state, Varka suddenly sees who is responsible for ruining her life. Her enemy is the screaming baby:

> The hallucination takes possession of Varka. She gets up from her stool, and with a broad smile on her face and wide unblinking

eyes, she walks up and down the room. She feels pleased and tickled at the thought that she will be rid directly of the baby that binds her hand and foot – kill the baby and then sleep, sleep, sleep. Laughing and winking, Varka steals up to the cradle and bends over the baby. When she has strangled him, she quickly lies down on the floor, laughs with delight that she can sleep, and in a minute is sleeping as sound as the dead.

Not one of Chekhov's jollier tales. The sad truth is that many cases of child abuse and even infanticide have stemmed from the corrosive effects of sleep deprivation. Shaking a baby is one of the leading causes of infant death – sometimes unintended, sometimes deliberate. Exhausted people cease to be rational. Sleep deprivation also fuels the postnatal depression experienced by some new mothers. At the very least, many people feel that tiredness detracts from their ability to be a good parent.

An infant's sleep patterns exert pervasive and complex influences on the parent–infant relationship, in ways that are not well understood by science. The way an infant sleeps (or fails to sleep) will affect the behaviour of its parents, which in turn will affect how the infant sleeps, and so on. Our individual sleep patterns emerge over time from this continual interplay; they are not wholly predestined or fixed at birth.

A mother's sleep problems can begin even before she becomes pregnant. The quality of women's sleep varies somewhat according to the menstrual cycle, with some women experiencing more noticeable fluctuations than others. Sleep tends to be at its best early in the cycle, just after menstruation, and worst during the premenstrual phase. Some women who suffer from premenstrual syndrome experience noticeable sleep disturbances during the premenstrual phase of their cycle. The symptoms can include more frequent awakenings during the night, heightened mental activity on awaking, nightmares, failure to wake at the expected time in the morning, and daytime tiredness. Oral contraceptives also alter sleep patterns. Women taking the pill have less slow-wave sleep and more of the shallower stage 2 sleep than naturally cycling women.

Sleep patterns change during pregnancy, and problems are fairly common among pregnant women. There is usually a marked reduction in deep, slow-wave sleep during pregnancy. Pregnant women typically spend only about 5 per cent of their total sleep time in slow-wave sleep, compared with around 25 per cent when they are not pregnant. This drop in slow-wave sleep is accompanied by an increase in REM sleep and therefore in dreaming. Dreams tend to be more vivid during pregnancy, in part because there are simply more of them. In the later stages of pregnancy, brute mechanical effects such as the need to urinate during the night can further disrupt sleep.

Women become more inclined to snore during pregnancy. Research found that 23 per cent of pregnant women were snoring every night by the final week of their pregnancy, whereas only 4 per cent reported snoring before they became pregnant. We shall be taking a closer look at snoring in chapter 15; suffice it here to say that snoring sometimes has a more serious side, especially when it is a symptom of a sleep-related breathing disorder.

Echoing this theme, the same research discovered that heavy snoring during pregnancy was associated with a heightened risk of developing high blood pressure and giving birth to a growth-retarded baby. High blood pressure was found in 14 per cent of women who frequently snored, compared with only 6 per cent of non-snorers. The pregnant women who snored were also found to be at greater risk of developing pre-eclampsia, a potentially dangerous form of pregnancy-induced high blood pressure. Pre-eclampsia is currently the main cause of death and illness in pregnant women and their foetuses, and of admissions to neonatal intensive care units. The data showed that pre-eclampsia occurred in 10 per cent of snorers, compared with only 4 per cent of non-snorers. Furthermore, the snorers were more likely to give birth to babies who were small for their gestational age: 7 per cent of babies born to snoring mothers were growth-retarded, compared with less than 3 per cent of infants born to non-snorers.

Similar findings emerged from another study that investigated a group of pregnant women with severe pre-eclampsia. Partial obstruction of the upper airway, which is the underlying cause of snoring and more serious disorders, was found in every one of the women with pre-eclampsia. When their breathing problem was treated, by giving them special air masks to assist their breathing during sleep, their blood pressure fell markedly. In chapter 15 we shall see why snoring and its big brother, sleep apnoea, are linked with high blood pressure and cardiovascular disease.

What of the baby? By about 4–6 weeks after birth, infants exhibit a distinct circadian rhythm in their activity, heart rate and body temperature. A basic circadian rhythm is present before birth, but this becomes more pronounced between one and three months of age. Sleep occurs in fewer but longer episodes as the infant grows older. The total time spent asleep falls during the first year, from around 14–16 hours a day to about 12 hours, while the number of separate episodes of sleep drops by half. By six months of age most infants are sleeping for 12–14 hours a day, mostly in one long block augmented by one or more substantial naps. The frequency of night-time awakenings usually falls between three and six months of age. Even so, a national survey of English mothers found that a quarter of one-year-olds were still waking during the night at least five nights a week.

The composition of sleep alters markedly during human development. One of the main changes is a continuous decline in the proportion of REM sleep. The older you are, the less REM sleep you get, both in absolute terms and as a proportion of your total sleep. Over the first two years of life, in particular, the total time spent asleep and the amount of REM sleep both fall. The drop in REM sleep is compensated for mainly by increases in stage 1 and stage 2 NREM sleep, with the proportion of slow-wave sleep (stages 3 and 4) remaining fairly stable.

By three months of age infants tend to be awake during the late afternoon and early evening. At this stage, REM sleep occurs mainly during the night, and the infant spends less of its daytime sleep in REM. By six months, REM sleep has dropped from 50 per cent of sleep to the adult level of around 25 per cent. The length of the sleep cycle also increases, from about 60 minutes at birth to 90 minutes. Strictly speaking, foetuses and very young babies do not display the fully-fledged EEG patterns of NREM sleep stages 1–4 and REM sleep. Instead, their sleep consists of two precursor forms known as Active Sleep, which develops into REM sleep, and Quiet Sleep, which develops into NREM sleep.

Other factors that influence our early sleep patterns include our physical size and food source. During the first few months of life, small infants sleep more on average than larger infants of the same age, and bottle-fed infants differ in their sleep patterns from breast-fed infants. A study that compared the sleep patterns of breast-fed and bottle-fed infants aged four months found that the breast-fed infants spent a higher proportion of their sleep time in deep sleep and had lower pulse rates during sleep. Breast is best, including when it comes to sleep.

Encouraging stroppy babies to go to sleep is a subject close to the hearts of sleep-deprived parents, and also an area of steaming controversy. There are no magic solutions and little consensus about what to do. Opinions differ, for example, as to whether babies should be placed in their cots (or cribs, if they are American) while they are still awake, or whether you should wait until they fall asleep. The nineteenth-century essayist Charles Lamb held strong views on this subject:

> Parents do not know what they do when they leave tender babes alone to go to sleep in the dark. The feeling about for a friendly arm – the hoping for a familiar voice – when they wake screaming – and find none to soothe them – what a terrible shaking it is to their poor nerves!

In practice, most parents seem to share Lamb's sympathy with the infant (at least, when it is very new). American research found that at three weeks of age most infants were not placed in their cots until they were already asleep. By three months, however, infants were often put into their cots while still awake. Perhaps the parents felt their babies could cope better when older, or perhaps they had just grown tired and fed up. The evidence also shows that infants who are allowed to fall asleep on their own are more likely to go back to sleep by themselves if they subsequently awake during the night. Many child-care experts recommend placing babies in their cots when they are sleepy but still awake, rather than waiting until they have fallen asleep.

Parents of older infants are usually advised that if their offspring wakes during the night they should respond in a minimalist way, neither lifting it out of its cot nor providing it with free entertainment. The thinking behind this advice is that you should avoid inadvertently training your infant to cry during the night by rewarding it whenever it does. Nearly everyone agrees that establishing a simple, predictable and relaxed bedtime routine helps.

How should parents deal with the bedtime tantrums of a toddler who really does not want to go to bed? Again, there is no magic key. However, controlled experiments have established that two strategies are both reasonably effective. The 'positive-routine' strategy involves initially moving the child's bedtime to the time when he or she naturally falls asleep, engaging in mutually enjoyable activities before the child is put to bed, and then gradually shifting the bedtime to an earlier time. The 'extinction' strategy involves putting the child to bed and ignoring its tantrums for progressively longer periods each time, so that its terror tactics are not rewarded. Both strategies are found to be more effective than simply waiting for the child to outgrow its tantrums. However, parents are often drawn towards the 'positive-routine' approach because they find it less distressing for themselves.

Though differing in their specifics, most 'toddler-management' regimes rely on one or more of the basic psychological principles of learning and behaviour modification derived from studying rats in mazes. These principles include rewarding the infant for behaving in the desired manner, not rewarding undesired behaviour, and associating the desired behaviour with predictable cues (such as a bedtime routine). Another common technique is gradually modifying the recalcitrant infant's bedtime behaviour through a series of small, incremental changes rather than an abrupt shift. Quick results usually generate the most tears. But perhaps the single most useful piece of advice for sleep-deprived new parents is to remember that it will not last for ever. (Just a few years.)

In times past, parents relied less on protracted negotiations and more on physical restraint. Swaddling – the practice of tightly wrapping infants in clothes or strips of cloth so that their movement is constrained – was commonplace throughout the world until the eighteenth century. Babies would be swaddled soon after birth and kept like that for anything from several weeks to more than six months. Perhaps the best-known image in Christian iconography is that of the baby Jesus in the manger, wrapped in swaddling clothes.

Swaddling is much less widespread nowadays in industrialised countries, although it is still commonly practised in non-industrialised societies, often for long periods of a child's early life. Swaddling is practised by many native Americans, including the Navajo of North America, the Cree peoples of Canada, the Aymara in Bolivia and the Quechua in Peru. It also remains common in some parts of Turkey and China. A fairly recent survey of child-rearing practices in the Yunnan province of southwestern China found that 52 per cent of rural mothers put their infants in swaddling clothes.

Swaddling does work, in that it pacifies babies and helps them to sleep. Swaddled babies spend more time asleep, even if they are only swaddled above the waist. And because they have much less scope to be troublesome when they are trussed up, swaddled infants also tend to spend more time in physical and social contact with adults.

They can be carried around and even hung on the nearest nail or branch when the parent is busy. A study of swaddling among Aymara Native Americans in Bolivia found that swaddled infants received more care from other family members than did non-swaddled infants. The simple explanation is that babies become better company when they are less bother. Swaddling may also help protect the infant from unsafe or unsanitary conditions.

The practice of swaddling does have its downsides, however. Prolonged swaddling may be a risk factor for congenital dislocation of the hip and for respiratory infections. Swaddling has also been found to increase the risk of Sudden Infant Death Syndrome, a subject we shall be returning to in chapter 16 – but only when the swaddling is combined with placing the infant in the face-down sleeping position. Anyway, whatever its potential attractions for the busy working parent, there seems little prospect that swaddling will become fashionable again in industrialised societies.

Another way of improving young children's sleep is through education. A two-year project in French nursery schools in the 1990s demonstrated that education could enhance the awareness of sleep among families with young children. The project used community doctors to distribute educational material and basic advice about good sleep practices to families with three-year-old children. The programme proved to be successful in improving children's sleep, especially among the urban poor. Tackling society's widespread ignorance about sleep ought to be a higher priority in education.

Bad children

> Where unbruisèd youth with unstuffed brain
> Doth couch his limbs, there golden sleep doth reign.
> William Shakespeare, *Romeo and Juliet* (1595)

In his 1693 treatise *Some Thoughts Concerning Education*, the philosopher John Locke argued that the foundations of a good education are play, physical exercise and plenty of sleep, which he described as 'the great cordial of nature'. Nothing contributes more to the growth and health of children, wrote Locke, than sleep. He recommended that children should be permitted as much sleep as they desire.

Things have changed since then, and not entirely for the better.

Modern education and parenting, with their single-minded focus on measurable attainment, pay less attention to unstructured play and physical exercise, and virtually ignore sleep. Partly as a consequence of this shift in attitudes, children spend less time sleeping now than they once did. In 1693 John Locke felt it necessary to warn the parents of children aged between seven and 14 years that if their child was too fond of sleep, they might consider gradually *reducing* its sleep to about eight hours a day, which Locke regarded as sufficient for most healthy adults. How many contemporary pre-adolescents and teenagers routinely sleep more than eight hours a day?

Sleep problems and daytime tiredness are alarmingly common among children. For example, an American survey of children aged between 4 and 12 years found that more than one in five were experiencing problems such as daytime sleepiness, snoring and difficulty falling asleep. Another study found that one in four children aged between 9 and 14 years felt tired when they were at school. A similar picture of tired, underperforming children applies in other countries.

It is not hard to see why many children are getting less sleep than they need. In many homes sleep is regarded as a maintenance activity, to be squeezed into the busy schedule when everything else has been done. In wealthy societies, children's bedrooms have increasingly become places of entertainment and stimulation rather than rest, with many having access to the phone, television, computer games or the Internet. Furthermore, working parents understandably wish to spend time with their offspring after they get home, perhaps late in the evening, and so bedtime has become a movable feast. Being sent to bed is a traditional punishment for children, not a reward. Some children learn the lesson that their parents have inadvertently taught them – that staying up late is a desirable part of adult life and going to bed is bad. Ironically, many of today's overworked and underslept adults would regard being sent to bed early as far more of a reward than a punishment.

Insufficient sleep may contribute to many contemporary childhood troubles. There is mounting evidence that what one scientist has called 'an epidemic of sleeplessness' is helping to fuel the rising incidence of behavioural and emotional problems among children. The connection between inadequate sleep, daytime tiredness and difficult behaviour in children is not always obvious, because lack of sleep affects adults and children in quite different ways. Tired adults lack energy and become less active. But tiredness can have the reverse effect on young children. Tired children try to resist their fatigue by becoming even more active. This revving up, combined with their underlying weariness, interferes with their ability to pay attention in the classroom and behave appropriately. The tired child becomes irritable, fidgety, inattentive, disruptive and generally bad company, but without necessarily looking obviously tired.

Scientists have been accumulating more and more evidence that some children who have been diagnosed with behavioural disorders are suffering from unrecognised sleep problems, some of which are treatable. For instance, an American study of pre-school children found that the individuals who got the least sleep were significantly more likely to display behaviour problems during the day. Similarly, when scientists assessed Dutch children aged 9–14 years, they found that the individuals who slept well had a more positive self-image and were more motivated at school, more receptive to their teachers, less bored and better able to control their aggression. Spending more time in bed will not necessarily transform your fractious offspring into a paragon of virtue, but it will probably help.

Teachers often encounter school-age children who are clearly sleep-deprived to the extent that they have noticeable difficulty staying alert in the classroom. These same children often revealingly refer to television programmes shown late at night. However, an experienced educational psychologist reported that whenever she had raised the issue of sleep with the parents of such children, the parents had invariably pooh-poohed the suggestion that their children might not be getting enough. Perhaps sleep just strikes parents as too mundane an explanation for their children's problems at school, or the suggestion of inadequate sleep may hint too strongly that the parent might be partly responsible for the child's problems. Or perhaps parents regard sending their child up to their bedroom as synonymous with ensuring they get enough sleep, which it clearly is not. Whatever the explanation, the practical experience is that many parents of tired schoolchildren have a blind spot about sleep.

One issue that definitely deserves closer scrutiny is the linkage between poor sleep and Attention Deficit/Hyperactivity Disorder, or ADHD, in children and young people. ADHD is diagnosed by the presence of impulsive, overactive or inattentive behaviour (or a combination of the three) to an extent that interferes with education, home life and social relationships. According to some authorities, ADHD affects between 5 and 10 per cent of school-aged children in the USA, making it the most commonly diagnosed childhood psychiatric disorder. Of course, being somewhat impulsive, highly active and inattentive is normal behaviour for a healthy child, so the problem is essentially one of degree. The incidence of diagnosed ADHD has rocketed in recent years. But there is growing evidence that some of this increase might stem from undiagnosed sleep problems, and that some children who are diagnosed with ADHD may be suffering primarily from lack of sleep.

Objective measurements have shown that children with ADHD have poorer quality sleep on average. A large proportion of children and adolescents who have ADHD, or the related Attention Deficit Disorder (ADD) or Asperger's syndrome, suffer from sleep problems including insomnia, early waking and excessive daytime sleepiness. Their sleep patterns tend to be less stable than normal and they are generally less consistent in the timing, duration and continuity of their sleep. A recent Canadian study confirmed the everyday experience of many parents of ADHD children, by demonstrating that ADHD is accompanied by significantly more bedtime conflict between child and parents. ADHD children were found to be more resistant to going to bed, and their behavioural interactions with their parents at bedtime were more challenging than normal.

One promising line of enquiry concerns factors that disrupt children's sleep and thereby trigger daytime behaviour problems. Part of the problem may lie with sleep-related breathing disorders. American scientists found that one third of children diagnosed with ADHD were habitual snorers, compared with about one in ten of non-ADHD children of the same age. As we shall see in chapter 15, habitual snoring is sometimes a sign of a sleep-related breathing disorder such as sleep apnoea, which can severely disrupt sleep. Moreover, the children with the most serious snoring were also the most inattentive and hyperactive during the day. Further evidence for a connection between behaviour problems, hyperactivity and poor sleep has emerged from research in China, where 14 per cent of elementary school-age children were reported to be getting insufficient sleep. The children with sleep problems were found to have poorer relationships with their parents and peers, worse social skills and lower achievement at school. Even more tellingly, the children with sleep problems were more likely to be described as hyperactive.

One interpretation of evidence like this is that undiagnosed sleep-related breathing disorders, such as sleep apnoea, disrupt some children's sleep, causing daytime fatigue, hyperactivity and inattention. In some severe cases, this leads to a diagnosis of ADHD. According to one estimate, a quarter of all children diagnosed with ADHD could have their behaviour problems cured by treating them for a sleep-related breathing disorder. Bad sleep is not going to emerge as *the* cause of ADHD, but it will probably turn out to be an important ingredient.

There is a further twist in this tale. As the number of children who are deemed to be suffering from ADHD has soared in the USA, UK and elsewhere, so too has the use of the standard treatment – namely, daily doses of the stimulant drug methylphenidate, better known as Ritalin. Some doctors, psychiatrists and parents regard Ritalin as little short of a miracle cure, and the drug has certainly provided enormous relief to many families. More children in the USA take Ritalin than any other mood-altering drug (except, perhaps, caffeine). Ironically, Ritalin is an amphetamine-like stimulant that disturbs sleep. In fact, it has been widely used since the 1950s to treat the serious sleep disorder narcolepsy. It is therefore possible that some children whose behaviour problems stem originally from poor sleep might find themselves locked in a vicious cycle. Because they get inadequate sleep they become fractious, inattentive, impulsive and hyperactive at school, and eventually acquire the diagnosis of ADHD. To deal with the ADHD they are put on Ritalin, which suppresses their daytime behavioural symptoms but further disrupts their sleep.

Governments that are really serious about improving the academic standards of the nation's children and tackling the problems of bad behaviour in schools could do worse than take children's sleep more seriously and encourage parents to ensure their offspring get enough of it.

Yawning youth

> We were very young. I think I never slept that year. But I had a
> friend who slept even less . . .
>
> Cesare Pavese, The Devil in the Hills (1954)

In our teens we are inclined to treat sleep as a disposable commodity, pretty near the bottom of life's long list of priorities. Think of the archetypal adolescent: irritable, moody, forgetful, often seemingly irrational, and inclined to lie in bed all hours at weekends. These are also the symptoms of sleep deprivation, and there are good grounds for suspecting a link. Research shows that chronic sleep deprivation is widespread among adolescents and teenagers, affecting their mood, behaviour and academic achievement. Once again, we see the spiral of disrupted nights followed by disrupted days.

The lifestyles of many teenagers in industrialised societies place conflicting demands on their time, resulting in less sleep and more daytime sleepiness. As adolescents get older they tend to go to bed later, spend less time asleep and consequently feel sleepier during the day. The need for sleep is, if anything, even greater during adolescence than at younger ages, but the actual amount of sleep generally declines. Many adolescents and teenagers end up with seven hours of sleep on weekday nights, when they probably need eight or nine. One American survey found that more than 60 per cent of high-school students experienced daytime fatigue that interfered with their studies.

The consequences of chronic sleep deprivation in teenagers include mood and behaviour problems, accidents, and increased vulnerability to drug and alcohol abuse. Evidence has begun to accumulate that sleep problems and tiredness heighten the risk of substance abuse in adolescents and teenagers. For example, research with college students found that those who fell asleep during the day from tiredness also tended to smoke more and drink more alcohol, while another large study discovered that adolescents who reported frequent sleep problems were more likely to use illicit drugs.

One reason why adolescents and teenagers are particularly prone to sleep deprivation is that our biological clocks shift to a more owl-like setting during adolescence. Getting up early in the morning therefore becomes harder, but it remains an absolute requirement. American school students making the transition from ninth grade to tenth grade usually start the school day an hour earlier, but in practice they go to bed no earlier and therefore get less sleep. Thereafter, sleep tends to be squeezed at the other end too, as older teenagers stay up later. A survey of American high-school students found that between the ages of 13 and 19 their total daily sleep duration fell by 40–50 minutes. Teenagers went to bed later as they got older, but the time at which they awoke remained roughly constant. They consequently got less sleep and the result was impaired daytime performance. The students who slept the least on school nights suffered the most from tiredness and depressed mood. They were also less successful academically: the students who performed poorly at school (gaining C grades or worse) went to bed on average 40 minutes later and got 25 minutes less sleep each night than the most successful students. They also tended to sleep in later at weekends, again indicating a sleep deficit.

In the 1990s, education researchers at the University of Minnesota concluded that American teenagers are chronically sleep-deprived and would benefit from an extra hour in bed on schooldays. In one school district of Minnesota, teenagers were allowed to start school an hour later than younger children. Their average sleep duly increased by about 45 minutes a night. The students who started school at this more civilised time felt less sleepy during the day and less depressed than students at similar schools where the early start was retained. They were less likely to arrive late for school and had significantly fewer days off school for sickness. Their teachers felt better too. Incidentally, British school students who envy that extra hour in bed should bear in mind that American schools generally start earlier in the morning. In the Minnesota experiment, the start-time was put back from the unspeakably early 7:30 to a more reasonable 8:30.

Insufficient sleep during the teenage years is a problem in many other countries too. When scientists assessed Israeli school children aged 10–12 years, they found that the early risers among them (those who started school just after 7 a.m. at least twice a week) got significantly less sleep than those who started school at the more civilised hour of 8 a.m. As well as sleeping less, the early risers complained more of daytime sleepiness and found it harder to concentrate at school. Similarly, psychologists who assessed Swiss high-school students aged 10–14 years discovered that between half and three quarters of them felt they were not getting enough sleep. And in a large French survey, four out of ten French high-school students said they got insufficient sleep or found it difficult to stay awake during the day.

Being part of a large family tends to make matters worse. A study of sleep patterns among 14-year-olds in England found that the more siblings an individual had, the later on average they went to bed and the less time they spent in bed. Younger siblings are often kept up by older siblings who go to bed later.

Teenagers develop characteristic strategies for coping with their chronic sleep deprivation. The two main coping strategies are sleeping in late at weekends and consuming prodigious amounts of caffeine. Many teenagers build up a substantial sleep deficit during the week, which they fend off with caffeine-laced soft drinks and coffee, before recouping some of their losses by sleeping longer over the weekend and during vacations. Again, this pattern seems to be widespread. For example, a study of Brazilian teenagers found that at weekends they woke on average three hours later and slept for between one and one and a half hours longer than on weekdays. Other studies have confirmed the common observation that adolescents and teenagers sleep considerably more during holidays than during the school term.

Sleep can be squeezed even harder when teenagers leave home to go to college or start work. Whatever positive influences their parents might have exerted soon vanish, while the social pressures to skimp on sleep become stronger. The upshot is even later bedtimes and even less sleep. And the problem appears to be growing. An analysis of several published studies found that over a 20-year period the average daily sleep duration of college students had dropped by one hour.

Old and grey and full of sleep

> When you are old and grey and full of sleep,
> And nodding by the fire, take down this book,
> and slowly read . . .
>
> W. B. Yeats, 'When You Are Old' (1893)

Once upon a time it was believed that old people needed more sleep. This commonly held view was reflected in a popular eighteenth-century guide to health for the older man, which declared this advice:

> Sleep recruits nature, and restores the wasted spirits. This is necessary to all persons; but to the aged most, because they can least bear the waste of them. They may therefore indulge in it longer than the young.

We actually sleep less in old age, mainly because our ability to sleep soundly diminishes rather than because we need less sleep. With advancing years, sleep becomes increasingly disrupted by changes in our circadian rhythms and by medication, sleep-related breathing disorders, other health problems and bad sleep habits.

As we progress from early adulthood into late middle age we spend less time on average in bed and less time asleep. People in their fifties and sixties get on average 46 minutes less sleep each night than adults in their twenties, mainly because they wake more often during the night and stay awake longer. The net result is that sleep becomes less satisfactory and more precious. 'The harassed middle aged,' wrote A. Alvarez, 'are in love with sleep in the same way as the young are in love with love; chastity is the torment of youth, insomnia of age, and at neither stage of life does it ever seem possible to get enough of what you want.'

The nature of our sleep changes throughout adulthood. With advancing years, a dwindling proportion of our dwindling sleep is spent in either REM sleep or deep, slow-wave sleep. Instead, more of our sleep is made up of the shallower stages 1 and 2 of NREM sleep. Slow-wave sleep declines from about 20–25 per cent of total sleep in early adulthood to less than 5 per cent by middle age. Elderly people get very little slow-wave sleep. This reduction in slow-wave sleep is accompanied by a substantial fall in the production of growth hormone. Older people are also slower to recover from lost sleep; following a period of sleep deprivation, they take longer to enter the most refreshing form of slow-wave sleep.

As old age advances we spend more time awake during the night; the amount increases by an average of almost half an hour for every additional ten years of age. The middle-aged and elderly spend twice as long awake at night as people in their twenties. The more frequent awakenings during the night explain why many people become more conscious of dreaming as they grow old, despite the fact that they actually spend less time doing it. When you are young you might dream for two hours a night but remember none of it because you sleep so soundly.

Age also brings a systematic shift in our circadian rhythms towards greater 'larkness'. We tend to go to bed earlier and wake up earlier as we get older. The slight consolation is that we feel more alert when we do wake early. This shift to a more lark-like sleep pattern reduces our capacity to cope with working at night. Elderly people are generally more suited to morning shifts than night shifts. Ernest Hemingway's *The Old Man and the Sea* captures the larkish trend in ageing in this exchange between the old man and his young assistant:

'Good night then. I will wake you in the morning.'
'You're my alarm clock,' the boy said.
  'Age is my alarm clock,' the old man said. 'Why do old men wake so early? Is it to have one longer day?'
  'I don't know,' the boy said. 'All I know is that young boys sleep late and hard.'

As well as drifting towards an earlier setting, our circadian rhythms also deteriorate with age. The sleep of older people is less neatly consolidated into a single block of unbroken sleep at night. They wake more often during the night and nap more during the day. The circadian fluctuations in core body temperature are 40 per cent smaller in people aged over 65 years, and the low point occurs nearly two hours earlier in the day than it does in younger people.

Insomnia becomes much more common as we grow older. The elderly consequently consume more sedatives and sleeping pills than any other age group. Insomnia is especially prevalent among middle-aged and elderly women. More than 40 per cent of healthy middle-aged American women suffer from some form of sleep disturbance. Those who experience the worst problems with their sleep are found to have higher levels of anxiety, stress and depression, and higher blood pressures. They also tend to be fatter, with higher waist-to-hip ratios. The menopause is particularly associated with deterioration in sleep.

Even after the menopause, women generally continue to experience poorer quality sleep than men. A Swedish study of people over 65 years of age found that 24 per cent of elderly women and 13 per cent of elderly men experienced difficulties in staying asleep, falling asleep or waking too early. Sleep problems were often associated with medical conditions in men and with depression in women. In another study, researchers found that women over the age of 50 took longer to fall asleep than men of the same age, woke up more often during the night, spent more time awake and had poorer quality sleep.

One reason why elderly people are more prone to insomnia is that they are prey to more sleep-disrupting influences such as chronic pain, the need to urinate during the night, sleep-related breathing disorders, and an increased sensitivity to noise. For example, an American study of people in their sixties and seventies found that more than one in three displayed frequent breathing disturbances that often disrupted their sleep. Many elderly men have an enlarged prostate gland, which makes them need to urinate during the night. This disrupts their sleep, sometimes with serious consequences. In one extreme case, a man suffering from benign prostate enlargement was so afraid of wetting his bed at night that he could not sleep. His sleep deprivation eventually became so severe that it triggered a form of dementia. His symptoms were wrongly diagnosed as a chronic brain disorder. But when the man received surgery to correct his prostate problem, he started sleeping normally again and his supposed 'brain disorder' vanished.

Another reason why old people often sleep badly is that they tend to be more sensitive to noise. The minimum amount of noise that is required to wake someone decreases substantially between early adulthood and later middle age. This greater sensitivity to noise is most apparent during slow-wave sleep, which is the type of sleep that declines most in old age.

As we saw in chapter 3, there are striking similarities between the effects of chronic sleep deprivation and many of the characteristics that we often associate with old age, including slowed reactions, impaired memory, depressed mood, irritability and somnolence. Some of the symptoms that are so readily dismissed as the unavoidable consequences of old age might in fact be the partially avoidable consequences of chronic sleep deprivation, in people whose sleep has become fragile and inadequate. Long-term insomnia is now known to be an independent risk factor for mental decline in the elderly. Conversely, a short afternoon nap of less than half an hour has been shown to produce significant improvements in the alertness and mental performance of elderly people.

From supposedly hyperactive children to supposedly senile old people, many of the seemingly intractable problems we encounter at different stages in our lives would benefit from greater attention to sleep.

12

The Reason of Sleep

Life is something to do when you can't get to sleep.
Fran Lebowitz, *Metropolitan Life* (1974)

Sleep, like play behaviour, has no self-evident biological purpose. When you see an animal eating or drinking or copulating or fighting, it is generally obvious why it is doing what it is doing. However, when you see an animal or a person curled up asleep, the ultimate explanation for their behaviour is less clear.

We assume that sleep must bring benefits, otherwise why would humans and animals all spend so much of their time doing it every day? Or, as a biologist would put it, since sleep has evolved through Darwinian selection it must confer adaptive advantages. If living organisms could survive and reproduce slightly better by not sleeping, then the world would by now be full of organisms that do not sleep. Instead, the world is notably full of organisms that dedicate large slices of their lives to sleeping. So, what are those adaptive advantages of sleep? Why did sleep evolve to become a universal characteristic of animals?

Throughout human history, scholars have mused on theories about the purpose of sleep. Most of those theories held sleep to be some form of recuperation, restoring the body and the mind after the rigours of the day. William Shakespeare memorably encapsulated this view in *Macbeth*:

> Sleep that knits up the ravelled sleeve of care,
> The death of each day's life, sore labour's bath,
> Balm of hurt minds, great nature's second course,
> Chief nourisher in life's feast.

In more recent times, various scientific theories have been advanced to explain the biological origins and functions of sleep. Meanwhile, more pragmatic souls have ignored the theorising and got on with the business of investigating the phenomenon. When Nathaniel Kleitman, the great pioneer of modern sleep science, was once asked to explain the role of sleep, he replied that he would explain the role of sleep if someone would first explain to him the role of wakefulness.

In 1624, John Donne expressed the view that sleep has two functions – to maintain the body and prepare it for the afterlife:

> Natural men have conceived a twofold use of sleep: that it is a refreshing of the body in this life; that it is a preparing of the soul for the next.

Modern science may not agree with the specifics of Donne's explanation, particularly the bit about preparing for the afterlife. But it does share his belief that sleep has more than one biological function or purpose. You will recall that two distinct states, REM sleep and NREM sleep, are lumped together under the heading of 'sleep'. In many respects, REM sleep and NREM sleep are as different from one another as both are from the waking state. We are therefore unlikely to discover the one true explanation for sleep because there are almost certainly at least two. But before we delve into the biological functions of sleep – that is, what sleep does for us now – let us consider where sleep came from.

The evolution of sleep

> Drinking when we are not thirsty and making love all the year round, madam; that is all there is to distinguish us from other animals.
>
> Pierre Beaumarchais, *The Marriage of Figaro* (1778)

To understand the current biological functions of sleep, it helps to know how sleep emerged over the course of evolutionary history. How did sleep evolve over millions of years into the various forms that we now observe around us in the natural world? Mammals and birds, together with reptiles that evolved more recently, come equipped with both NREM sleep and REM sleep, whereas primitive reptilian species that originally evolved a very long time ago have only NREM sleep. From this fact we can infer that REM sleep evolved more recently in biological history than NREM sleep.

Further clues about the evolutionary origins of sleep may be gleaned from existing primitive mammals that are known to have remained relatively unchanged for a very long time. One such beast is the echidna, or spiny anteater, an egg-laying mammal that lives in Australia. The echidna belongs to a class of mammals that branched off from all other mammals (which, like us, do not lay eggs) about 120 million years ago, early in mammalian evolutionary history. For a long time, scientists believed that the echidna had no REM sleep. However, more recent research has overturned this belief. The echidna does have REM sleep after all. Scientists discovered that these animals do display the characteristic EEG patterns and other physiological signs of REM sleep when they are sleeping at moderate temperatures, although the signs disappear if the temperature is too low or too high. Another primitive egg-laying mammal, the platypus, has lots of REM sleep. In fact, the platypus has more REM sleep than any other animal, and several times more than humans. It sleeps for 12–16 hours a day, of which 6–8 hours are spent in REM sleep.

The fact that primitive mammals like the echidna and platypus exhibit REM sleep implies that REM sleep appeared earlier in evolutionary history than was previously thought – quite possibly in the reptilian ancestors of all mammals, who lived around 250 million years ago. If REM sleep did emerge that long ago, it may well have evolved originally to perform some basic maintenance role, rather than anything concerned with honing the higher intellectual faculties. Once it had evolved, however, REM sleep could then have been co-opted by Darwinian selection and exploited in new ways. Dreaming might have been grafted on to REM sleep later in evolutionary history, in much the same way that the capacity for speech was grafted on to the capacity for breathing during human evolutionary history. Whether or not this theory is correct, the point to remember is that REM sleep and dreaming, although often coincident, are not the same thing. But what is sleep for *now*?

What is sleep for?

> O Sleep, in whom all things find rest, most peaceful of the gods,
> you who calm the mind, put cares to flight, soothe limbs wearied
> by harsh tasks and refresh them for their toil . . .
>
> Ovid, *Metamorphoses*, Book XI (AD 2)

One of the oldest theories about the purpose of sleep, dating back at
least as far as Aristotle, is that it is a mechanism for physical recuperation. However, the scientific evidence provides only lukewarm
support for this view.

We saw in chapter 8 that slow-wave sleep tends to increase after
vigorous physical exercise, perhaps indicating that slow-wave sleep
has some sort of restorative function. Both sleep deprivation and
intense stimulation during wakefulness provoke transient increases
in slow-wave sleep. Moreover, the synthesis of protein and nucleic
acids (the building blocks of DNA) in the brain increase during sleep.
Again, these observations fit with some restorative role for sleep. On
the other hand, many parts of the body, notably the brain and heart,
remain active even when we are asleep. Indeed, some parts of the
brain are more active during REM sleep than in the waking state. So
it cannot be true that the whole body rests during sleep. Moreover,
people who spend all day lying in bed still sleep as much as (if not
more than) people who live highly active and strenuous lives. Sleep
is not simply an automatic consequence of physical exertion.

A more specific and plausible theory is that sleep originally evolved
as a way of conserving energy. In warm-blooded animals like us,
maintaining a constant body temperature requires lots of energy,
especially if you are small or live in a cold climate. Physical activity
adds substantially to that cost. Periods of enforced inactivity therefore
reduce the requirement for food, especially if the animal rests somewhere sheltered from the cold. The energy consumption of the brain
also drops substantially during NREM sleep (though not in REM
sleep) producing a further saving. The brain is a major consumer of
energy. In line with this energy conservation theory, the evolutionary
evidence suggests that sleep and warm-bloodedness both evolved at
around the same time. The energy-conservation theory also fits the
observation that small mammals, particularly those inhabiting cold
climates, tend to sleep a lot.

Again, though, the theory does not easily account for all the facts. Sleeping – as opposed to just resting – does not conserve very much energy in humans. We can reduce our energy expenditure almost as much just by lying still and relaxing; going to sleep further reduces our metabolic rate by only about 5–10 per cent. When aquatic mammals such as porpoises sleep they carry on swimming, so sleeping cannot reduce their energy expenditure very much. And there are other ways of saving energy, notably by hibernating.

Hibernation, which is quite different from plain sleep, is accompanied by long periods of NREM sleep punctuated by periods of wakefulness. During normal NREM sleep the body temperature and metabolic rate both decrease slightly, whereas in hibernation they drop markedly. For instance, when ground squirrels hibernate, their body temperature falls to 23 degrees Celsius. As their temperature falls, so does the amount of REM sleep; by the time it has dropped to 23 degrees, REM sleep has vanished and the hibernating animal spends its entire time in slow-wave NREM sleep.

In the 1970s the British biologist Ray Meddis proposed that sleep evolved as a way of keeping animals inactive during periods when being inactive is simply the best thing to do. For most species, this means being inactive during the hours of darkness, when it would be hazardous and wasteful to blunder around. By sleeping during the night, preferably somewhere sheltered, animals avoid predators, conserve energy and protect themselves from the vagaries of their environment. Some species specialise in doing the reverse, by being nocturnal and hunting while everyone else is asleep, but the basic principle is the same. According to this theory, sleep might have no recuperative or other functions: it is simply a mechanism to ensure that animals stay still when they have nothing better to do with their time. The inactivity theory also offers an explanation of why babies sleep much more than adults. Small babies are entirely cared for by their parents and cannot usefully spend their time doing much else besides sleeping.

Theories in this mould do at least help to make sense of the enormous variations in sleep patterns between species. Among mammals, for example, there is a general correlation between an animal's physical security and the duration of its sleep. Predators tend to sleep a lot, while the vulnerable species they prey on tend to sleep less. For instance, jaguars and other large cats sleep for ten or more hours a day, whereas sheep, deer and many other animals that graze in the open sleep for less than half that amount. Species that sleep in safe places also tend to spend longer asleep, even if they are potential meals for predators. For example, the common ground squirrel sleeps 14 hours a day in underground burrows where it is relatively well protected.

It is easy to imagine how Darwinian selection might have arrived at some optimal trade-off between the biological benefits of sleeping and the conflicting pressure to remain awake and vigilant for danger. An animal is at its most vulnerable during sleep – especially in REM sleep, when it is virtually blind, deaf, dumb and paralysed. Most large animals sleep standing up, but they have to lie down for REM sleep or they would fall over. Elephants, for example, sleep for only a few hours a day; they take their NREM sleep standing up, but lie down for short bursts of REM sleep. Cows similarly sleep standing up but lie down to dream.

A reasonable conclusion is that NREM sleep, which emerged first, originally evolved as a way of ensuring that animals are inactive during periods in the 24-hour cycle (usually the dark hours) when inactivity is the optimal strategy. What this theory fails to explain, however, is why we all come equipped with two different forms of sleep. REM sleep evolved more recently than NREM sleep, and the evidence suggests that it has quite different biological functions.

What is REM sleep for?

> Every person may derive advantage from dreams.
> William Smellie, *The Philosophy of Natural History* (1790–99)

One of the more unusual theories about the biological function of REM sleep was proposed by David Maurice of Columbia University. He suggested that the rapid eye movements during REM sleep serve to stir up the aqueous humour (the fluid inside the eyeball).

This theory is not as fanciful as it might seem. The aqueous humour has to circulate continuously in order to transport oxygen from blood vessels in the iris to the surface of the cornea, which is poorly supplied with blood. Without adequate circulation, the cornea would be starved of oxygen. When we are awake, the aqueous humour is stirred up by continual small eye movements and by convection currents that are generated by air cooling the outer surface of the eyeball. But when the eyes are closed, the fluid inside them becomes almost static. Something needs to keep it circulating, and Maurice argued that rapid eye movements fit the bill. His theory would account for the fact that successive episodes of REM sleep become progressively longer during the course of a night's sleep, as the eyes become more in need of oxygenation. But it obviously does not explain the really interesting aspects of REM sleep such as the distinctive pattern of brain activity, the dreaming and the muscular paralysis.

Most of the evidence about REM sleep points firmly towards the conclusion that it is concerned with maintaining the brain in some way. During REM sleep the brain is highly active, but in a quite different way from the waking state. Moreover, REM sleep is a universal characteristic of large-brained animals like birds and mammals.

Young animals spend much more of their time in REM sleep than adults, and one theory is that REM sleep promotes the development of the brain in early life. The idea is that during REM sleep the immature brain generates activity patterns that stimulate the brain and assist in its own self-assembly. Sensory stimuli from the eyes, ears and other organs contribute to brain development, but these sources of stimulation are largely cut off during sleep. And therefore, according to the theory, the brain generates its own stimuli in the form of REM sleep. The importance of appropriate stimulation for normal brain development is demonstrated by the fact that if a young animal is deprived of, say, visual experience during a crucial phase in its early life, the regions of its brain that process visual information do not develop properly. The stimuli generated within the brain during REM sleep could complement the role of sensory stimuli during the waking state. If so, REM sleep would be especially important before birth, when the developing brain receives less sensory stimulation from the outside world.

The development theory would explain why less well-developed brains tend to spend more time in REM sleep. This correlation between brain immaturity and REM sleep is true both within species and across species: so, for example, kittens have more REM sleep than adult cats, and babies have more REM sleep than adult humans. Similarly, species whose young are born with relatively immature brains (such as cats and humans) have more REM sleep than species whose brains are relatively mature at birth (such as sheep and guinea pigs). The theory is also supported by experiments on the development of vision. If a kitten is deprived of sight in one of its eyes, the brain cells involved in processing visual information fail to develop normally; this impairment in brain development is amplified if the kitten is also deprived of REM sleep.

If REM sleep were exclusively concerned with the construction of the developing brain then there would be no obvious reason why it should continue to occur every night in mature adults with fully developed brains. But it does. Although REM sleep accounts for less of our total sleep as we grow older, it still makes up 20–25 per cent of all sleep during most of our adult life. This suggests that it must perform a valuable biological function in the adult brain as well. Over the past few years, scientists have unearthed compelling evidence that this function is concerned with learning and memory.

Reverse learning

> Sleep is a relaxation of the conscious guard, the sorter. Sleep is when all the unsorted stuff comes flying out as from a dustbin upset in a high wind.
>
> William Golding, *Pincher Martin* (1956)

In the early 1980s the British scientists Francis Crick (of DNA double-helix and Nobel Prize fame) and Graeme Mitchison put forward their 'reverse-learning' theory of REM sleep. This proposed that the biological function of REM sleep is to eliminate undesirable, 'parasitic' modes of electrical activity from the brain – roughly the opposite of learning. According to their theory, dreams are merely reflections of neural dross that is being erased from the brain during REM sleep. Freud would not have approved.

Crick and Mitchison argued that the brain stores information in a fundamentally different way from an electronic computer. Memories are stored in the brain in a distributed and overlapping format, with each memory spread over many synapses (the junctions between nerve cells) and each synapse involved in storing many different memories. That is why our memories are associative: a single thought or stimulus can evoke a whole string of related memories, because they all reside in the same network of neurons. Information-storage networks of this nature are inherently easy to overload, they argued, because stimulating one part of the system can trigger mixed outputs. Such systems need continual pruning to avoid confusion. Crick and

Mitchison proposed that 'reverse learning' during REM sleep counteracts this inherent tendency to overload by weakening the memory traces that are already weak, so that the stronger traces can become fixed. They further suggested that this memory-pruning mechanism has enabled evolution to construct brains that are much smaller in size than they would otherwise have to be without REM sleep.

The Crick-Mitchison theory was put to the test by using a computer to simulate a highly simplified network of nerve cells. The model neural network was able to store many memory traces, each of which could be evoked by stimulating only part of that trace. In that sense it behaved like real associative memory. The model network had other interesting properties as well. When new memories were stored, spurious memory traces were created as by-products. More importantly, an 'unlearning' process, analogous to the Crick-Mitchison reverse learning, had the beneficial effect of reducing spurious memory traces, making the network more efficient at recalling genuine memories. The model system behaved roughly as predicted by the Crick-Mitchison hypothesis.

The reverse-learning theory of REM sleep echoes a much older strain of thought, which regarded dreams as the waste products of the mind. The English satirist and playwright Thomas Nashe espoused this nihilistic view of dreaming in his 1594 treatise *The Terrors of the Night*:

> A dream is nothing else but a bubbling scum or froth of the fancy, which the day hath left undigested; or an afterfeast made of the fragments of idle imaginations . . . You must give a wounded man leave to groan while he is in dressing. Dreaming is no other than groaning, while sleep our surgeon hath us in cure.

Is the Crick-Mitchison theory right? The evidence is not wholly convincing. The theory says that we dream about the rubbish our minds are trying to forget, not the important memories that our minds want to retain. This makes some sense, in that most dreams are composed of mundane elements drawn from everyday experiences. But there are important anomalies. The contents of dreams might be mostly mundane, but they are not entirely random. Dreams do sometimes focus on matters of obvious personal significance, and they are almost always bizarre and exotic. During dreams the brain constructs a highly selective but distorted model of the world, in which some elements of waking reality are systematically over-represented while others are under-represented. Dreams neither resemble the random rubbish left on the brain's cutting-room floor nor the crucial memories we are specifically trying to retain.

Another snag with the Crick-Mitchison theory is its implication that we should not try to remember our dreams, because these are the very traces our brains have been trying to delete. In fact, there is no convincing evidence that simply remembering dreams is harmful. People who practise remembering their dreams become adept at it and seem to benefit from doing so. Recalling and discussing dreams is common in many human cultures and appears to do more good than harm.

The Crick-Mitchison theory is probably wrong in its specifics. But it does hit the mark by focusing on how the sleeping brain processes information and stores memories. More recently, scientists have been accumulating a convincing mass of evidence that the key function of REM sleep lies with the consolidation of memory.

To sleep, perchance to learn

> Weary with toil, I haste me to my bed,
> The dear repose for limbs with travel tired;
> But then begins a journey in my head,
> To work my mind when body's work's expired.
>> William Shakespeare, Sonnet 27 (1609)

Something happens to our thoughts, anxieties, preoccupations and feelings while we sleep. Common experience suggests that when we are confronted with some knotty problem or emotional dilemma, it often helps if we 'sleep on it'.

Most languages have a saying or proverb that conveys the notion of 'sleeping on it'. In Russian, for example, there is a saying that the morning is the time to make up one's mind. German, French and Italian have proverbs to the effect that the night brings counsel and you should always sleep on an important decision. Folk wisdom and science have yet again converged. Currently, the most powerful scientific theories about the biological function of REM sleep revolve around the idea that it is a special state in which the brain processes newly acquired information and consolidates memories.

Many interlocking strands point to a connection between REM sleep and learning. People and animals display a temporary increase in REM sleep shortly after they have been exposed to new stimuli or learnt a new task. Conversely, if animals or people who have just learnt something new are selectively deprived of REM sleep, their ability to recall and use that new information is impaired. For example, rats that have just learned how to navigate around a novel maze spend more time in REM sleep when they next sleep, whereas depriving them of REM sleep after training mars their ability to get round the maze the following day. Moreover, naturally occurring differences in the sleep patterns of individual rats are correlated with their learning abilities. When researchers measured the abilities of individual rats to learn a new task, they found that the rats that had learned quickly had more REM sleep than those that learned slowly.

The learning of new information is followed by a transient rise in REM sleep in humans too. In one experiment, men were taught Morse code shortly before they went to bed. Monitoring of their subsequent sleep revealed an increase in the number and duration of REM sleep episodes. Another study, which assessed the sleep of English-speaking students taking an intensive French course, also recorded an increase in REM sleep after learning. The students who were most efficient at learning the new language exhibited the biggest increases in REM sleep. Similarly, researchers logged an increase in REM sleep in people taking a course in trampolining, which required them to learn complex new patterns of movement. A comparison group who engaged in physical activities that did not require them to learn new skills (dancing and soccer) showed no such increase in REM sleep.

These and other experiments suggest that new information acquired during the day is processed and stored during the ensuing night's sleep. The experimental evidence also implies that this consolidation of memory occurs only during a limited time window following the learning experience. If an animal or human is deprived of REM sleep during this critical period, their long-term memory for the new information is impaired. The importance of timing was demonstrated by an experiment in which rats were trained during daily sessions to find food by navigating around an eight-armed radial maze. After each training session the rats were selectively deprived of REM sleep. The REM sleep deprivation began either immediately, or four hours or eight hours after the training session ended. Only the rats that were deprived of REM sleep starting four hours after the training sessions showed a deterioration in their long-term ability to remember the task.

REM sleep deprivation interferes with the consolidation of memory in humans as well. An experiment in the 1970s found that volunteers who were selectively deprived of REM sleep were subsequently less accurate in their ability to recall a story than control subjects who had been selectively deprived of slow-wave sleep. In a more recent experiment, volunteers learned a new skill that required them to discriminate between different visual stimuli. Their performance improved after a normal night's sleep even though they had no further practice, implying that their sleeping brains were continuing to process the newly acquired information. However, this overnight improvement was negated if they were selectively deprived of REM sleep. Disrupting their slow-wave sleep had no such effect. Furthermore, depriving them of REM sleep did not reduce their ability to perform a similar task that they had already learned some time before. REM sleep only made a difference when it came to learning a *new* skill.

The timing of REM sleep in relation to learning is similarly important in humans. REM sleep must occur within 24 hours of the learning experience if it is to enhance memory and performance. People who learn a new skill during the day and then stay up all night display no further improvement in their performance, even if they go on to have unlimited amounts of sleep the following night.

Further evidence for a link between REM sleep and learning has emerged from studies of brain activity. These have revealed remarkable similarities between specific patterns of brain activity during the initial learning process and hours later, during REM sleep. In one study, researchers compared the electrical activity patterns in rats' brains while the rats were learning new tasks, and later while they slept. Clusters of neurons in a particular region of the brain (known as the hippocampus) fired synchronously when the rats were in particular locations during the learning task. These same clusters of neurons also fired synchronously when the rats subsequently slept, repeating the precise pattern that had accompanied the original learning. This pattern of neural activity did not appear when the rats were sleeping before learning the task, only afterwards.

More recent work has reinforced the intriguing idea that the sleeping brain rehearses new experiences from earlier in the day during REM sleep. Scientists recorded the activity patterns in rats' brains while the rats learned to run around a circular track, where they were rewarded with food at particular checkpoints. Recordings from the hippocampus region of the rats' brains revealed highly specific patterns of neural activity that corresponded with specific aspects of the rats' experience. When the rats were in particular regions of the running track, their brains generated distinctive patterns of electrical activity. The rats' brain activity was then monitored while they subsequently slept. During REM sleep the rats' brains displayed patterns of activity that precisely matched those shown when they were previously learning about the running track. The correspondence was so precise that the scientists could deduce exactly where the dreaming rat was on the dream running track, and whether it was dreaming of running or standing still. It was as though a recording of the rat's experiences from earlier in the day was being replayed off-line in its brain while it slept.

A similar correspondence between brain activity during learning and during subsequent REM sleep has been recorded in humans. Researchers used brain scanning to monitor activity patterns in the brains of volunteers while they learned a new skill, which involved pressing buttons in response to symbols appearing on a computer screen. The volunteers were then monitored while they slept. The brain scans showed that the same brain areas became active again during REM sleep, as though the sleeping brains were running an action replay of the day's events. These patterns of brain activity were absent in control subjects who had not learned the new skill.

Parallel strands of evidence linking sleep and learning have emerged from investigations of how songbirds (specifically, zebra finches) learn to sing. In their natural environments, young male songbirds learn to sing the courtship song that is typical of their species by copying the songs of adult birds. Each individual then adds a few idiosyncratic flourishes to make his song unique, while retaining the essential characteristics needed to attract a mate of the right species. Scientists discovered that part of the song-learning process occurs when the birds are asleep. When they analysed the activity patterns in regions of the birds' brains that are known to be involved in song learning and production, they found remarkable similarities between the waking and sleeping states, suggesting that songs are processed off-line while the bird sleeps. When a bird sang during the day the complex patterns of muscular activity that are required to produce the song were mirrored by equally complex patterns of electrical activity in its brain. Later, when the bird was asleep, the same specific patterns of brain activity were repeated. In effect, the bird was practising the song in its dreams. What is more, the fine details of the dream song were then modified, with small variations in tempo, as though the sleeping bird was experimenting with minor variations on the basic song.

The relationship between REM sleep and learning is not simple, however. Individuals who spend more time in REM sleep do not appear to be consistently better at learning than those who get less sleep. Moreover, research suggests that learning is not exclusively the province of REM sleep: NREM sleep seems to play a role too. For certain types of learning, the overnight consolidation of memory occurs best when individuals get plenty of slow-wave sleep during the first part of the night and plenty of REM sleep during the final part.

One plausible theory is that NREM sleep and REM sleep are involved in the storage and consolidation of two different sorts of memory, which are known as declarative memory and procedural memory, or 'knowing that' and 'knowing how'. Declarative memory ('knowing that') refers to the recognition and conscious recall of factual knowledge and events – for example, knowing that a bike has two wheels, or that I rode a bike last Tuesday, or that Beethoven composed the Moonlight Sonata. This corresponds to the colloquial sense of the word 'memory'. Procedural memory ('knowing how') refers to knowledge of skills, habits, relationships and procedures – for example, knowing how to ride a bike or how to play the Moonlight Sonata. Our daily behaviour is heavily dependent upon procedural memory, even though we do not necessarily recall any conscious information. Different parts of the brain are involved in serving these two types of memory.

The evidence indicates that slow-wave (NREM) sleep is more important for the consolidation of declarative memory, while REM sleep helps to consolidate procedural memory. Learning and recalling a new skill such as trampolining, Morse Code, a mating song (if you are a bird) or navigating a maze (if you are a rat), all involve procedural memory and are therefore followed – as observed in experiments – by an increase in REM sleep. The storage of new facts – the kind of thing that students are supposed to worry about – should therefore benefit more from NREM sleep. Despite continuing disputes over the details, most scientists now agree that the brain does carry out some form of off-line memory processing during sleep, and that lack of sleep interferes with learning and memory.

One practical implication is that students preparing for exams are better off getting a good night's sleep than staying up all night cramming. Indeed, sleeping should produce better results than pulling an all-nighter for at least two reasons. First, sleeping after a period of intensive learning should help to embed the new information in memory. Secondly, and more obviously, someone who has stayed up all night will suffer the inevitable consequences of sleep deprivation the next day: their concentration, reactions, thinking, judgment, mood, creativity and communication skills will all deteriorate. Getting up very early in the morning to revise is probably not a good idea either, because the early morning is when most REM sleep occurs. So go to bed and stay there, asleep, for several hours.

Should machines sleep?

Do Androids Dream of Electric Sheep?
Philip K. Dick, title of novel (1969)

Biological theories such as these about sleep, learning and memory have interesting implications for the designers of intelligent machines. Sleep is evidently a necessary state for living brains. All complex organisms sleep – even if, as in the case of dolphins, they have to do it with only one half of their brain at a time. If sleep were dispensable, or if some superior alternative existed, then evolution would by now have found it. The strong implication is that when you build a small but immensely powerful organic computer, it must spend a substantial portion of its time asleep, focusing inwards rather than outwards.

The lesson here from nature may be that intelligent machines should be explicitly designed to undergo the computer equivalent of sleep, in which information and memories are processed and consolidated off-line in ways that are not possible or efficient during the conventional 'waking' state, when the machine is being bombarded with new information about the outside world. Intelligent machines might even be able to enhance their creativity by spending an hour or two dreaming every night (though goodness knows what the computer equivalent of a nocturnal erection would entail).

PART VI

Problems

13

Bad Sleepers

And as for those who persist in finding fault with me, I hereby declare them to be bad sleepers. *Bad sleepers!* This is a new insult, and I intend to take out a patent for it, being the first to have discovered that it is an excommunication in itself.

Jean-Anthelme Brillat-Savarin, *The Physiology of Taste* (1825)

While millions of people routinely deprive themselves of sleep because they are too busy working or being entertained, millions more are trying to sleep but cannot. Sleep is prey to many problems and the most common of these is insomnia, which just means difficulty falling asleep or staying asleep. Younger insomniacs tend to have problems falling asleep, while older insomniacs are more likely to have problems staying asleep. We shall be looking at more exotic sleep disorders in the next two chapters.

Insomnia is a symptom, not a disease. It can affect any of us, and usually does at some point in our lives. Insomnia is an egalitarian disorder from which no one is entirely exempt, regardless of health, wealth or age. On average, one person in three experiences inadequate or unsatisfactory sleep at least once a year. The occasional night of bad sleep is normal, and nothing to lose sleep over. The problems arise when insomnia becomes persistent. The usual responses are either to ignore it, or to treat the symptom without identifying its cause. About one in four of the adult population, if asked, complains of insomnia in some form. However, only one in 20 out of this vast army of self-proclaimed insomniacs ever seeks medical advice, and only a small proportion of those receive any treatment for their problem. The words 'tip' and 'iceberg' spring to mind.

An intolerable lucidity

> All that night and all day, the intolerable lucidity of insomnia
> weighed upon him.
>
> Jorge Luis Borges, *Labyrinths* (1964)

William Shakespeare well understood the burden of insomnia, which plagued several of his characters. Macbeth, for instance, could not sleep because he was haunted by his crimes. And in *Henry IV Part 2*, the king muses on the irony that his most lowly subjects are sleeping soundly while he, weighed down with the responsibilities of state, is stricken with insomnia:

> How many thousand of my poorest subjects
> Are at this hour asleep! O sleep, O gentle sleep,
> Nature's soft nurse, how have I frighted thee,
> That thou no more wilt weigh my eyelids down,
> And steep my senses in forgetfulness?

Charles Dickens was plagued by insomnia. In an essay called 'Lying Awake' he wrote that 'something in me was as desirous to go to sleep as it possibly could be, but something else in me *would not* go to sleep, and was as obstinate as George the Third'. The Lord Chancellor in Gilbert and Sullivan's *Iolanthe* sings at great length about his infernal insomnia:

> When you're lying awake with a dismal headache, and repose is
>   taboo'd by anxiety,
> I conceive you may use any language you choose to indulge in,
>   without impropriety;
> For your brain is on fire – the bedclothes conspire of usual
>   slumber to plunder you:
> First your counterpane goes, and uncovers your toes, and your
>   sheet slips demurely from under you . . ., etc, etc.

Insomnia is common in most countries. But if it does have a homeland, that homeland appears to be the USA. About one in three Americans experience sleep problems in the course of a typical year. In one large and fairly representative study, the US National Sleep Foundation and Gallup surveyed a thousand randomly selected Americans. One third reported some problem with their sleep, a quarter experienced occasional insomnia, and for 9 per cent of Americans insomnia was a nightly experience. The most common manifestations, in descending order of frequency, were waking in the morning feeling tired, waking in the middle of the night, difficulty returning to sleep after waking, and difficulty falling asleep. Relatively few of these self-proclaimed insomniacs had consulted a doctor about their problem, but 40 per cent had treated themselves with nonprescription medicines. Another American study found that about one in three people had current problems with their sleep. Those with sleep problems reported poorer general health, poorer performance at work, less satisfaction with their job and greater absenteeism.

The prevalence of insomnia appears to be lower in many other countries: more like one person in five. But this apparent difference may well stem mainly from different research methods and cultural expectations, rather than some peculiar American inability to sleep. In the UK, for example, a large survey found that 21 per cent of adults aged between 20 and 45 were dissatisfied with their sleep and 8 per cent were taking medication to help them sleep. Among a large sample of Canadians of all ages, 22 per cent displayed insomnia in some form. A survey of French people found that 20 per cent were unsatisfied or anxious about their sleep and 10 per cent were taking medication for it, while a study of Japanese men found that 28 per cent suffered from difficulty in falling asleep, frequent interruption of sleep or waking too early. Slightly lower estimates emerged from surveys in Iceland, Sweden and Belgium, where 10–15 per cent of adults reported persistent difficulties in getting to sleep or waking up too early in the mornings.

Insomnia increases markedly with age, and as many as 40 per cent or more of people over the age of 60 experience difficulties sleeping. But insomnia is certainly not confined to adults: children and adolescents can also be affected. Many pre-school children find it hard to get to sleep and stay asleep, to the frustration of their parents. Belgian research discovered that 43 per cent of primary school children had some sleep difficulties. The worst sleepers were more likely to perform badly at school, and more than one in five of them had failed one or more years. Sleep problems are particularly prevalent among children with learning difficulties: a British survey found that more than half of all children with severe learning difficulties had persistent problems with their sleep. Even teenagers, notorious for their ability to sleep all day, are not immune. When scientists surveyed teenagers in France, Germany, Italy and the UK, they found that about one in four had some symptoms of insomnia. Four per cent of these teenagers displayed the full-blown symptoms that qualified them for a formal diagnosis of a sleep disorder.

Anyone who has experienced insomnia for more than the odd night will know how unpleasant and debilitating it can be. Apart from the frustration of being unable to get a decent night's sleep, insomniacs experience a range of knock-on problems that erode the quality of their waking life. These include deteriorations in concentration, memory, personal relationships and the ability to accomplish the tasks of everyday life. A nationwide investigation in Germany found that 22 per cent of insomniacs rated their overall quality of life as bad (compared with only 3 per cent of people without sleep problems). Insomnia also has large economic costs. In France, where the consumption of sleep-promoting drugs is greater than most other European countries, insomnia is big business. When the money spent on sleep medications in France was added to the costs of medical care, the estimated annual bill (at 1995 rates) came out at over two billion US dollars.

Why can't you sleep?

How do people go to sleep? I'm afraid I've lost the knack.
Dorothy Parker, 'The Little Hours' (1944)

Insomnia is a slippery beast and tricky to pin down objectively. As a general rule, we are all poor at judging the quantity and quality of our own sleep. Our subjective feelings about our sleep can differ substantially from the objective reality as measured in a sleep laboratory. This gap between perception and reality is particularly apparent in many insomniacs. Bertrand Russell went so far as to assert, rather harshly, that:

Men who are unhappy, like men who sleep badly, are always proud of the fact.

The truth is that some insomniacs seriously overestimate how long it takes them to fall asleep and underestimate how long they actually spend asleep. They genuinely believe they have been awake much of the night, even though EEG measurements prove that they have actually been asleep. That is not to say, of course, that insomnia is all in the mind: most people who judge themselves to have sleep problems do indeed sleep badly. But for some, their sleep feels much worse than it really is.

Whatever the most convinced insomniac may believe, no person has ever gone completely without sleep for more than a few days, other than with the active assistance of a scientist studying sleep deprivation or a torturer. Even by the distorted perceptions of some insomniacs, Evelyn Waugh's character Professor Silenius was grossly exaggerating when he proclaimed:

> I haven't been to sleep for over a year. That's why I go to bed early. One needs more rest if one doesn't sleep.

According to one theory, the mismatch between perception and reality in chronic insomniacs might stem from a fault in the mechanisms that normally erase our memories during the transition from wakefulness to sleep. An abnormally good retention of memory from the time of sleep onset could blur the distinction between sleep and wakefulness, creating a false perception of having been awake. This theory is supported by experimental data. Compared with good sleepers, chronic insomniacs are better able to recognise and recall information that has been presented to them while they are falling asleep. If you remember most of what happens to you during the night, it will seem like it lasted much longer.

Insomnia sometimes has an obvious cause, such as shift work or illness. Working at strange hours of the day and night is a major source of sleep disturbances in industrialised societies, as are chronic respiratory problems such as coughs or wheezes. It should come as no surprise (but often does) that people who cannot breathe properly have difficulty sleeping. Chronic pain is another frequent cause of insomnia.

Like many generalised and vaguely indeterminate disorders, insomnia is often assumed to have psychological or emotional origins. This may indeed be true in some cases. Insomniacs are found to differ statistically in certain respects from normal sleepers, both psychologically and physically. On average they are more anxious. Insomniacs also tend to have a slightly higher body temperature, a higher metabolic rate and higher levels of the stress hormones cortisol and adrenaline. The more disturbed their sleep, the higher their cortisol and adrenaline levels. However, these psychological and physiological differences may well be consequences of insomnia rather than causes. As we saw in earlier chapters, loss of sleep has all sorts of unwelcome effects on our mental and physical states.

Some cases of insomnia stem from bad sleep habits or learned associations. If you unconsciously learn to associate the act of going to bed each night with feelings of anxiety then you are even less likely to sleep well. Most self-help books sensibly advise that bedrooms should be associated with only two activities: sleep and sex. Televisions, computers, office desks and the angst-inducing paraphernalia of work should ideally be banished from the bedroom, since they are all firmly associated with being awake.

A few individuals develop curious obsessions or neurotic rituals that can interfere with their sleep. Marcel Proust, for example, became unable to go to sleep unless his underpants were circling him tightly around his stomach and fastened with a special pin. The loss of his special pin on one occasion caused Proust great distress and even greater sleeplessness. It kept him awake all day.

Persistent insomnia can be a reflection of a psychiatric disorder such as depression. Severe depression is generally accompanied by insomnia, often in the form of early-morning waking. In *The Anatomy of Melancholy*, the seventeenth-century philosopher Robert Burton wrote that:

> Waking, by reason of their continual cares, fears, sorrows, dry brains, is a symptom that much crucifies melancholy men, and must therefore be speedily helped, and sleep by all means procured, which sometimes is a sufficient remedy of itself without any other physic.

(We shall take a closer look at the intimate relationship between depression and troubled sleep in the next chapter.) Conversely, sleep disorders can generate complex symptoms that are occasionally mistaken for mental illness. Some individuals who seek medical help for what appear to be psychiatric disorders turn out to have undiagnosed sleep disorders. Unfortunately, most of the medical profession still has a blind spot when it comes to recognising the sleep disorders that lie at the heart of many vague medical problems.

Storm and stress and sleep

> Methought I heard a voice cry, 'Sleep no more:
> Macbeth does murder sleep.'
>
> William Shakespeare, *Macbeth* (1606)

Psychological stress is a powerful and regrettably commonplace disrupter of sleep. However, individuals vary considerably in their reactivity to stress, and hence its impact on their sleep. Under demanding conditions where the individual can comfortably cope with the pressures, there is often an increase in sleep. Moderately stressed people who remain in control of their situation tend to sleep more. However, if the physical or psychological demands placed on someone exceed their capacity to cope, then a biological stress response is triggered. This stress response disrupts sleep, particularly slow-wave sleep.

Everyday life is teeming with potential sources of emotional and psychological stress, but most of us manage to cope with most of them most of the time. Sometimes, however, the pressures become too great or our attitudes are wrong. Anxieties about personal relationships or work can provoke sleeplessness, especially if we allow them to prey on our minds. Research has shown that people who regard themselves as overinvolved with their jobs are at greater risk of insomnia. But the anxiety may be entirely rational if the problem is real and potentially serious.

Unemployment, or the threat of unemployment, is a frequent creator of anxiety and insomnia. Extensive research has uncovered firm statistical links between unemployment and insomnia. For instance, a lengthy study of British civil servants found that those who faced the prospect of unemployment experienced more sleep problems and poorer health than those whose jobs were secure. Insomnia is not the only adverse effect of unemployment, of course. A mass of compelling evidence from several countries confirms that unemployed people have poorer mental and physical health than people in work, and die younger on average. Sleep disruption might play a role in this pervasive link between unemployment and bad health. For example, a long-term study of Swedish shipyard workers whose jobs were under threat found an increase in their blood-cholesterol levels – a risk factor for cardiovascular disease. The rise in cholesterol was greatest in those individuals whose sleep was the most disturbed.

Turbulent personal relationships, hardly a rarity, add to the stock of everyday stressors. American research discovered that people whose marriages were breaking up had significantly less slow-wave sleep than married people of the same age. A year later, those who had completed their divorces were once again getting more slow-wave sleep, presumably reflecting a reduction in their stress levels.

War is a prolific generator of trauma and insomnia. During the 1991 Gulf War the threat of attacks from Iraqi Scud missiles provoked widespread insomnia among the Israeli population. A survey carried out during the third week of the war found that 28 per cent of Israelis were experiencing significant problems with their sleep. The proportion of people affected by insomnia was higher in those areas of the country that were most at risk from attack, such as Tel Aviv and Haifa.

Severe psychological stress can cause Post-Traumatic Stress Disorder, or PTSD. One of the hallmarks of PTSD is disrupted sleep, often accompanied by recurrent nightmares. As a general rule, the more traumatic a sufferer's experience, the worse their sleep problems are likely to be. Survivors of torture, for example, almost invariably experience long-lasting sleep disturbances and nightmares. An investigation of PTSD sufferers who had served in the Vietnam War discovered that nightmares and insomnia were more frequent among individuals who had experienced the greatest exposure to combat conditions. PTSD can also be very persistent. Some Holocaust survivors and combat veterans from World War Two, Korea and Vietnam were still experiencing sleep disturbances and nightmares decades later.

In severe cases of PTSD there can be a large drop in total sleep duration, a big deficit in REM sleep and a virtual absence of deep, slow-wave sleep. When PTSD sufferers do sleep, their sleep is mostly light NREM sleep and they consequently wake the next morning feeling as though they have hardly slept at all. When they do eventually sink into REM sleep, PTSD sufferers are more difficult to wake up. Researchers measured the 'awaking thresholds' of PTSD sufferers by determining the minimum noise level required to wake them from REM sleep. The PTSD sufferers were generally more difficult to wake than healthy people, and the most depressed and anxious individuals were the most difficult of all to wake. Combat-related PTSD, or 'combat fatigue' as it was once known, often arises in physically demanding situations where the individual is already badly sleep-deprived because of the combat conditions. This sleep deprivation probably plays a role in triggering or amplifying the ensuing PTSD.

What to do?

> That ideal reader suffering from an ideal insomnia
>> James Joyce, *Finnegans Wake* (1939)

Curing persistent insomnia is seldom easy. Many sages have promised relief but few have delivered it, and only a charlatan or fool would claim to have the complete answer. Nonetheless, sleep scientists and physicians do at least agree on several basic principles of how to tackle the problem.

Anyone who is dissatisfied with their sleep should at least start by getting the basic elements of their 'sleep hygiene' right. This means, among other things, sticking to a regular routine of bedtime and waking time, having a comfortable sleeping environment free from disturbances, and avoiding too much caffeine. Simple relaxation techniques can help. Conversely, fretting about not sleeping is guaranteed to make matters worse. The fabled remedy for insomnia, of course, is counting sheep. William Wordsworth tried this, apparently to no avail:

> A flock of sheep that leisurely pass by,
> One after one; the sound of rain and bees
> Murmuring; the fall of rivers, winds and seas,
> Smooth fields, white sheets of water, and pure sky;
> I have thought of all by turns, and yet do lie
> Sleepless!

Dylan Thomas in *Under Milk Wood* wrote of the sleepless Mr Utah Watkins who 'counts, all night, the wife-faced sheep as they leap the fences on the hill, smiling and knitting and bleating just like Mrs Utah Watkins'. Dorothy Parker characteristically refused even to try the old sheep-counting ploy:

> Well isn't that nice. Isn't that simply ideal. Twenty minutes past four, sharp, and here's Baby wide-eyed as a marigold . . . I really

can't be expected to drop everything and start counting sheep, at my age. I hate sheep.

Groucho Marx unhelpfully recommended *subtracting* sheep as a tactic for those who wish to stay awake. Not every folk remedy for insomnia involves sheep. When Marcel Proust could not sleep, which was often, he liked to read a train timetable. Charles Dickens dealt with his insomnia by going on long nocturnal walks.

Many insomniacs reach for drugs or alcohol to ease their problem. We looked at various sleep-inducing substances in chapter 8. The ideal sleep-inducing drug, or hypnotic, should have several characteristics: it should work rapidly and provide immediate relief soon after it is taken; it should promote sleep throughout the night but cease to be active by morning, so you do not wake up feeling drugged; it should have no adverse side effects; and it should not induce tolerance such that you need to keep increasing the dose.

The latest generations of hypnotics come much closer to attaining these ideals, and offer substantial improvements over their predecessors. Drugs such as zolpidem and zaleplon are generally safer and more effective than the barbiturate and benzodiazepine sleeping drugs that were once the only options. Zolpidem, for example, is effective even when used intermittently; insomniacs can take it one or two nights a week, as and when they feel the need. The older drugs tended to produce a rebound effect, in which users would suffer even greater insomnia if they failed to take the drug one night. Modern hypnotics are eliminated faster from the body, which means they can be taken during the night without causing residual sleepiness the next day. They are also relatively free from the tolerance problems that affected long-term users of benzodiazepine drugs. The newer hypnotics are particularly helpful for treating elderly people, who experience a high incidence of insomnia and consume large quantities of sleeping aids. The earlier generations of hypnotics were more persistent, causing residual sleepiness the following day. Their use was therefore accompanied by a heightened risk of falls and hip fractures in elderly people.

Drugs are definitely not the only effective way of tackling insomnia. Various psychological treatments also provide relief. These generally work by modifying bad sleep habits or removing the stimuli that disrupt sleep. A review of 59 published scientific studies concluded that behavioural interventions, involving on average only five hours of treatment, produced reliable and enduring improvements in most cases. American researchers separately found that after completing at least four sessions of cognitive behavioural therapy, insomniacs displayed a 34 per cent reduction in the time taken to fall asleep, a 29 per cent increase in the total time asleep and a 56 per cent reduction in the time spent awake during the night. These are big improvements that compare well with hypnotic drugs.

A more exotic approach, called 'brain music', has been developed by scientists in Moscow. This involves recording the insomniac's EEG brain-wave patterns during sleep and converting the electrical waveforms into audible music. The insomniac then listens to their own 'brain music' before going to sleep. A 15-day treatment with 'brain music' was shown to help more than 80 per cent of insomniacs, who slept better and felt better after the treatment.

Staying awake

> And miles to go before I sleep,
> And miles to go before I sleep.
>> Robert Frost, 'Stopping By Woods
>> on a Snowy Evening' (1923)

What if you *want* to be a bad sleeper? What if, for example, you are in the middle of fighting a war or, more mundanely, partway through a long car journey? What should you do if you need to stay awake for some reason and must resist sleep at all costs? Chemical stimulants such as caffeine can prop you up temporarily, but the only proper advice for someone intending to drive a car when sleep-deprived is: don't.

If you are planning to go without sleep for a while, it will certainly help if you are not already sleep-deprived when you start, as Charles Lindbergh nearly found to his cost during his transatlantic flight. A good strategy is to sleep well the night before, or at least take a prophylactic nap. In Shakespeare's *Richard III*, the future King Henry decides to nap in preparation for the following day's battle with King Richard:

> I'll strive with troubled thoughts to take a nap,
> Lest leaden slumbers peise [weigh] me down tomorrow.

Military personnel are often required to function effectively for long periods with little or no sleep. Take B-2 bomber crews, for example. The US Air Force B-2 stealth bomber is so jam-packed with technological secrets and so fiendishly expensive, costing at the very least 1.3 billion dollars each, that all of its missions are flown from the Whiteman Air Force Base in Missouri. During the 1999 Balkan conflict, B-2 bombers flew numerous combat missions over former Yugoslavia. Each mission involved around 30 hours of non-stop flying. The planes would take off from Missouri early one morning and not return until the afternoon of the following day. In missions flown over Afghanistan in October 2001, B-2s stretched the record for non-stop combat sorties to an astonishing 44 hours in the air – 10 hours longer than Charles Lindbergh's exhausting flight across the Atlantic.

The B-2 bomber has a crew of only two, so long rest periods are not an option on combat missions. During these marathon sorties the two crew members therefore accumulate large sleep deficits. To mitigate this, they take short naps on folding beds behind their seats (though presumably not both together at the same time). Longer naps are avoided because waking from deep sleep would leave them feeling groggy from sleep inertia. Preparation and training are also important. B-2 crews are trained to fly on simulators for up to 50 hours. Before each mission they pack in lots of extra sleep during an 18-hour rest period, erasing any pre-existing sleep deficit before they take off. They are also trained in the techniques of power napping. Even so, flying one of those 44-hour sorties must be exhausting.

Chemical stimulants provide another way to stay awake. Caffeine is by far the most widely used remedy for sleepiness, as we saw in chapter 8. If you have to cope with insufficient sleep then caffeine will bolster your performance – or rather, it will make the decline in your performance less severe than it would otherwise have been. Large doses, equivalent to several cups of coffee, can maintain alertness at almost pre-deprivation levels for a while. But caffeine cannot eliminate the corrosive effects of sleep deprivation.

Caffeine is most helpful in the early hours of the morning when alertness is normally at its lowest, thanks to the circadian rhythm. Experiments using driving simulators have demonstrated that a 200-milligram dose of caffeine (equivalent to two or three cups of instant coffee) can reduce the sleepiness and improve the performance of sleep-deprived drivers when they are behind the wheel in the early morning. These stimulant effects of caffeine are not long-lasting, however; they persist for only about two hours in a driver who has had only a few hours' sleep the night before, and for barely half an hour if the driver has been up all night. Exposure to bright light can boost the ability of caffeine to keep you awake. Experiments with sleep-deprived volunteers have demonstrated that a combination of caffeine and very bright light (brighter than that produced by ordinary electric lighting) is effective at offsetting the usual deterioration in alertness and performance. Again, the treatment works best in the small hours of the morning, when alertness and performance are naturally at their lowest.

So-called 'energy drinks' can also provide some temporary relief. They also rely on caffeine as their main active ingredient, but offer an alternative to coffee or soft drinks as the vehicle for delivering it. A typical can of 'energy drink' contains about the same amount of caffeine as a cup of strong coffee. A British motoring organisation has marketed one variety containing 124 milligrams of caffeine, which is intended to be equivalent to two cups of coffee.

Another reasonably good countermeasure against sleepiness is a combination of napping and caffeine. If you are tired but nevertheless insist on driving, a good way of improving your chances of staying alive is to take a nap and ingest caffeine. A prophylactic nap combined with a dose of caffeine can maintain alertness and performance at near-normal levels for up to 24 hours in healthy young adults. In one experiment, sleepy volunteers were tested on a simulated driving task lasting two hours. Before each test they took a nap of up to 15 minutes, or 200 milligrams of caffeine, or both (or a placebo). The combination of caffeine plus nap reduced the usual mid-afternoon peak in simulated driving 'accidents' to less than one tenth the frequency with placebo only. Caffeine takes about 15–30 minutes to start working after you ingest it, so if you drink your cup of coffee just before you start your nap, the caffeine should be just kicking in as you wake up. Listening to music can perk you up a little, if you are only mildly fatigued. Experiments have confirmed that background music lifts the heart rate and alleviates sleepiness in sleep-deprived people. The boost is not enough, however, to transform a tired driver into a safe driver.

If you are willing and able to reach for more exotic chemicals, then a new generation of 'wakefulness-promoting' drugs awaits you. One of these drugs is called modafinil; it was developed in the 1970s and has been used successfully to treat the symptoms of narcolepsy.

Another is called pemoline. These drugs have the advantage of maintaining wakefulness without causing nasty side effects. They are more specific in their actions and generally less hazardous than the traditional blunderbuss stimulants like amphetamine. The military love them. A combination of modafinil and napping has proved to be highly effective at maintaining alertness in the face of prolonged sleep deprivation. Physical exercise helps too. Ultimately, though, the only way to stop feeling sleepy is to get plenty of sleep.

14

Dark Night

Care-charmer Sleep, son of the sable Night,
Brother to Death, in silent darkness born
Samuel Daniel, Delia (1592)

The large slice of each human life that sleep occupies is not always
serene and untroubled. Sleep is a nightly stage upon which unattractive characters sometimes strut. The cast ranges from sleepwalking, migraine, bed-wetting and night terrors to stress-induced
insomnia, narcolepsy and sudden death. In this chapter we will
sample the less restful side of sleep.

Walking and talking

I have seen her rise from her bed, throw her night-gown upon
her, unlock her closet, take forth paper, fold it, write upon't, read
it, afterwards seal it, and again return to bed, yet all this while
in a most fast sleep.
William Shakespeare, Macbeth (1606)

In *Macbeth*, Shakespeare paints a vivid picture of three nocturnal
phenomena: sleepwalking, sleeptalking and nightmares. Lady Macbeth, tortured by anxiety, has taken to walking in her sleep. Macbeth's
doctor remarks on the mental anguish that has produced her nocturnal
wonderings:

> unnatural deeds
> Do breed unnatural troubles; infected minds
> To their deaf pillows will discharge their secrets.

The doctor is wise enough to know his limitations, however. He tells Macbeth that he cannot cure Lady Macbeth's disorder, which manifests itself in 'thick-coming fancies', or frequent nightmares, that disrupt her sleep.

Sleepwalking and sleeptalking are products of slow-wave NREM sleep, not REM sleep. The muscular paralysis that accompanies REM sleep would prevent the sleeper from walking or talking. Sleepwalking and sleeptalking, together with a range of other phenomena, including night terrors and nocturnal tooth-grinding, are known collectively as parasomnias. They occur mostly during the first half of the night, when NREM sleep dominates. Parasomnias result from a glitch in the mechanisms controlling arousal; in effect, the mind and body fail to wake up in synchrony.

Sleepwalking is relatively common and generally benign in children, but it can signify psychological problems if it persists into adulthood. Many children sleepwalk occasionally, although only a minority do it repeatedly. Sleepwalking peaks in pre-adolescence and becomes relatively rare by adulthood. A study of more than eleven thousand adults found that about 9 per cent had sleepwalked in childhood and about 3 per cent had sleepwalked as adults. The great majority of those who were sleepwalking as adults had also sleepwalked as children. You are more likely to sleepwalk if you are very tired. For that reason, chronic stress, anxiety and other factors that disrupt sleep tend to exacerbate sleepwalking. A large survey found that sleepwalking was more prevalent among people who had suffered a stressful event such as a road accident within the previous year. Regular sleepwalkers tend to have more disturbed sleep, containing more microarousals, during the first sleep cycle of the night.

A propensity to sleepwalk can run in families. In 1930 a scientist documented a family of six, comprising a husband, his wife (who was also his cousin) and their four children, all of whom were somnambulists. According to the account, the entire family arose one night at about three in the morning and gathered in the servants' hall around the tea table. They only awoke when one of the children accidentally knocked over a chair. William Dement relays an anecdote told to him by a somnambulist patient whose entire family were sleepwalkers. While attending a Christmas family reunion, the man awoke one night in his grandfather's dining room to find himself surrounded by all his sleeping relatives. According to another tale, a lady awoke one night with the disagreeable sense of not being alone. It soon became apparent that someone was moving about in the darkness of her bedroom. There was a thud on her bed. The lady was terrified and fainted. When she regained consciousness in the morning she found that her butler, in a fit of sleepwalking, had laid the table for 14 people upon her bed.

The vast majority of sleepwalkers do not seek medical help and do not need to. In extreme cases, however, there is a risk of physical injury and treatment is needed. This may involve drugs or sleeping in a physically secure environment where the sleepwalker cannot come to harm or cause harm to others. Violence during sleep is rare but not unknown. In a large survey, 2 per cent of British people reported that they had behaved violently during sleep, usually while sleepwalking during the first few hours of sleep. The survey also revealed that these sleepwalkers consumed above-average amounts of caffeine, tobacco and alcohol, and had a higher incidence of mood and anxiety disorders.

A tiny minority of sleepwalkers are capable of severely injuring themselves or others while asleep, yet retain no memory of their actions when they awake. In one case, a 14-year-old boy rose from his bed in the small hours of the morning and stabbed his five-year-old cousin, causing serious injuries. Several people have been acquitted of murder on the grounds that they were asleep at the time and therefore not responsible for their actions. In one famous case in 1987, a young Canadian man called Kenneth Parks got up in the middle of the night and drove 14 miles from his home in Toronto to the house of his sleeping parents-in-law, where he proceeded to stab his mother-in-law to death in a violent, unprovoked and apparently motiveless attack. Parks was charged with murder, but he was acquitted on the grounds that he had been asleep at the time he did the deed.

Sleeptalking, like sleepwalking, is common and benign in children, but may indicate psychological problems if it persists into adulthood. An analysis of twins found that the propensity for sleeptalking in adulthood has a substantial inherited component.

If you speak two or more languages fluently, which one do you use when you talk in your sleep? To answer this question, Spanish scientists enlisted several hundred parents and children from bilingual schools in the Basque region of northern Spain, where two completely different languages, Spanish and Basque, are spoken. Sleeptalking was reported in 56 per cent of the children. The great majority used their dominant, native language when they talked in their sleep. However, a small minority (less than 4 per cent) consistently used their non-dominant language instead. Balanced bilinguals, who were equally proficient in both languages, would sleeptalk in either language.

Nightmares, night terrors, sleep paralysis and the Old Hag

> It was during the horror of a deep night.
>
> Jean Racine, *Athalie* (1691)

The night terror is another form of parasomnia associated with deep NREM sleep. In a night terror, the sleeper suddenly awakes with a feeling of, well, terror, which is often accompanied by an oppressive sense of some alien presence. This presence is variously described as a monster, an incubus or an extraterrestrial. Unlike nightmares, night terrors are rarely remembered in any detail the next morning.

Sleep paralysis is another very unpleasant nocturnal experience. This description of sleep paralysis, taken from Herman Melville's *Moby-Dick*, predates the first clinical account of the phenomenon by a quarter of a century:

> At last I must have fallen into a troubled nightmare of a doze; and slowly waking from it – half steeped in dreams – I opened my eyes, and the before sun-lit room was now wrapped in outer darkness. Instantly I felt a shock running through all my frame; nothing was to be seen, and nothing was to be heard; but a supernatural hand seemed placed in mine. My arm hung over the counterpane, and the nameless, unimaginable, silent form or phantom, to which the hand belonged, seemed closely seated by my bedside. For what seemed ages piled on ages, I lay there, frozen with the most awful fears, not daring to drag away my hand; yet ever thinking that if I could but stir it one single inch, the horrid spell would be broken.

Erasmus Darwin, writing more than two hundred years ago, attributed sleep paralysis to a combination of unusually deep sleep and 'uneasy sensations' arising in the mind. He noted that:

Where the sleep is uncommonly profound, and those uneasy sensations great, the disease called the incubus, or nightmare, is produced. Here the desire of moving the body is painfully exerted, but the power of moving it, or volition, is incapable of action, till we awake.

Scientists have since established that sleep paralysis and night terrors result from an anomalous partial awakening from slow-wave sleep. Like sleepwalking and sleeptalking, they usually occur during the first episode of slow-wave sleep, in the first hour or so after the onset of sleep. You are more likely to experience night terrors or sleep paralysis if you are tired. For that reason, sleep paralysis is particularly common among shift workers, and has even been dubbed 'night-nurse paralysis'. Night terrors are relatively common in young children and rarely a cause for concern, although in adults they may be symptomatic of anxiety, stress or depression.

Sleep paralysis appears to be a worldwide phenomenon, but the language used to describe it varies enormously from culture to culture. A large survey in Germany and Italy found that 6 per cent of healthy adults had experienced at least one instance of sleep paralysis. Among Hong Kong Chinese, sleep paralysis is known as 'ghost oppression', and more than a third of young adults have had it at least once. In Japan, a form of sleep paralysis known as *kanashibari* is remarkably common: four out of ten healthy young adults have experienced it. The first episode of *kanashibari* often occurs in adolescence and is usually associated with stress or disturbed sleep. In Newfoundland, where it has been the subject of folklore for generations, sleep paralysis is referred to as an attack of the Old Hag.

Language and cultural influences have an important bearing on whether people notice and remember the experience of sleep paralysis, and how they then interpret it. Although sleep paralysis is equally prevalent in Canada and Japan, far fewer Canadians are conscious of having experienced it. Canadians are much more likely than Japanese to describe their sleep paralysis as a dream. The probable explanation is that the Japanese have a commonly used word in their language for sleep paralysis, whereas Canadians do not. Having a specific word for the phenomenon, and hence a mental pigeonhole for the experience, seems to make people more likely to register it in the first place. (For the same reason, connoisseurs of wine like to use elaborate language to capture and describe a wine's complex characteristics.)

The basic features of sleep paralysis are remarkably consistent around the world, but different cultures interpret and describe essentially the same experience in very different ways. In societies where sleep paralysis is not widely acknowledged or discussed, such as the UK, USA and many other Western nations, the hallucinations and other dramatic symptoms are sometimes mistaken for evidence of psychotic illness or supernatural phenomena. Sleep paralysis probably lies behind many reported cases of paranormal experiences and alien abductions. People who genuinely believe that they have encountered extraterrestrial aliens often have these experiences during the night. Many accounts of alien encounters or abductions also contain other telltale features of sleep paralysis, including an overwhelming sense of dread, an inability to move, a threatening presence and a crushing weight on the chest.

Night terrors and sleep paralysis are quite distinct from nightmares, which are bad dreams and therefore generally occur during REM sleep. Nightmares tend to happen during the second half of a night's sleep, when REM sleep predominates. Like the parasomnias, nightmares are more common in children than adults. The word nightmare, incidentally, derives from the Anglo Saxon *mare*, meaning 'demon', which in turn derives from the Sanskrit word *mara*, meaning 'destroyer'. Bad nightmares can be a real nightmare. Hamlet, for example, is plagued by them:

> I could be bounded in a nutshell, and count myself a king of infinite space, were it not that I have bad dreams.

Samuel Taylor Coleridge's notebooks are packed with vivid descriptions of his grotesque nightmares, in which he saw giant pigs leaping on his legs and claw-like hands grasping his stomach. 'With Sleep my Horrors commence,' wrote Coleridge, 'Dreams are no shadows with me; but the real, substantial miseries of Life.' Coleridge's nightmarish visions were amplified by his consumption of opium. By the age of 31, his nightmares had become so ghastly that he was afraid to go to sleep – a common characteristic of opium eaters. Coleridge's contemporary Thomas de Quincey described the drug-fuelled horrors that he experienced nightly in his *Confessions of an English Opium Eater*:

> I seemed every night to descend, not metaphorically, but literally to descend, into chasms and sunless abysses, depths below

depths, from which it seemed hopeless that I could ever reascend. Nor did I, by waking, feel that I *had* reascended. This I do not dwell upon; because the state of gloom which attended these gorgeous spectacles, amounting at least to utter darkness, as of some suicidal despondency, cannot be approached by words . . . I sometimes seemed to have lived for 70 or 100 years in one night.

The sixteenth-century writer Thomas Nashe was another martyr to nightmares, which he gloomily attributed to sin:

As touching the terrors of the night, they are as many as our sins. The night is the devil's Black Book, wherein he recordeth all our transgressions . . . The rest we take in our beds is such another kind of rest as the weary traveller taketh in the cool soft grass in summer, who thinking there to lie at ease and refresh his tired limbs, layeth his fainting head unawares on a loathsome nest of snakes.

Research has found that adults who experience frequent nightmares are significantly more likely to require hospital treatment for mental illness at some stage in their lives. And for reasons that are far from obvious, nightmare sufferers are also easier to hypnotise. A large study of students who had taken part in tests of hypnotism discovered that those who frequently experienced nightmares scored higher on measures of hypnotisability. Inherited factors have some bearing on an individual's propensity to suffer from nightmares. An investigation of twins found that if one identical twin had frequent nightmares then the other twin had a 45 per cent chance of suffering too, whereas with non-identical twins there was only a 20 per cent chance.

What goes on in a nightmare? Much the same as in an ordinary dream, it would seem, but with more unpleasant contents. One theory is that during REM sleep the brain is exploring and resolving emotional problems. If the problem is too intense to resolve, however, the process can become stuck and give rise to a recurring dream or nightmare. Fortunately, we can exert some influence on our dreams, and it is possible to ameliorate or even eliminate recurrent nightmares by consciously thinking about how to change them. If you promise yourself during the day that your recurring nightmare will get better or go away, there is a reasonable chance that it will. If this do-it-yourself approach fails, chronic nightmare sufferers can be trained to render their nightmares less distressing. A cognitive-behavioural technique called imagery rehearsal involves learning to visualise and then change the nightmare images while awake. Sufferers who learn this technique usually end up with fewer nightmares and better sleep.

Moving sleep

Our nature consists in movement; absolute rest is death.
Blaise Pascal, Pensées (1670)

Sleepwalking aside, you normally stay in the same place when you are asleep. But you do not stay still. Unless you are very tired, very drunk, drugged or tied up (or, perish the thought, all four) you will normally change your body posture every 15–20 minutes during sleep, usually at the end of an episode of REM sleep or slow-wave sleep. This is just as well, because lying in the same position for too long could restrict your circulation, damage nerves and even cause pressure sores. Prolonged pressure on a nerve can cause temporary numbness and loss of movement in the limb. If someone drinks too much alcohol and falls asleep in a chair with their arm over the back of the chair, they may discover in the morning that, in addition to a hangover, they have a numb arm and cannot flex their wrist or straighten their fingers. This sobering phenomenon, which is known to some doctors as 'Saturday night palsy', is caused by sustained pressure on the radial nerve in the arm.

Your limbs may also make occasional jerking movements during sleep, known as periodic limb movements (or nocturnal myoclonus). These usually occur during light NREM sleep. We tend to notice them most during the transition from waking to sleep because they sometimes wake us up again. Limb movements during sleep are remarkably frequent. One study of elderly people recorded an average of 26 per hour. But no correlation was found between the frequency of limb movements and the quality of sleep or daytime sleepiness, suggesting that the movements rarely disrupted sleep. However, some people experience more violent limb movements that do occasionally jerk them out of sleep (usually slow-wave sleep). These sudden awakenings can be mildly distressing and, if they are frequent, may cause insomnia.

If you really start thrashing about in your sleep you might be suffering from something called REM sleep behaviour disorder, or RSBD. As its name suggests, RSBD occurs during REM sleep. This unpleasant malady stems from an intermittent failure in the mechanisms that normally suppress muscle activity during REM sleep, thereby freeing the dreamer to move about during their dreams and perhaps cause injury. The dreams associated with RSBD are often violent or alarming. RSBD mostly affects older men, but some cases have been reported in children. It is often misdiagnosed as sleepwalking or night terrors. A few unfortunate individuals suffer from a combination of RSBD, severe sleepwalking and night terrors – a distressing and confusing condition that is known as parasomnia-overlap disorder.

Midnight feasting

Last night I dreamt I ate a ten-pound marshmallow. When I woke up the pillow was gone.

Tommy Cooper, British comedian (1921–84)

Sleep can be fattening. A few people regularly wake up during the night with an irresistible urge to eat or drink. Night-eating syndrome, as this disorder is known, often results in obesity and it can require medical intervention.

Severely afflicted nocturnal eaters wake up several times a night with an urgent need to swallow food, despite the absence of real hunger. They cannot get back to sleep until they have eaten. Not surprisingly, they tend to become fat and tired. One of the few systematic investigations discovered that nocturnal eaters were consuming more than half their total daily energy intake at night. They typically woke three or four times and ate food twice a night. It is a night thing, however: nocturnal eaters do not eat compulsively during the day. They always awake from NREM sleep and are fully conscious when they eat. They are not sleepwalking. Night-eating syndrome is often associated with psychological problems and with other sleep disorders such as sleepwalking or periodic leg movements. The origins of this strange disorder remain obscure. Nocturnal eaters are known to have unusual hormone patterns, including high levels of the stress hormone cortisol, a subnormal night-time rise in melatonin, and unusually low levels of the appetite-suppressing hormone leptin. This altered pattern of hormonal changes makes them feel hungrier and less sleepy at night, as though they were jet-lagged.

A leading expert on the subject has speculated that Elvis Presley might have been a victim of nocturnal eating syndrome. Presley died in 1977 of a heart attack, which was attributed at the time to obesity and abuse of prescription drugs. This posthumous diagnosis would account for Presley's legendary obesity, together with his nocturnal lifestyle and late-night gorging on greasy foods. Nocturnal grazing seems to have been a frequent temptation for our ancestors too. A common item of bedroom furniture in the fifteenth century was the aumbry – a small bedside cupboard intended to hold food and drink in case of 'night starvation'.

Soggy sheets

> The nights of sleep seemed to him deep, dark pools in which he could submerge himself.
>
> Jorge Luis Borges, *Labyrinths* (1964)

Nocturnal enuresis, better known as bed-wetting, is common among children. It is estimated that between five and seven million children in the USA wet their beds. A survey of Finns found that almost 4 per cent had frequently wet their bed and 7 per cent had suffered the occasional mishap. By adulthood, however, it is rare: the same study found that only two people in a thousand admitted to wetting their bed as adults. Even so, the occasional grown-up accident is not unknown. In his diary for 28 May 1660, Samuel Pepys recorded how he fell foul of a freak episode of nocturnal enuresis:

> This night I had a strange dream of bepissing myself, which I really did; and having kicked the clothes off, I got cold and found myself all muck-wet in the morning and had a great deal of pain in making water, which made me very melancholy.

The reported incidence of bed-wetting varies considerably between countries, although this might be partly an artefact of methodology. A study of children in Saudi Arabia found a surprisingly high incidence. Among a randomly selected sample of several hundred school children aged from 6 to 16 years, 15 per cent (more than one child in seven) were wetting their beds. Thirty per cent of these bed-wetters were wetting their beds during the day, reminding us that the problem is not confined to night-time sleep. The risk factors associated with bed-wetting among the Saudi Arabian children included stressful life events before the age of six, acute psychosocial problems in the family, and constipation. Children from stable families and those who had been breast-fed were less likely to be bed-wetters. Recent work in Turkey found that 12 per cent of primary school children aged four to 12 years were wetting their bed at least once a week. More than 40 per cent of the children affected came from families with a history of bed-wetting. In China, however, only 4.5 per cent of primary school children are reported to wet their beds – a lower prevalence than in most Western countries.

Boys are more prone to leaking at night than girls. A large British study of children aged between 5 and 11 years found that bed-wetting was more common in boys than girls and declined markedly with age in both sexes. Bed-wetting was also found to be more common in children whose mothers smoked, Afro-Caribbean children, later-born children, children whose sleep was disturbed, and children with young mothers. Far more obscure associations have also been unearthed, including a link between bed-wetting and left-handedness. (You might recall from chapter 6 that statistical associations have been uncovered elsewhere between left-handedness and a baffling array of things, including allergies, asthma, dyslexia, autism, alcoholism, birth complications and accidents.)

Many families take bed-wetting in their stride, stoically regarding it as an inconvenient but common aspect of growing up. Only half the parents of bed-wetters in England and Scotland ever consult a doctor about their child's problem. When they do, bed-wetting is usually attributed to physical causes such as a delay in the maturation of bladder control, decreased bladder capacity, constipation or increased urine production at night. Inherited factors also play a role in some forms of nocturnal enuresis, which tend to run (as it were) in families. The problem is seldom attributed exclusively to psychological and emotional problems. The standard treatments include drugs, training procedures and urine alarms, which wake the bed-wetter at the first sign of seepage. In most cases the problem simply goes away with age. Not every bed-wetter dries out with age, however, and soggy sheets remain an occasional hazard for around one in 200 unlucky adults.

Aching heads

> What the will and reason are powerless to remove, sleep melts
> like snow in water.
>
> Walter de la Mare, *Behold, This Dreamer.* (1939)

Sleep and headaches: there is a connection. Headaches can disturb sleep and sleep problems can evoke headaches. Fatigue is a common trigger for headaches. The relationship between sleep and headaches is at its deepest and most complex when it comes to migraine, which affects about one in ten people in the USA and UK. Many who suffer from this unpleasant affliction find that overwork and lack of sleep can precipitate a migraine attack. Studies of migraine sufferers (or 'migraineurs') consistently find fatigue to be one of the most frequent triggers for an attack. Some sufferers are repeatedly hit by migraine at the end of the working week, giving rise to the notorious 'weekend-migraine' pattern. Sleep plays a noticeable role in more than three quarters of all migraine cases.

Most classic migraine attacks are accompanied by an unpleasant lethargy, with an overwhelming desire to lie down, be alone and sleep. A short, deep sleep, taken as soon as the early warning signs are noticed, can sometimes stave off an impending attack. Failing that, most sufferers find that sleep brings some relief, and many attacks end naturally with a long, deep sleep. Migraine attacks sometimes occur during sleep. Indeed, some sufferers only ever have attacks during sleep, often accompanied by nightmares. But no one really understands what lies at the root of this intimate, two-way relationship between migraine and sleep.

If instead of a headache you awake most mornings with an ache in your jaw, neck or ear, you may have a very different sort of problem known as nocturnal tooth-grinding, or sleep bruxism. This phenomenon involves repeated contraction of the jaw muscles for several seconds at a time, forcing the teeth to grind together and make a toe-curling noise. Nocturnal tooth-grinding is surprisingly common. When scientists surveyed many thousands of people in the UK, Germany and Italy, they found that 8 per cent ground their teeth during sleep at least once a week. The risk factors associated with sleep bruxism were found to include the usual sleep-unfriendly tribe of loud snoring, heavy alcohol consumption, caffeine, smoking, stress and anxiety. Once again we are reminded that anything likely to disturb your sleep may bring other unwelcome problems trailing in its wake.

The biting force exerted during nocturnal tooth-grinding can be more than 40 kilograms force, which exceeds the maximum biting force you would normally exert when awake. The dreadful grinding noise this makes would set any listener's teeth on edge. Naturally, all that gnashing and clamping does the jaw muscles and teeth no good, and muscle pain in the jaws is a common symptom. Persistent bruxism can wear down the teeth and cause problems with speaking, chewing and swallowing. Researchers have even been evaluating a treatment for severe cases that involves injecting botulinum toxin into the jaw muscles.

Troubled guts

> The winds come to me from the fields of sleep.
> William Wordsworth, 'Ode: Intimations of Immortality' (1807)

The myriad tentacles of sleep even reach into our guts. A subtle relationship exists between sleep and the behaviour of our digestive tracts. Once again, this relationship works in both directions: gut disorders can disrupt sleep, and sleep problems can exacerbate gut disorders.

You might have noticed that a late or sleepless night is often followed by a day of mild intestinal discomfort or indigestion. Shift workers, and especially those working nights, notoriously suffer from a much higher prevalence of gastrointestinal problems such as indigestion, flatulence, bloating, constipation, abdominal pain and heartburn. But what, you may be thinking, could sleep possibly have to do with flatulence or gut ache?

Scientists have uncovered at least one possible explanation for the link between disturbed sleep and digestive discomfort. The healthy functioning of the digestive system depends, among other things, on the production of a special protein called trefoil peptide TFF2. This protein activates protective systems in the stomach and gut that repair minor abrasions and inflammations. If you swallow a small fragment of bone, for example, it can scratch your gut lining and cause local damage. When this happens, trefoil peptide TFF2 triggers biochemical mechanisms that repair the lining, preventing a minor abrasion from turning into anything more serious. The level of TFF2 in your gastric juices varies dramatically according to a circadian rhythm. The low point of the cycle normally occurs in the early evening; TFF2 levels then rise sharply during your sleep, reaching a peak of more than 27 times the early-evening level. Anything that disrupts your normal sleep cycle will also disrupt your TFF2 system. The amount of TFF2 in your gastric juices drops dramatically after a night of partial or complete sleep deprivation, making your stomach and gut more vulnerable to irritation. So, if you eat hot curry, swallow bits of bone and gristle and then go to bed very late, do not be surprised when your entrails grumble and gripe the next day.

Your gut has a life of its own, and its movements are heavily influenced by sleep. The gut is generally less mobile during sleep, especially deep NREM sleep. But during REM sleep, the frequency of colonic contractions and the pressure in your colon both rise sharply. Colonic contractions also occur during transient arousals from sleep during the night and when you wake up in the morning (as you always suspected).

Scientists have discovered an intriguing link between sleep and irritable bowel syndrome (IBS) – a chronic condition involving recurrent abdominal pain accompanied by diarrhoea or constipation. It has become clear that IBS involves complex interactions between the gut and the brain, overturning the traditional view that it is purely a disorder of the digestive system. Psychological stress can induce the abnormal gut movements and other symptoms of IBS. Conversely, IBS is accompanied by changes in sleep patterns, especially in REM sleep. IBS sufferers spend substantially more time in REM sleep than healthy people, both in absolute terms and as a proportion of their total sleep time. Altogether, they have about double the average amount of REM sleep. They therefore presumably dream more than healthy people, though this aspect of IBS does not seem to have been explored.

As you might expect, IBS sufferers experience more disturbed sleep than healthy people. However, the causal connection between IBS symptoms and disturbed sleep does not operate in the way you might expect. The IBS symptoms appear to be more a consequence than a cause of disrupted sleep. Research has found that IBS symptoms wax and wane in line with the quality of the previous night's sleep, with poor sleep preceding worse symptoms. There are good reasons why scientists searching for ways of treating IBS should look beyond the gut itself and think about the sleeping brain. And the brain is where we go next.

Troubled minds

> In a real dark night of the soul it is always three o'clock in the morning.
>
> F. Scott Fitzgerald, *Esquire* (1936)

Competent physicians and healers have understood for millennia that depression and sleep disorders go hand in hand. In *The Anatomy of Melancholy*, published in 1621, Robert Burton noted the tendency of depressed people to lie in bed, and counselled against it:

> Seven or eight hours is a competent time for a melancholy man to rest; but as some do, to lie in bed and not sleep, a day, or half a day together, to give assent to pleasing conceits and vain imagination, is many ways pernicious.

Over the past century, hundreds of scientific studies have cast light on the relationship between depression and sleep. Once again, the relationship is complex and two-way. Disturbed sleep is a common consequence of depression and, conversely, it can fuel depression. The debilitating fatigue that results from chronic insomnia can erode a person's quality of life and impair their ability to function effectively, making them vulnerable to feelings of depression. Some of the drugs that have been used to treat depression elicit marked changes in sleep patterns and reduce the quality of sleep, potentially sparking a vicious cycle of insomnia and depression.

Sleep problems often precede depression, implying that they are not simply a side effect of depressive illness. For instance, one long-term study of elderly people living in inner London found that insomnia was linked with future depression. The best single predictor of whether an elderly person would subsequently become depressed was whether they were currently suffering from insomnia. Similar findings have emerged from the USA. In one study, for example, symptoms of sleep disturbance and depression were tracked over time among a large sample of apparently healthy Californians aged 50 or more. Almost one in four of them was suffering from insomnia and some went on to become depressed. The individuals who had problems with their sleep were significantly more likely to develop major depression over the following year. Once again, the sleep problems preceded rather than followed the depression.

Once depression has taken hold, whatever its origins, sleep patterns almost always change. Depression is usually accompanied by marked alterations in REM sleep – typically an increase in the proportion of REM sleep, a reduction in the time taken to enter REM sleep after falling asleep (REM latency) and a higher frequency of rapid eye movements. In severe cases of bipolar (or manic) depression, the reduced REM latency tends to rise back up towards normal as the patient's mood improves. The higher frequency of rapid eye movements is a curious hallmark; besides being associated with depression, it is also predictive of poor outcomes in other psychiatric disorders including suicide in schizophrenics and relapse in abstinent alcoholics.

We saw earlier that REM sleep may play a role in helping to regulate mood, through a form of off-line processing of emotions. If so, then it is understandable that someone whose REM sleep is disturbed might suffer from mood alterations the next day, and perhaps even depression. But the relationship between REM sleep and depression turns out to be more complicated than that.

Psychiatrists discovered decades ago that acute sleep deprivation can actually *relieve* clinical depression. When severely depressed people are deprived of sleep, about half of them stop feeling depressed almost immediately – at least, until they sleep again. For the other half, sleep deprivation makes no difference. Severe depression can be temporarily alleviated by as little as one night of sleep deprivation. Unfortunately, the majority of depressed patients relapse to their former state after another night of sleep. Depriving depressed patients of sleep for only part of the night has similar therapeutic benefits to total deprivation, so partial sleep deprivation is now the preferred method of treatment. The usual procedure is to wake the patient in the early hours of the morning, depriving them of half-a-night's sleep; this produces temporary remission in 40–60 per cent of cases of major depression.

The mechanism by which sleep deprivation relieves depression in some patients appears to have something to do with correcting unusual patterns of brain activity. Brain scanning has revealed that in those patients who respond favourably to it, sleep deprivation has the effect of reducing abnormally high levels of activity in certain regions of the brain. How and why this happens remains unclear. But it works.

Sudden nocturnal death

> How wonderful is Death,
> Death and his brother Sleep!
>
> Percy Bysshe Shelley,
> 'The Daemon of the World' (1816)

Mark Twain once said 'Don't go to sleep: so many people die there.' He was right – many people do die in their sleep. The phenomenon of sudden unexplained death during sleep, or SUDS, is well documented. Some cultures even have a special word for it. In Japan, for example, SUDS is called *pokkuri*, while in the Philippines it is known as *bangungut* (which literally means 'to rise and moan during sleep'). In north-eastern Thailand it is referred to as *laitai*.

For reasons that are unknown, SUDS is particularly prevalent among young Asian men. For example, an analysis of autopsy reports and death certificates filed in Manila uncovered many hundreds of cases. The victims were virtually all men, with an average age of 33 years, and the most common time of death was three o'clock in the morning. A similar picture has emerged from several other Asian countries. For instance, a survey of villages in north-east Thailand discovered many verified cases. The victims were all relatively young men, and a family history of SUDS was reported in more than a third of cases. The annual mortality rate from SUDS was 26 deaths per 100,000 person-years, making it one of the leading causes of death among young men in that region. SUDS is also a significant cause of death in Korea, and many cases are recorded each year in the USA among refugees from south-east Asia, particularly Laos and Kampuchea. As elsewhere, the victims are usually men in their early thirties and they nearly all die at night.

The most likely explanation for SUDS is cardiac arrest, brought on by heightened nervous system activity resulting from nightmares or night terrors. REM sleep is a highly active state in which the heart rate and breathing become faster and more irregular. Dreams or nightmares can be highly emotional, provoking marked physiological arousal. In individuals who have pre-existing abnormalities in the heart or coronary arteries, this physiological ferment may be sufficient to trigger a catastrophic event such as a heart attack, or ventricular fibrillation leading to cardiac arrest and death. So much for the belief that dreams are never of any consequence.

Narcolepsy

Sleep hath seized me wholly.

William Shakespeare, *Cymbeline* (1609-10)

Finally, we come to a truly nasty sleep disorder: narcolepsy. In narcolepsy, defects in the brain mechanisms that control sleep and wakefulness allow REM sleep to break through into waking consciousness. The resulting symptoms include excessive daytime sleepiness, sudden 'drop' attacks of muscular paralysis known as cataplexy, sleep paralysis and hypnagogic hallucinations.

Narcoleptics cannot sleep properly and consequently suffer from debilitating daytime sleepiness. Many also experience cataplectic attacks, caused by a sudden loss of muscle tone. They crumple or fall down, as though catapulted into sleep, but remain conscious while this is happening. These distressing attacks combine elements of both the waking state and REM sleep; the sufferer is both awake and asleep at the same time. In their less extreme form, cataplectic attacks may involve only a brief limpness of the arms or legs. They last for seconds or minutes at a time, and some narcoleptics can have several a day. The attacks are often triggered by emotions such as anger, fear, amusement or sexual arousal, so narcoleptics are apt to collapse during sex or after hearing a joke. They tend to avoid situations in which their emotions are likely to be aroused.

A notable feature of narcolepsy is the abnormally rapid onset of REM sleep. Whereas normally REM sleep does not appear until an hour or more after falling asleep, narcoleptics enter REM sleep very rapidly – in some cases almost immediately after going to sleep. They therefore get less NREM sleep. The daytime sleepiness stems mainly from this lack of NREM sleep. Narcoleptics also experience attacks of sleep paralysis, usually at the beginning or the end of the night's sleep. The sleep paralysis is often accompanied by hallucinations, making a formidably awful combination.

Narcolepsy is not as rare as its exotic symptoms might suggest.

The prevalence in the general population is around 0.05 per cent, or one case for every two thousand people, and it seems to be rising. This makes it about as common as multiple sclerosis or Parkinson's disease, and five times more common than leukaemia. The number of narcoleptics in the UK is about the same as the population of a small town – say, Stratford-upon-Avon.

Narcolepsy was first categorised and named as a distinct disorder in 1880, by a former naval doctor called Jean-Baptiste Gélineau. But scientific explanations for the bizarre symptoms did not begin to develop until the second half of the twentieth century. Until then, various psychoanalytic theories sought to explain the cataplexy and other symptoms in terms of repressed, guilt-laden sexual desires. Such theories bore only a tenuous relationship to reality and did little to help sufferers. Fortunately, we now know better. Although narcolepsy has been recognised as a distinct medical disorder for more than a century, it remains notoriously difficult to diagnose.

In recent years, intensive research into the genetics and neurobiology of narcolepsy has made tremendous progress in unravelling its biological mechanisms. It appears now that a class of neuropeptide molecules called hypocretins (also known as orexins) lies at the heart of the disorder. Abnormalities involving hypocretins have been discovered in narcoleptic dogs, mice and humans. Stanford University researchers have been breeding narcoleptic dogs since the 1970s, in order to analyse the neural and genetic mechanisms. In these dogs, narcolepsy is transmitted from one generation to the next by a single recessive gene, which codes for the hypocretin-receptor molecule. Mice that have a defect in the hypocretin gene suffer from a sleep disorder that closely resembles narcolepsy. The majority of people with narcolepsy also have very low hypocretin levels, and evidence is accumulating that hypocretin anomalies are a key mechanism in human narcolepsy as well. If so, then it might one day be possible to use molecules that mimic hypocretins to treat narcolepsy. Meanwhile, there are no truly effective remedies, although antihypnotic drugs such as modafinil (Provigil) can at least alleviate the daytime sleepiness. It is notable that modafinil activates nerve cells that contain hypocretins.

In humans, unlike the Stanford narcoleptic dogs, there is no single gene 'for' narcolepsy, and the disorder is not inherited in any simple manner. However, combinations of several genes in specific regions of the human genome do predispose individuals to develop the disorder.

Since the early 1980s it has been known that human narcolepsy is associated with a particular family of genes that code for protein-marker molecules found on the surfaces of most cells. Individuals who inherit particular combinations of genes in this region of the genome are at substantially greater risk of developing narcolepsy. But environmental factors also play a major role in its development. Even if one identical twin has narcolepsy, the other twin will still have only a 25–30 per cent risk of developing it, despite having exactly the same genes. Inheriting a particular combination of genes does not condemn that individual to developing narcolepsy.

As yet, there is no scientific consensus about the precise chain of events that triggers the development of narcolepsy. According to one theory it may be an autoimmune disease – that is, a disease in which the immune system mistakenly attacks healthy cells of its own body, rather than alien 'non-self' cells such as bacteria, virus-infected cells or cancer cells. Autoimmunity has been likened to 'friendly fire' in a war, when one side's gunfire or bombing accidentally kills some of its own troops. Many complex diseases are now known to arise from autoimmune reactions, including rheumatoid arthritis, multiple sclerosis, diabetes, pernicious anaemia and several thyroid disorders. Multiple sclerosis, for example, results from the autoimmune destruction of the myelin sheaths covering nerve cells. Autoimmune diseases can appear at almost any age, but they often appear in adolescence or early adulthood, as does narcolepsy. More specifically, scientists have suggested that narcolepsy may be the outcome of autoimmune reactions that destroy hypocretin-producing cells in the brain. Evidence supporting this theory emerged when researchers autopsied the brains of dead narcoleptics. The narcoleptics' brains contained only a small fraction of the hypocretin cells found in normal brains. Nearly all of the hypocretin cells were missing, leaving extensive scarring where they should have been. Brain cells of other types were present in normal quantities. It appeared as though these cells had been destroyed by an autoimmune reaction, perhaps triggered by a prior viral infection. That, at least, is the theory. And now we turn from brains to breathing.

15

Pickwickian Problems

The Trick is to Keep Breathing.
Janice Galloway, title of novel (1989)

We generally take our breathing for granted, especially when settling down for a nice, peaceful night's sleep. It seems reasonable to assume that we will continue to breathe, smoothly and without interruption, during those next several hours while we slumber. Breathing is just one of those things that happens automatically, doesn't it? But our confidence may be misplaced. In fact, our breathing may be disrupted during sleep in ways that range from the relatively benign (if anti-social) phenomenon of snoring to something far more serious and potentially life threatening. Let us start at the least alarming end of the spectrum.

The wonderful world of snoring

Laugh and the world laughs with you; snore and you sleep alone.
Anthony Burgess, *Inside Mr Enderby* (1963)

When you snore – and most of us do on occasion – you emit an annoying racket capable of keeping anyone within earshot awake. The ears closest to the source of this drain-like noise are your own. But you will be the last person to hear it. Mark Twain, in *Tom Sawyer Abroad*, remarked upon the injustice of the fact that snorers cannot hear themselves snore:

Jim begun to snore – soft and blubbery at first, then a long rasp, then a stronger one, then a half a dozen horrible ones, like the last water sucking down the plug-hole of a bath-tub, then the same with more power to it, and some big coughs and snorts flung in, the way a cow does that is choking to death . . . Now there was Jim alarming the whole desert, and yanking the animals out, for miles and miles around, to see what in the nation was going on up there; there warn't nobody nor nothing that was as close to the noise as *he* was, and yet he was the only cretur that wasn't disturbed by it.

An engineer would describe snoring as a periodic resonant noise generated in the upper respiratory tract, in the frequency range 140–1,200 hertz and with an intensity of 70–90 decibels. As you might know from sharing a bed with one, a snorer in full voice can emit a loud and immensely irritating sound. A very bad snorer can produce a noise that is, literally, deafening. In a snoring competition (sic) held in England in 1992, one Melvin Switzer gave birth to a 92-decibel snore, which is louder than a motorbike revving at close quarters and enough to induce hearing loss in anyone exposed to it for too long. According to one account, Mr Switzer's wife is deaf in one ear.

The curiously aggravating sound of snoring is caused by vibrations in the soft palate, the base of the tongue and other wobbling flesh in the back of the throat. A prime contributor is the uvula – the flap of tissue you will see dangling at the back of your throat if you look in a mirror and open your mouth wide. Its normal function is to direct food into your throat and away from the cavity behind your nose. The principal sources of noise, however, are the edges of the soft palate.

We snore because the soft tissues of the nose and throat are an engineering compromise. If their only function in life were eating or talking, they would be more flexible. If, on the other hand, their only function were to support breathing, they would be rigid. In fact they must do both, and so their mechanical properties are somewhere in between. The limited rigidity the tissues do have is partly maintained by muscle tension rather than the inherent physical properties of the tissue. Therefore when the supporting muscles relax – as in REM sleep, for example – the tissues become floppier and the upper airway narrows. This narrowing of the upper airway means that a greater effort is required to pull air through it. The air travels faster through the smaller cavity, causing the elastic walls of the upper airway to vibrate. The effect is somewhat like a flag flapping in the wind. The point to remember is that snoring can be provoked by anything that constricts the upper airway or relaxes the surrounding muscle tissue, including nasal congestion, unusually deep sleep, obesity, enlarged tonsils, a receding chin or alcohol.

Gravity contributes to snoring by helping the tongue and other tissues to sag back into the upper airway. Reducing gravity should therefore reduce snoring, and recent research with Space Shuttle astronauts has confirmed that this is indeed the case. Scientists monitored the sleep of five extremely healthy astronauts before, during and after Space Shuttle missions. When they were on earth, and subject to normal gravity, the astronauts spent a surprisingly large 16 per cent of their total sleep time snoring. (Comfort yourself with the thought that even astronauts snore.) However, when they were in space and therefore subject to a greatly reduced gravitational field, their snoring virtually disappeared, dropping to less than 1 per cent of total sleep time. The frequency with which their breathing was interrupted during sleep also fell by more than half. Unfortunately for those of us who would love to curb our snoring, blasting off into space does not offer a particularly practical remedy. Probably the next best thing to being in space, for these purposes, is sleeping on your side or front rather than on your back. That should at least help to stop your tongue from lolling back and restricting the airflow.

Snoring can sound much worse (or, strangely, better) than it really is. A listener's subjective perception of someone else's snoring bears only a loose relationship to objective measures of the actual sound intensity. Measurements have proved that there is at best only a modest concordance between people's subjective ratings of recorded snoring and the actual sound intensity. Even snoring of relatively modest sound intensity can still induce apoplexy among bystanders when it is repeated at slightly varying and unpredictable intervals. According to legend, the notorious American gunfighter John Wesley Hardin was so infuriated by a man snoring in an adjoining hotel bedroom that he fired his gun through the wall and killed the unfortunate snorer.

Snoring is very common. In the USA and UK, between 30 and 40 per cent of adults confess to snoring regularly. And they will only know they snore because someone else has told them. Much the same is true of France, where about one in three men snore habitually. The town of Uppsala in Sweden seems to be a better bet for a quiet night's sleep because, according to a study of its residents, only 15 to 20 per cent of men there are habitual snorers. Children and teenagers snore too. A recent survey in Vienna high schools established that 21 per cent of school students admitted to snoring, while research in Portugal found that a remarkable 39 per cent of primary school-age children snored occasionally, with 9 per cent snoring loudly and frequently. Overall, it is safe to conclude that at least one adult in five snores regularly, and probably more like one in three. To find out how badly you snore, if you have no sleeping partner to ask, you could try placing a tape recorder next to your bed at night. You might be surprised.

Certain sections of the population contribute more than others to the nocturnal cacophony. Regular snorers are more likely to be male than female, older rather than younger, and fatter rather than thinner. In Finland, for example, between 28 and 44 per cent of healthy, middle-aged men snore habitually, compared with only 6–19 per cent of middle-aged women. The archetypal snorer is an overweight, middle-aged man. Why are men more prone to snoring than women? The answer may lie in the distribution of fat and the size of muscles in their necks. The volume of soft tissue in the neck region is one third higher in men than in women. Much of this difference is accounted for by muscle. Men also have more fatty tissue in the tongue and soft palate – the prime villains in snoring.

Humans are not the only species to snore, as many owners of elderly, obese dogs will attest. But snoring is not particularly prevalent elsewhere in the animal world, and it is unclear whether it serves any biological purpose. Some hard-line proponents of the view that all biological characteristics must have specific adaptive advantages have suggested that the sound of snoring might have helped our hunter-gatherer ancestors to survive by deterring nocturnal predators. But this imaginative hypothesis fails to explain why the individuals who snore the most and loudest tend to be old and fat. A simpler explanation is that snoring is an unfortunate by-product of other evolutionary pressures, combined with recent historical changes in human lifestyles. Obesity, fatigue and alcohol have a lot to answer for.

Snoring can obviously be undignified and bothersome, but is it harmful? Although a large proportion of adults snore, the vast majority of snorers do not consult their doctors or feel they have a medical problem. Whether snoring is in itself a health hazard or danger sign remains a matter of controversy. We shall shortly be looking at obstructive sleep apnoea, the big brother to snoring and a much more serious disorder. But the jury is still out on whether plain, simple snoring is by itself bad for you.

Some studies have uncovered apparent connections between snoring, high blood pressure and heart disease. But the evidence is equivocal, and more than half of all published studies have failed to find any independent link between snoring and high blood pressure. Researchers discovered that sudden death from cardiovascular causes is somewhat more common among men who have habitually snored. Habitual snorers are more likely to die during their sleep or in the morning. In many cases, however, their snoring may be a sign of a more serious but undiagnosed disorder, namely obstructive sleep apnoea. When snoring is accompanied by sleep apnoea it is definitely associated with high blood pressure and cardiovascular disease. Habitual snoring can have undesirable side effects, even if it may not be an independent risk factor for heart disease. It can impair sleep quality and cause daytime sleepiness. Heavy snorers tend to feel sleepier during the day than non-snorers. More worryingly, they also report more instances of narrowly avoiding car accidents – presumably a consequence of their daytime sleepiness. There is also some evidence that people who snore badly have slightly impaired cognitive performance. One study of heavy snorers (none of whom had sleep apnoea) recorded deficits in their manual dexterity, attentiveness and early-morning alertness.

The suggestion that habitual heavy snoring can take the edge off your mental skills is borne out by research showing that medical students who snore are more likely to fail their exams. A study of trainee doctors taking their final exams found that those who snored frequently had lower exam scores and were more likely to fail than those who snored occasionally or never. The differences were substantial: 42 per cent of the frequent snorers failed their medical exams, compared with only 13 per cent of non-snorers. Snoring is also linked to headaches. Healthy men who snored were found to be 50 per cent more likely to have headaches than non-snorers, even after taking account of factors such as alcohol consumption. Child snorers are affected too. Portuguese research discovered that school-age children who snored loudly and frequently at night were sleepier and more irritable during the day, exhibited more behavioural disturbances and were more likely to suffer from recurrent infections of the respiratory tract.

Silence is golden

A barbarous noise environs me.

John Milton, Sonnet 12 (1673)

Whether or not snoring is bad for your health, few would disagree that it is a pointless and unattractive trait with few redeeming features. So, how do you stop it? Snoring is notoriously impervious to remedies. Many have tried to cure it, but few have succeeded. There is little to recommend the following harsh method, espoused by a clergyman called Prichard in a letter to *The Times* many years ago:

> No cure for snoring? We had a most reliable one at Sherborne forty-five years ago – dropping specially kept pellets of soap into the offending mouth. Further treatment was seldom needed.

Anything that relaxes the muscles in the upper airway will exacerbate snoring, so snorers should try to avoid alcohol and other sedatives in the evenings. Tiredness is another offender, because it increases the depth of sleep and hence the degree of muscle relaxation. If you are prone to snoring and deprive yourself of sleep, rest assured that you will snore louder. Allergies, colds and other upper respiratory disorders make snoring worse because they narrow the upper airways; nasal-decongestant sprays often provide some relief. Obesity narrows the upper airways through the deposition of fat. Losing weight therefore often helps to reduce snoring. Finally, many snorers find that they snore less if they sleep on their sides rather than on their backs. For the snorer who is truly determined to cut down the volume, a combination of shedding fat, sleeping on the side and a nasal-decongestant spray should help in most cases.

Another simple and harmless remedy that does the trick for some snorers is the nasal-dilator strip – a springy, self-adhesive strip that sticks over the nose. It works by dilating the nostrils and reducing the resistance to airflow. Snorers who have evaluated the strip have reported pleasing reductions in mouth dryness and daytime sleepiness, while their long-suffering sleeping partners have noticed a welcome reduction in snoring. The nasal-dilator strip works best in mild cases of snoring. A study of snoring men found that using the nasal dilator for a month made a substantial difference: the men's snoring (as rated by their partners) improved significantly, the men felt less tired during the day, and their self-rated quality of life also improved. Their partners benefited too: they slept better and felt better in the mornings. Peace and happiness reigned.

Obesity is a big risk factor for snoring. The usual remedy for excess fat, sagging muscles and flabby tissues elsewhere in the body is physical exercise. However, surprisingly little attention seems to have been paid to the possibility that exercising the muscles of the upper airway might alleviate snoring. One of the few exceptions was some work done in the 1960s by a British doctor, who developed a series of exercises to prevent snoring by toning up the muscles in the throat, jaw and tongue. The regime comprised a two-week programme of nightly exercises that included pressing the tongue against the lower teeth for 3–4 minutes, pressing the fingers firmly against the chin and holding the jaw against the pressure for 2–3 minutes, and holding a pencil or some other object firmly between the teeth for ten minutes. The exercises were tested on snorers and had measurable benefits. Exercises like these are presumably harmless (provided you do not swallow the pencil), cost nothing and might be worth a try. If all else fails, you could always issue your bed partner with earplugs.

Breathless in bed

> Blow, winds, and crack your cheeks!
>> William Shakespeare, *King Lear* (1605–6)

Snoring lies at the benign end of a spectrum of sleep-related breathing problems. At the other end lurks an altogether more insidious disorder called sleep apnoea. The name apnoea derives from the Greek *apnoia*, meaning 'without breath'. That about sums it up.

Sleep apnoea occurs when the upper airway is temporarily obstructed during sleep, despite efforts by the respiratory muscles to inhale. The sleeper stops breathing for anything from ten seconds to more than a minute at a time. Physiological alarm mechanisms eventually trigger a partial reawakening, allowing the airways to reopen. The sleeper then gasps in some air, with what is chillingly known as a resuscitative snort, before the airways close up again. This disturbing sequence can happen more than a hundred times during the course of a night's sleep. Sleep apnoea has been likened to having your head held under water, over and over again, throughout the night.

Although sleep apnoea looks and sounds dramatic to an observer, the sufferer rarely wakes up during these brief asphyxiations and therefore remains oblivious of the fact that they have repeatedly stopped breathing. What they do notice, however, is feeling the next day that they have not had a refreshing night's sleep. Sleep apnoea can be like waking up with a hangover every morning. The repeated arousals throughout the night severely disrupt sleep, causing daytime sleepiness, irritability and other signs of sleep deprivation.

What snoring and sleep apnoea have in common is an increased resistance to airflow in the upper airway during sleep. Not everyone who snores has sleep apnoea, but anyone who snores loudly and persistently should at least consider the possibility that they might have sleep apnoea.

The intimate relationship between good sleep and good breathing was discovered a long time ago. Ancient Greek writings describe a condition reminiscent of sleep apnoea, and European works dating from the sixteenth century note that breathing through the mouth can interfere with sleep. In 1861 one George Catlin devoted a whole book to the subject of nasal breathing and its superiority over breathing through the mouth, entitled *The Breath of Life; or Mal-respiration, and its Effects upon the Enjoyments & Life of Man*. A definitive connection between obstruction of the upper airway, snoring and daytime sleepiness had been made by the second half of the nineteenth century.

Charles Dickens was once again ahead of his time. The term Pickwickian was coined in 1889 to categorise obese, sleepy patients with severe sleep apnoea; it was derived from Dickens's character Joe the Fat Boy in *The Pickwick Papers*, which was written more than 50 years earlier. Joe the Fat Boy presents an authentic clinical portrait of sleep apnoea:

> On the box sat a fat and red-faced boy, in a state of somnolency . . .
> 'Damn that boy,' said the old gentleman, 'he's gone to sleep again.'
> 'Very extraordinary boy, that,' said Mr Pickwick, 'does he always sleep in this way?'
> 'Sleep!' said the old gentleman, 'he's always asleep. Goes on errands fast asleep, and snores as he waits at table.'
> 'How very odd!' said Mr Pickwick.

'Ah! odd indeed,' returned the old gentleman; 'I'm proud of that boy – wouldn't part with him on any account – he's a natural curiosity!'

The horses were put in – the driver mounted – the fat boy clambered up by his side – farewells were exchanged – and the carriage rattled off. As the Pickwickians turned round to take a last glimpse of it, the setting sun cast a rich glow on the faces of their entertainers, and fell upon the form of the fat boy. His head was sunk upon his bosom; and he slumbered again.

Despite the perceptiveness of Dickens's observations, a further 140 years elapsed before medicine could adequately explain the nature of the Fat Boy's condition. Sleep apnoea was overlooked for much of the twentieth century because of general scepticism about the clinical significance of daytime sleepiness, and because of confusion between the symptoms of sleep apnoea and narcolepsy. It was not properly recognised as a serious medical problem in its own right until the 1960s.

Occasional, brief interruptions in breathing during sleep are common and normal. The real problem arises when these interruptions are prolonged and occur repeatedly throughout the night. The number of individual episodes of apnoea (interruption in breathing lasting at least ten seconds) per hour of sleep is known as the sleep apnoea index. So, for example, someone who had 160 episodes of apnoea during eight hours of sleep would have a sleep apnoea index of 20. Anyone with a sleep apnoea index of more than ten is usually diagnosed as having sleep apnoea syndrome. A sleep apnoea index of more than 20 would usually imply that the sufferer needs some form of medical attention. It is not unusual to find sleep apnoea patients who have hundreds of episodes each night, some lasting more than a minute.

Sleep apnoea is actually of two different types. The predominant form, which we are focusing on here, is called obstructive sleep apnoea. This occurs when mechanical obstructions in the upper airway impede the flow of air. There is another, much less common, form called central sleep apnoea, in which the sleeper makes no effort to breathe in the first place. This rare condition stems from a defect in the central brain mechanisms controlling breathing. Central sleep apnoea was once known as Ondine's curse, after a German legend in which a maiden named Ondine punished her former lover by placing a curse on him. Ondine's curse rendered him unable to breathe except through conscious will; if he forgot to breathe, he stopped breathing. Central sleep apnoea causes serious complications, including high blood pressure and growth retardation.

Obstructive sleep apnoea (or, more simply, sleep apnoea) is an alarmingly common and widely undiagnosed disorder. The main symptoms are habitual snoring, which affects 70–95 per cent of sufferers, daytime sleepiness, morning headaches, depression and reduced libido. Sleep apnoea is estimated to affect around 4 per cent of the general population, making it almost as common as asthma. Among middle-aged and older men the prevalence rises to more than 8 per cent. One large study of Americans aged 65 and older found sleep apnoea in 13 per cent of elderly men. In high-risk groups – notably the severely obese and individuals with abnormalities in their facial bone structures – the prevalence of sleep apnoea may exceed 30 or 40 per cent.

Most people who suffer from sleep apnoea remain oblivious of their problem. They often assume that their chronic daytime sleepiness is normal, and ignore it. Others ascribe their symptoms to depression. The disabling tiredness is usually what drives a few sufferers to seek medical help. The only way of accurately diagnosing sleep apnoea is by overnight polysomnograph recordings in a sleep laboratory: clinical examination alone is an unreliable method of diagnosis. Most cases of sleep apnoea in the general population therefore go undiagnosed. According to one published estimate, almost 90 per cent of the middle-aged Americans who are suffering from moderate or severe sleep apnoea have not been clinically diagnosed as such. Sleep apnoea is even more widely underdiagnosed in the UK. A 1993 report by the Royal College of Physicians of London conceded that the UK had lagged behind other countries, notably the USA, Australia and continental Europe, in recognising sleep apnoea as an important clinical problem. The report concluded that there was 'an urgent need to increase awareness and knowledge of sleep apnoea'. Its recommendations included the very modest target of establishing one specialist respiratory sleep centre for every three million of the British population (which works out at around one for every 120,000 apnoea sufferers).

Although sleep apnoea is primarily a disorder of middle age, it is not confined to adults. Children can suffer from sleep-related breathing disorders that vary along a spectrum from slightly increased resistance in the upper airway to fully-fledged sleep apnoea. The prevalence of these disorders among children is not known with any accuracy, but some estimates have put the figure as high as 11 per cent. In the 1990s, scientists at the University of Iceland conducted a major study of sleep-related breathing disorders among pre-school children. They surveyed all the children between six months and six years of age living in the town of Gardabaer, ten kilometres south of Reykjavik. Initial questionnaires were followed up with overnight monitoring of those individuals who reported symptoms of sleep apnoea. Snoring and sleep apnoea turned out to be surprisingly widespread: 17 per cent of the children snored occasionally, 3 per cent snored frequently and 3 per cent had sleep apnoea.

The signs and symptoms of sleep-related breathing disorders in children can include irritability, hyperactivity, daytime sleepiness, morning headaches and bed-wetting. Young children who have difficulties breathing through their nose often end up habitually breathing through their mouth; this in turn can affect the physical development of their face, further worsening the original breathing problem. The consequences of severe, untreated sleep apnoea in children may include stunted growth, behaviour problems and underachievement at school.

What causes sleep apnoea? One factor that has a major bearing on the likelihood of developing sleep apnoea is age, and one reason why older people are at greater risk is that the soft tissues in the upper airways become slacker and floppier as we grow older. Regrettably, the daytime sleepiness, mild depression and befuddlement that can result from sleep apnoea in the elderly are all too easily ignored on the grounds that they are merely the signs of 'old age'. Being male also raises your risk. Other risk factors include smoking, alcohol and poor physical fitness. But the single most powerful predictor of sleep apnoea in adults is obesity, especially of the upper body. Clinical research has consistently found a firm correlation between the severity of sleep apnoea and the degree of obesity.

Obesity of the upper body is linked to sleep apnoea because fat deposits in the neck constrict the upper airway. The archetypal sleep apnoea victim – like the archetypal snorer – is an overweight, middle-aged man with a short, fat neck. More than three quarters of sleep apnoea patients who receive medical treatment are men, and more than half of them are obese. Measurements reveal a strong association between individuals' external neck circumference and the severity of their sleep apnoea. One telltale sign of the overweight middle-aged man is the undone shirt-collar button.

To add to the brew, take one vicious cycle. Obesity can trigger sleep apnoea, but sleep apnoea can also worsen obesity. Sleep apnoea is associated with abnormalities in metabolism that promote weight gain, creating a cycle in which obesity and sleep apnoea feed off each other. To make matters worse, some apnoea sufferers turn to sugar-rich foods, chocolate and caffeinated soft drinks to perk themselves up and help them resist their daytime sleepiness. Ultimately, of course, guzzling excess calories only makes both problems worse. The victim just gets progressively fatter and sleepier. Think of Dickens's Joe the Fat Boy.

The anatomical configuration of your upper airway, palate and jaw can also have a bearing on whether you will develop the disorder. On average, people with obstructive sleep apnoea have longer faces, longer soft palates and narrower airways than non-sufferers. A receding jaw is a definite risk factor, since it usually produces a smaller aperture where the upper airway meets the back of the tongue. The English upper-class chinless wonders, so lovingly portrayed in the books of P. G. Wodehouse, would have been prime candidates for sleep apnoea. (Perhaps that is why many of them were so dim-witted.) Because facial anatomy makes a difference, sleep apnoea sometimes occurs in several members of the same family. (Think of long lines of chinless wonders.) In one documented case, all ten members of an American family, spanning three generations, were habitual snorers and half of them had sleep apnoea. However, before we leap to the spurious conclusion that sleep apnoea must be 'all in the genes', we should remember that genes are not the only things that run in families. Lifestyles and environments are inherited too. Close relatives often share other risk factors for sleep apnoea, such as a tendency to become obese or to drink too much alcohol at bedtime.

The role of facial anatomy is illustrated by the pedigree English bulldog. This snuffly beast, with its flattened face and abnormal upper airway structure, is the loving product of many generations of artificial selection by human breeders. It is also a natural animal model of obstructive sleep apnoea. A bulldog's breathing is severely disordered during sleep, especially in REM sleep, and its blood oxygen drops to abnormally low levels for prolonged periods. When bulldogs are awake they often display signs of sleepiness, and they tend to fall asleep faster than other breeds.

A blocked or runny nose is another potential trigger, which is why people with allergic rhinitis (hay fever) are more prone to sleep apnoea. The nasal discharge generated by their allergic reaction increases the resistance to airflow in the upper airway. In people suffering from a seasonal allergy to certain plants, measurements have revealed an increase in upper airway resistance during the allergy season, accompanied by an increase in the severity of their sleep apnoea. Outside the allergy season there is a drop in both nasal resistance and sleep apnoea, accompanied by a rise in the amount of slow-wave sleep. They sleep more when they can breathe more easily. Strangely, and for reasons that remain unclear, there is also a statistical association between sleep apnoea and left-handedness. (Yet another curious connection between left-handedness and something apparently quite unconnected.) A large study discovered that left-handers had significantly more severe apnoea than right-handers. The explanation might lie with allergies and asthma, which are also statistically associated with left-handedness.

Consequences – mostly dire

> That he who many a year with toil of breath
> Found death in life . . .
> > Samuel Taylor Coleridge, 'Epitaph for Himself' (1834)

So much for the causes: what does sleep apnoea do to us and does it really matter? Sleep apnoea is potentially dangerous for two main reasons. First, full-blown sleep apnoea severely disrupts sleep, causing chronic daytime sleepiness and thereby heightening the risk of having an accident. If you repeatedly stop breathing throughout the night, you will not be on peak form when you crouch behind the wheel of your car the next day. Secondly, sleep apnoea raises blood pressure and hence worsens the risk of heart disease and stroke.

The most noticeable symptom of sleep apnoea is feeling excessively sleepy during the day. Sleep apnoea disrupts sleep because it provokes numerous brief awakenings, or microarousals. The sufferer rarely wakes up, but their sleep becomes highly fragmented. A side effect of this repeated disruption is that people with sleep apnoea remember far fewer dreams; their REM sleep is just too fragmented. Sleep apnoea produces a form of chronic sleep deprivation, which impairs mental performance. Apnoea sufferers display the classic signs: they find it difficult to concentrate, learn new tasks or perform monotonous tasks, their reactions are slower, and they fall asleep easily. Experiments have found that people with moderate sleep apnoea have slower reactions than healthy people with blood-alcohol levels that would make it illegal for them to drive.

The sleepiness and impaired reactions do not make for safe driving. American research found that almost one in four drivers with sleep apnoea admitted falling asleep while driving at least once a week. Not surprisingly, they have far more vehicle accidents than other drivers. At a conservative estimate, sleep apnoea more than doubles the risk of having a vehicle accident, even among experienced professional drivers such as long-distance truck drivers. Some estimates have put the relative risk much higher. Spanish investigators, for example, found that drivers with sleep apnoea were six times more likely to have had a serious accident. In another study, apnoea sufferers whose driving ability was assessed on a computer simulator had three times as many 'accidents' as healthy people. Moreover, their driving performance improved after their apnoea was treated, confirming that there had indeed been a direct causal relationship between the sleep apnoea and the bad driving.

Worryingly, most drivers who have sleep apnoea are completely unaware of the fact. When researchers analysed the history of vehicle accidents among supposedly healthy adults, they found that a disproportionately large number of drivers who had been involved in two or more accidents were suffering from previously undiagnosed sleep apnoea. Sleep apnoea is sufficiently widespread to affect large numbers of drivers, making it a significant cause of vehicle accidents. For example, Swedish scientists estimated that 100,000 car drivers in Sweden (2 per cent of the population) were suffering from excessive daytime sleepiness caused by sleep apnoea, making them potentially unsafe drivers.

Here is just one illustration of what happens as a consequence of undiagnosed sleep apnoea. On a sunny summer morning in 1995, a small car slowed down as it approached road works on a motorway in northern England. A large truck travelling behind the car failed to slow down. It ploughed into the car at 55 miles an hour, crushing the car and killing the two men inside it. The truck driver, who had been driving professionally for more than 20 years, had fallen asleep at the wheel. He later told police that he had been woken by the sound of the crash but remembered nothing of what happened immediately before the crash. It transpired that he was suffering from obstructive sleep apnoea. Sadly – but not unusually – his sleep apnoea was not diagnosed until seven months after the crash. This was despite the fact that over the previous six years he had complained to doctors on several occasions about his troubling tendency to fall asleep involuntarily. His illness was not recognised until two people had died.

The driving performance of people with sleep apnoea improves substantially after their condition is treated. A large Canadian study found that drivers diagnosed with sleep apnoea had three times the normal rate of collisions (18 accidents per 100 drivers per year, compared with the national average of six). But after their sleep apnoea had been successfully treated, their accident rate dropped to the average for the population as a whole. It is hard to escape the conclusion that identifying and treating more cases of sleep apnoea would substantially reduce the numbers of accidents, injuries and deaths on our roads.

But even if you stay inside, bolt the doors and never drive a car again, sleep apnoea is still bad for your health and may hasten your death. A ten-year investigation established that mortality rates were significantly higher among men who snored and suffered from excessive daytime sleepiness, which are the two main symptoms of sleep apnoea. Men under the age of 60 with both symptoms had a mortality rate that was between two and three times higher than average. Other research has reinforced the conclusion that people with sleep apnoea are at greater risk of dying prematurely.

Sleep apnoea is strongly associated with cardiovascular disease, for reasons that are not hard to understand. Each time an apnoeic sleeper stops breathing, their heart suddenly has to work much harder and their blood pressure rises sharply. The interruptions in breathing reduce the amount of oxygen in their blood and increase the level of carbon dioxide. This, in turn, makes their blood more acidic, which can trigger abnormal heart rhythms. A specific connection between sleep apnoea and high blood pressure was uncovered in the 1970s, and subsequent research has consistently shown that sleep apnoea is an independent risk factor for high blood pressure. Blood pressure increases in proportion to the severity of sleep apnoea, with each additional episode of apnoea per hour of sleep increasing the risk of developing high blood pressure by about one per cent. As you might expect, sleep apnoea is also a risk factor for heart attack and stroke.

A large American investigation uncovered firm links between sleep apnoea, heart failure and stroke, even among people whose apnoea symptoms were mild.

In children, sleep apnoea can result in stunted growth and failure to thrive. As we saw in chapter 6, disrupting someone's sleep will raise their metabolic rate. In other words, they burn more energy. When children with sleep apnoea were treated surgically for the disorder, their overnight energy expenditure fell and their growth rate improved, even though their calorie intake remained unchanged.

Sleep apnoea has pervasive and malign effects on children's behaviour and academic performance at school. American research discovered an unusually high proportion of apnoea sufferers among the first-grade children who were ranked in the bottom 10 per cent for school performance. Sufferers who later received surgical treatment for their apnoea improved their academic performance during the second grade, whereas those whose parents declined treatment showed no improvement in their grades. We saw in chapter 11 that some childhood behaviour problems might stem from undiagnosed sleep disorders. Evidence is mounting that sleep apnoea might contribute to some cases of Attention Deficit/Hyperactivity Disorder (ADHD). Some of the symptoms of ADHD resemble those of sleep deprivation in children, including hyperactivity, poor performance at school, impaired social skills, reduced attention span and impulsiveness. Conceivably, many children diagnosed with ADHD are suffering from undiagnosed sleep apnoea, in which case their behaviour problems should improve if they were treated for their sleep apnoea.

In adults, sleep apnoea may be a risk factor for depression. A study of sleep apnoea patients found that 45 per cent of them scored highly on an index of depression. Those with the highest depression scores had the most severe sleep apnoea. Moreover, their depression improved after their sleep apnoea was treated.

Sleep apnoea even changes the way people talk. It alters the acoustic properties of the vocal tract, producing distinctive differences in sound quality, particularly for the vowel sounds /i/ (as in 'see') and /e/ (as in 'get'). The top frequency of the harmonics of both vowel sounds is much lower in people with sleep apnoea, probably because they have a longer vocal tract. Their /i/ vowel sound also has fewer harmonics. One day, analysing the acoustic properties of voices might provide clinicians with a non-intrusive method for diagnosing sleep apnoea.

The partners of apnoea sufferers suffer too. Night after night they share their bed with someone who is restless, snores loudly, and repeatedly stops breathing in a hair-raising way. Not surprisingly, the partners often have poor quality sleep, with the added bonus that they are all too aware of the fact.

We are still not done with our catalogue of woes. To round it off, sleep apnoea can give you headaches and make you impotent. Heavy snoring and sleep apnoea are both linked with headaches, especially first thing in the morning. One in five people with sleep apnoea frequently experience headaches when they wake in the mornings, compared with only one in 20 of the general population. Their headaches stem from the repeated surges in blood pressure during the night and the dilation of blood vessels around the brain. Sleep apnoea can also induce impotence. Men with sleep apnoea are more likely to experience problems producing and maintaining erections. When treated for their sleep apnoea, around a third find that their erections are restored to their former glory.

The overall picture, then, is moderately alarming. Sleep apnoea has an insidious and damaging impact on health. It makes people feel chronically tired, erodes their quality of life and impairs their ability to perform safety-critical tasks like driving a car. It makes children behave badly and perform badly at school. And it heightens the risk of high blood pressure, heart attacks and strokes. Redeeming features? None. The disorder is neither easy nor cheap to diagnose and treat. But the economic costs of neglecting it are far greater. Untreated sleep apnoea causes accidents and illness, undermines people's effectiveness in the workplace and imposes large burdens on the healthcare system. An economic analysis in Germany concluded that the benefits of diagnosing and treating sleep apnoea outweighed the costs to the tune of several hundred million pounds a year. Paying more attention to sleep apnoea would improve the wealth of nations, as well as their health and happiness.

Unblocking those tubes

. . . a sleep
Full of sweet dreams, and health, and quiet breathing.
John Keats, *Endymion* (1818)

How is sleep apnoea treated? The approaches to treatment are many and varied, differing both in their effectiveness and their intrusiveness. They range from not sleeping on your back to radical surgery, and include lifestyle changes such as losing weight and drinking less alcohol. Of course, anyone who suspects that they might have sleep apnoea should seek expert advice from a qualified doctor, not a book. Remember that sleep apnoea can only be diagnosed following overnight monitoring. What follows is a brief survey of the current state of play.

The simplest and safest place to start in mild cases is by not sleeping on your back. Body posture affects our breathing during sleep. Snoring and obstructive sleep apnoea are more likely to occur when lying than when sitting, and more likely when lying on the back than on the side. Lying on the side can help because the uvula and tongue are less likely to obstruct the throat. The upper airway also becomes more collapsible when lying on the back. Many people find that learning to sleep on their side alleviates snoring and mild sleep apnoea. The tactic works best in sufferers who are not obese. One study found that sleeping on the side produced a big improvement in more than 90 per cent of non-obese apnoea sufferers, but helped less than 60 per cent of obese sufferers.

Sufferers can train themselves to sleep on their sides by using a positional alarm – an electronic device that makes a loud noise if the sleeper lies face-up for more than a few seconds. In one experiment, men with sleep apnoea who wore a chest-mounted alarm for one night had fewer episodes of apnoea and the symptoms abated in more than half the subjects. As an added benefit, sleeping on your side also reduces the likelihood of nocturnal tooth-grinding. A crude but effective method of stopping yourself sleeping on your back is to use 'snore balls', consisting of two or three tennis balls sewn along the spine of your pyjama top. The balls are certainly more convenient than relying on your bed partner to elbow you in the ribs every time you lie back and snore. Posture can also be improved with the help of various anti-snoring pillows, which cradle the head on one side. If you are using an ordinary pillow, it is a good idea to use a single, fairly thin one, since this will help to keep your neck straight and your airway open. Propping your head up on two or three big, fat pillows, so that your neck is bent at a sharp angle, is not recommended if you snore or suspect that you might have sleep apnoea.

Sleeping on your side is not a panacea, however. Most sleep apnoea sufferers require more. The next notch up in terms of intrusiveness is to avoid alcohol and sedatives in the evenings. We saw earlier that alcohol helps you to fall asleep but then disrupts your sleep. In the case of sleep apnoea, the sedative effect of alcohol is the problem, because it allows the upper airways to sag. If you snore heavily or have sleep apnoea, the conventional advice is to abstain from alcohol in the evenings.

A common strategy for tackling sleep apnoea is to lose weight. Obesity, especially in the upper body, is a major risk factor for obstructive sleep apnoea, and shedding excess fat cracks the problem for many moderate apnoea sufferers. But losing weight is not easy, as any dieter will know. Many apnoea sufferers, like the vast majority of dieters, fail to sustain the dietary regime needed to lower their weight and keep it there. And even among the determined few who do manage to remain slimmer, sleep apnoea can still reappear.

Various oral appliances have also been developed to treat sleep apnoea. These come in many shapes and sizes, but they all work by altering the relative positions of the lower jaw and tongue so as to increase the aperture of the upper airway. They can reduce or even eliminate snoring. They can also reduce the severity of sleep apnoea in many patients, with resulting improvements in sleep quality and daytime alertness. But, like all the other methods, they do not work for everybody.

More intrusive techniques may be called for when sleep apnoea is more severe. One of the most effective and widely used treatments is called continuous positive airway pressure, or CPAP ('see-pap'). This works by gently pumping air at a slightly increased pressure through the nostrils via a comfortable nose mask. The air acts like a pneumatic splint, holding the saggy upper airway open and preventing it from becoming blocked during sleep. CPAP, which was developed in the early 1980s, is the most common treatment for obstructive sleep apnoea, and is used by hundreds of thousands of patients around the world. For some sufferers, CPAP comes like a miracle cure, suddenly relieving them from years of chronic sleep deprivation. They become quieter and more restful in bed, so their partners benefit too. CPAP is relatively simple, safe and effective, but it is not a permanent cure. The sleep apnoea will return if the treatment is stopped. There are other pitfalls too. Many patients find CPAP difficult to live with in the long term: after all, it does require them to go to bed each night with a mask over their nose, connected to a gently humming box by a tube. About 30 per cent of patients find it hard to comply with CPAP for long. In more severe cases, where CPAP fails to work, surgery might have to be considered.

A standard surgical treatment for obstructive sleep apnoea is to remove the uvula, together with part of the soft palate and the tonsils, thereby enlarging the throat. This procedure is known in the trade as uvulopalatopharyngoplasty, or UPPP (or, less reverently, a throat facelift). It is said to be easier to perform than pronounce. Newer variations on the theme incorporate additional surgical flourishes, designed to pull the tongue muscles forward and away from the back of the throat. In its current form, UPPP was developed in the early 1980s. Not much under the medical sun is entirely novel, however, and certainly not the practice of removing uvulas. Aetios, a Greek scholar who wrote about medical matters in the late fifth or early sixth century, advised how to amputate uvulas using an instrument known as a 'uvula crusher'. The British Museum houses an example of what is thought to be a Roman uvula crusher, dating from the first or second century AD. This daunting instrument consists of bronze forceps with fine-toothed jaws.

UPPP is effective in about half of all cases and completely cures only a minority of patients. All forms of surgery for sleep apnoea depend on accurately identifying where the obstruction is occurring, and UPPP is fully effective only when the uvula and palate are the main sites of obstruction. In some cases, however, the obstruction lies elsewhere. Very obese patients tend to have sites of obstruction that are lower down their respiratory tracts, and they therefore respond less well to surgery in the upper airway. Moreover, all surgery carries risks. Surgery on the roof of the mouth causes serious complications in a small minority of cases. Less grave side effects, such as making the voice sound nasal or difficulty in swallowing, are fairly common. That is why alternatives to surgery, such as weight loss and CPAP, are given careful consideration before reaching for the knife.

Other surgical techniques use lasers, microwaves or radiofrequency heating in place of the traditional scalpel to remove or shrink the offending tissue. These methods have the advantage that they can be performed quickly under local anaesthetic. But even they are not without their problems. A study of patients who had received laser-assisted surgery for snoring found that some ended up worse off. One in five developed mild sleep apnoea following surgery, having originally complained only of bad snoring. And although most of them experienced an initial reduction in snoring, this improvement declined over time and snoring tended to return. The lesson seems to be the old one: that we should not rush to have surgery unless we are reasonably confident that it is necessary and that the benefits are likely to outweigh the risks.

Almost a century ago, George Bernard Shaw's play *The Doctor's Dilemma* satirised the medical profession for the way in which financial incentives seduced some doctors into performing unnecessary surgery. The removal of the uvula was alleged to be a case in point, even then:

> They've found out that a man's body's full of bits and scraps of old organs he has no mortal use for. Thanks to chloroform, you can cut half a dozen of them out without leaving him any the worse, except for the illness and the guineas it costs him . . . The father used to snip off the ends of people's uvulas for fifty guineas, and paint throats with caustic every day for a year at two guineas a time. His brother-in-law extirpated tonsils at two hundred guineas until he took up women's cases at double the fee.

The most radical treatment of all for obstructive sleep apnoea is tracheotomy, which means cutting a hole through the neck into the windpipe so that the patient can breathe through a tube. (Obviously, this can make swimming tricky.) Tracheostomy is highly effective and it saves lives in extreme cases. But radical surgery of this sort is always risky and can cause real problems. Fortunately, it has become rare since the advent of UPPP.

We have dwelt long enough on the darker side of sleep. Now it is time to explore its rich pleasures. Meanwhile, sleep well and breathe easily, preferably through your nose.

PART VII

Pleasures

16

And So to Bed

When Stephen opened his eyes he was lying on a bed ... All
the worst and the best things that had ever happened to him
had happened here. This was where he belonged.

Ian McEwan, *The Child in Time* (1987)

'The bed,' wrote Guy de Maupassant, 'is our whole life. It is there
that we are born, it is there that we love, it is there that we die.' Most
of us aim to love and die in bed even if we were not conceived there,
and we will spend a good third of our existence in one. Humans have
striven for millennia to ensure that this large portion of our lives is
as comfortable and pleasurable as possible. In short, the bed has a
huge significance in human experience, even if nowadays we rather
take it for granted.

The bed has been the setting for many other fine activities besides
sleep, sex, birth and death. It has been the birthplace of some of the
finest writing, music and art. Cicero, Horace, Milton, Swift, Rousseau,
Voltaire, Anthony Trollope, Mark Twain, Robert Louis Stevenson,
Marcel Proust, Colette and Winston Churchill all wrote in bed. Glinka,
Puccini and Rossini composed music in bed. Matisse painted in bed.
Samuel Pepys even discussed Royal Navy business in bed with the
Deputy Treasurer of the Fleet. Bed is simply the best place to lie and
think. The only other location that comes close is the bath, but baths
go cold and beds stay warm.

John Milton wrote *Paradise Lost* in his bed, sending for one of his
daughters to transcribe each day's composition. (He was blind.) Someone unkindly suggested that bed is also the best place to read *Paradise
Lost*, since Milton's tome is a sure-fire cure for insomnia. Rossini
composed many of his operas from the comfort of his bed. It is said

that on one occasion the manuscript of a duet he had just written fell to the floor, but rather than get out of bed to retrieve it Rossini wrote another duet, quite different from the first version. G. K. Chesterton imagined other peculiar pleasures that could be taken while in bed:

> Lying in bed would be an altogether perfect and supreme experience if only one had a coloured pencil long enough to draw on the ceiling.

Marcel Proust adored his bed. He transformed it into his office and wrote *Remembrance of Things Past* there. The last three years of his life were spent there. Proust even aspired to live on a yacht so that he could travel around without having to get out of bed. The author and academic John Bayley maintained that the secret of writing books is to stay in your pyjamas and not get dressed until you have produced your daily quota of words; that way, he advised, you cannot be tempted into displacement activities such as going to the shops or meeting friends for lunch.

The writer Thomas Mann, a great sleeper all his life, was passionate about his bed. He claimed to have preserved a clear and fond memory of every bed he had ever slept in for any length of time:

> It is a sweet-odoured linen shrine, where we lie unconscious, our knees drawn up as once in the darkness of the womb, attached anew, as it were, to nature's umbilical cord and by mysterious ways drawing in nourishment and renewal; it is a magical cockle shell, standing covered and unheeded by day in its corner, wherein by night we rock out upon the sea of forgetfulness and infinity.

Winston Churchill worked from his bed most mornings, even when he was prime minister. His working day began at eight, when the morning newspapers were brought in. He would read them thoroughly while propped up in his enormous bed, consuming a hearty breakfast. Churchill would remain in bed all morning, receiving official visitors and continuing with his work while secretaries took dictation and officials hovered in attendance. His bed became an office, and he was more productive there than at his desk. Only a national crisis would force him out of it before lunchtime. Anyone who came to see Churchill in the morning, whoever they were, was taken to see him in his bed. Churchill had a huge lust for life and packed an enormous amount into every day, even in his old age, but much of it was conducted from the comforting environment of his bed.

A brief history of beds

> This bed thy centre is, these walls thy sphere.
>> John Donne (1572–1631) *Songs and Sonnets*, 'The Sun Rising'

A comfortable bed can make the difference between a night of serene slumber and the grating frustration of insomnia. Charles Dickens observed of one bed in *The Pickwick Papers* that 'it would make anyone go to sleep, that bedstead would, whether they wanted to or not . . . poppies was nothing to it'. Here in similar vein is the American writer Damon Runyon rhapsodising on the joy of a good bed:

> We will take all the luxury with which a bed can possibly be surrounded – a gentle, yielding mattress, and quiet, cushiony springs, and soft, downy pillows, and snowy linen and the richest of coverings. A fellow gets little enough out of life under any circumstances without making his hours of rest too tough.

If beds reflect a society's attitudes towards sleep, then our ancestors appear to have taken their sleep fairly seriously. The beds of the ancient civilisations of Egypt, Greece and Rome were graceful and civilised. The Egyptians were constructing elegant beds more than five thousand years ago, displaying a standard of craftsmanship that was not seen in Europe until the Renaissance. Rectangular wooden beds with legs shaped like the legs of bulls or lions are recorded as far back as the First Dynasty (around 2925 to 2775 BC).

When the tomb of King Tutankhamen, who reigned in the fourteenth century BC, was discovered in 1922, it contained several fine beds. One was made from ebony, ivory and gold, with legs shaped like a cat's. Another, built of white-painted wood, had folding legs and was light enough to be carried by one man. Tutankhamen's beds were probably upholstered with leather and matting, though these materials did not survive the millennia. On dying, wealthy Egyptians often had their bed placed in their tomb with them, supporting the coffin. A headrest would be placed inside the coffin by the corpse's head. This desire to make the deceased comfortable reflected the Egyptians' belief in death as merely a transition between this life and the afterlife.

Beds in ancient Greece looked more like couches and they could be moved around easily. They were made of wood and were often elaborately decorated. The Greeks liked to eat while lying on their beds. The Romans later developed the Greek design, adding a headboard, footboard and turned legs. Some Greek and Roman beds were so high they needed steps to reach them. The Persians were connoisseurs of the bed and placed great emphasis on the luxury of their sleeping arrangements. The water bed originated in Persia three thousand years ago, when Persian nomads discovered the delights of sleeping on goatskins filled with water. Raised beds with canopies were being made in China more than two thousand years ago.

In northern Europe, however, beds remained primitive affairs until much more recently. In Saxon England, beds were just crude bags of straw, and whole families slept together on the floor. The expression to 'make a bed' stems from those times, when people picked up a sack and some straw each night and did just that. Dr Marie Carmichael Stopes, the eminent authority on birth control and sometime sleep expert, was still arguing in the 1950s that 'the finest, springiest and most comfortable mattress in the world is dry summer heather'.

From around the twelfth century onwards, English beds gradually became more comfortable and ornate. In the English medieval castle the whole household ate and slept together in the great hall. They slept in their day clothes, or naked if it was warm enough: specialised nightwear was still centuries away. As people increasingly sought a little privacy, a trend developed for sleeping in alcoves created by suspending material from the walls or ceiling. The earliest English bedsteads were made of four low wooden rails with a lattice of ropes threaded through holes to support the mattresses. However, the really important feature of a fine medieval bed was not the bedstead but the draperies that were used to provide privacy and exclude draughts. In wealthy households these draperies were often elaborate and expensive. The thirteenth century saw the advent of the tester: a fabric canopy hung from the rafters. Curtains were often hung from the tester. These could be pulled back, allowing the bed to be used as a couch during the day. In the fourteenth and fifteenth centuries, an era when the palest complexions were considered the most alluring, wealthy women in continental Europe favoured black satin sheets that contrasted with the pallor of their skin.

The type of bed favoured by the wealthy in England until the mid-nineteenth century was the four-poster. It reached its zenith in the Tudor and Stuart periods, when beds were more like works of architecture than items of furniture. Beds grew heavier and more elaborate, with elaborate carvings considered *de rigueur*. Such beds were limited to wealthy homes in Elizabethan England, but became more widespread by the seventeenth century. In winter, the chill was taken off with a warming pan filled with charcoal or hot embers. A servant often slept on a truckle (or trundle) bed – a low bed on casters that could be tidied away under the main bed during the day.

This was a time of great beds that were symbols of wealth and power. An account dating from 1598 describes the beds of Kings Henry VII, Henry VIII and Edward VI as being 11 feet (3.4 metres) square. The marriage bed of Isabella of Portugal was 5.5 metres (18 feet) long by 3.7 metres (12 feet) wide. In humbler homes the bed became one of a family's most prized possessions, to be bequeathed in wills and passed down through the generations. William Shakespeare's last will and testament famously contained this instruction: 'Item, I give unto my wife my second best bed, with the furniture.' This was not as churlish as it might sound. Shakespeare knew that his daughter Susannah and her husband would inherit his house in Stratford, where they would occupy the master bedroom. They would therefore need the master bed. Shakespeare's widow Ann would naturally receive the dowager bed, the second grandest in the household.

In the seventeenth and early eighteenth centuries the bed became even more a status symbol in Europe. Cardinal Richelieu, who was Chief Minister of France from 1624 until 1642, was devoted to his bed and seldom left it in his final years. This was a cause of some inconvenience to the many people he visited during his official travels around France, since his bed was enormous and often required the demolition of walls to allow its entry. One contemporary account describes a state visit that Richelieu made to the town of Viviers. He arrived by boat, still in his bed. A wooden jetty was built to enable the bed to be brought ashore, carried by six strong men. The townspeople were astonished to discover that Richelieu's chosen method of entry into their homes was through their walls or windows in his bed, rather than through their doors on foot. Before his arrival at a house,

Richelieu's personal masons would knock a hole in the wall of the room he was to occupy, or break open the window casements. A wooden bridge would be constructed, connecting the room to the street outside. The cardinal, temporarily decanted into a portable bed, would then be carried into the room where another, grander bed would be awaiting him. Only then would the interview begin.

Ceilings became taller in the seventeenth and early eighteenth centuries, and so did the beds. Men of great wealth would have state beds up to five and a half metres (18 feet) tall. When Parliament was in the throes of restoring the exiled King Charles II to the throne as king of Great Britain and Ireland in 1660, it issued a list of 'necessaries' to be provided for the new king's household. This list included various beds, among which was 'a rich bed to be of velvet, either embroidered with gold or laced [ie, trimmed with gold braid], lined with cloth of silver or satin'. An upholsterer's bill for one bed in late seventeenth-century England showed that the bed required 294 metres of velvet, 74 metres of satin, 307 metres of broad lace and 368 metres of narrow lace.

In Restoration London of 1660–70, a typical wealthy man's bed would have a wooden framework, with either wooden boards or a lattice of ropes to support the mattress. Where ropes were used, the bed had to be tightened from time to time by twisting the ropes with a key – hence the saying 'sleep tight'. Floor-length curtains could be pulled around the bed to keep out draughts in unheated rooms and create some illusion of privacy. A thick undermattress filled with wool or straw was topped by a thinner mattress filled with wool or feathers (or, if you were very rich, swansdown). None of these materials could be washed, so bedbugs were an insoluble problem for rich and poor alike. They remained a fact of life until the advent in the late eighteenth century of cheap cotton bedclothes that could be boiled, and cheap metal bedsteads in which the bedbugs could not 'nestle and breed'.

The seventeenth century also saw the advent of the day bed, or *lit de repos*, designed specifically for the daytime nap or siesta. The oldest surviving example in England is a rather splendid oak bed, measuring a less-than-compact 2.2 metres (7 feet 3 inches) in length. The leisured classes regularly used them for napping. It is said that on his death in 1673, the French playwright Molière was the proud owner of 100 day beds. King Louis XIV of France owned a mere 48 day beds, but these comprised only a small proportion of his total collection of 413 beds of at least 25 different types. He often held court from one of them.

The fashion for great beds lasted in Europe from Renaissance times until the early nineteenth century. From then on, beds went into decline all over Europe. The British, who were the first to enter the industrial revolution, reverted to lumpen, functional simplicity. Iron or brass bedsteads, which were cheaper and recommended on health grounds, became increasingly common. The spring mattress appeared in the 1850s. Meanwhile, the rest of Europe witnessed a final outburst of extravagant ornamentation, exemplified by two extraordinary beds that were installed in the Austrian palace of Furstenberg in 1820. These vulgar temples of slumber were built to resemble rockeries. They were richly decorated with bats and toads of gilded wood and carved lizards the size of a man's hand. The beds stood in an alcove lit by a lamp in the shape of an owl, with beams of light shining from its eyes.

Although the fashion for extravagant beds fizzled out as industrial-isation took hold, there continued to be some glorious and self-indulgent exceptions. King Ludwig II of Bavaria, who ascended to the throne in 1864 and was declared insane in 1886, had an outrageously expensive bed in the shape of a cathedral. Charles de l'Orme, a French physician, slept in a bed made of bricks with an integral lavatory. An Indian maharajah had a bed weighing more than a ton made for him in Paris. Four remarkably realistic and life-size models of nude women guarded the corners of the bed. The weight of the maharajah's body on the bed automatically triggered a musical box and set the naked houris in motion, two fanning his face and two whisking the flies from his feet.

Even the utilitarian twentieth century saw the occasional attempt to rediscover the concept of the grand bed. The eccentric American millionaire Howard Hughes had an outrageous hospital bed con-structed for him after he was injured in a plane crash. Hughes's bed was self-propelled, powered by thirty electric motors and controlled from an elaborate aircraft-style cockpit. From the comfort of this mobile sleeping machine, Hughes could tour the hospital wards, pos-ition his bed wherever he fancied, and summon up creature comforts such as music and hot and cold running water, all at the touch of a button. A British double bed exhibited in 1959 was described as a 'machine for living'. It comprised two mattresses that could be separ-ately adjusted at the touch of a button and individually heated to each sleeper's preferred temperature. On the man's side (sic) of the bed was a unit providing for a man's needs: a telephone, electric shaver and dictation machine. On the woman's side were an automatic tea-maker, a silver tea service and an electric massage machine. Each occupant had a personal radio and a half-share of a television at the foot of the bed. The curtains, lights and television were controlled from a switchboard. Mink coverlets (a very fifties touch) provided the final embellishment.

Elsewhere in the world, people have their own ideas about what constitutes a good bed. The Jivaro Indians of Amazonia sleep in rectangular bedsteads made from palm wood or bamboo. Their beds are enclosed on three sides, allowing them some privacy within their communal dwelling houses. The Jivaro believe that you will never be cold if your feet are warm, so they sleep with their feet projecting over the end of the bed, suspended above a smouldering hearth. This curious arrangement relies on someone getting up at regular intervals during the night to keep the embers glowing.

Pillows and other forms of headrest are important artefacts in many cultures. The ancient Egyptians used headrests made of wood, ivory or alabaster instead of soft pillows, presumably in deference to the heat. These headrests were often elaborately decorated. The wealthy preferred folding headrests in the shape of a folding stool. These were often carved with symbols designed to ward off the malign nocturnal forces that would threaten the sleeper with illness or nightmares. The Shona people of eastern Zimbabwe have a tradition of making highly distinctive wooden headrests, elaborately carved with geometric motifs symbolising ancestors. These headrests are regarded as a source of knowledge and prosperity. They are used mainly by men, who are believed to visit their ancestors during sleep. When the owner dies, his headrest is often buried with him or passed on to his surviving relatives.

Beds nowadays are usually accompanied (and marred) by the baleful presence of an alarm clock. Many of us get insufficient sleep and need the assistance of a machine to wake us up. In 2000, an Internet-based global survey of 1.2 million people in 251 countries found that only one person in three usually woke unaided, while half said they used an alarm clock. The extent of alarm-clock dependency is almost certainly higher in industrialised countries. Of the remainder, 12 per cent relied on other people to wake them, 4 per cent were woken by pets, while 1 per cent of the planet's population said they relied on a rooster (though presumably not the same rooster).

The oldest mechanical alarm clocks date back to the early fifteenth century. They became more elaborate over time and eventually evolved into the faintly absurd. In 1781 a Monsieur Morgues of Marseilles built an alarm clock that automatically lit a candle at the appointed hour of awakening, drew the bedroom curtains and opened the window. For the inveterate slugabed of the mid-nineteenth century there was Mr Robert Watson Savage's 'Alarum Bedstead', which was first unveiled at the Crystal Palace exhibition of 1851. This device would start off in a reasonably civilised fashion by sounding a bell. If the bell was ignored, it automatically stripped off the bedclothes. And if that failed to dislodge the sleeper, the mattress slowly tilted sideways to 45 degrees until gravity deposited him or her on the floor. Modern technology is helping to soften the brutal behaviour of alarm clocks, even if their very existence remains regrettable. One friendlier device is designed to simulate the dawn, by gradually increasing the intensity of a light over a period of half an hour before the chosen time of awakening. Our eyelids are surprisingly translucent, so enough light gets through them to inform the brain that morning has arrived and it is time to wake up. The result is an altogether gentler awakening, free from the harsh and malevolent beeping of the conventional alarm clock.

Other species have beds too. Elephants prepare a pillow of vegetation to sleep on. Parrotfish secrete a mucous envelope in which they sleep. Monkeys and apes spend about half their lives at sleeping sites, which provide comfort, safety from predators and endless opportunities for social interactions with other members of their species. Gorillas and chimpanzees construct nests of branches and leaves. Indeed, chimpanzees build separate and structurally distinct night nests, where they sleep, and day nests, where they rest and feed. Most other primates also sleep upstairs, like us. White-handed gibbons, for instance, use sleeping sites in tall trees to reduce the risk from pythons and other predators. They usually sleep alone in different trees, the only sharing being done by females with their infants. The gibbons arrive at the trees where they are going to sleep several hours before dusk and spend between 14 and 17 hours a day there. Black-and-white colobus monkeys in Kenya also sleep in tall trees, usually selecting trees in areas where food is available. Incidentally, we humans are recent converts to the practice of sleeping upstairs. Beds did not migrate upstairs until the seventeenth century, and downstairs bedrooms were still common in the eighteenth century.

Sleeping partners

> My brother Toby, quoth she, is going to be married to Mrs Wadman. Then he will never, quoth my father, lie *diagonally* in his bed again as long as he lives.
>
> Laurence Sterne, *Tristram Shandy* (1759–67)

The majority of adults (around two thirds) do not sleep alone. Sharing your bed with a partner is an arrangement that can bring trouble as well as joy. If you snore loudly, have frequent nightmares, sleepwalk, or talk in your sleep, your partner's sleep will probably suffer more than yours. In Shakespeare's *Richard III*, Richard's wife bemoans having to share a bed with a man who has bad dreams:

> For never yet one hour in his bed
> Did I enjoy the golden dew of sleep,
> But with his timorous dreams was still awaked.

Another pitfall with bed sharing is that sleep deprivation is contagious. If you are starving yourself of sleep by staying up late every night and getting up early every morning, the chances are that your partner will not be getting enough sleep either. Conflicts can also arise when an owl and a lark share the same bed, since they will rarely want to enter or leave it at the same hours. Dutch research found that couples whose sleep–wake patterns were mismatched tended to have less well-adjusted marriages than couples whose sleep patterns chimed. Bed sharing can even be physically hazardous, as this entry from Samuel Pepys's diary for 1 January 1662 reveals:

> Waking this morning out of my sleep on a sudden, I did with my elbow hit my wife a great blow over her face and nose, which waked her with pain – at which I was sorry. And so to sleep again.

Pepys and Mrs Pepys did not let this little mishap come between them. A few months later Pepys recorded this gorgeous little anthem to the delights of the shared bed:

> Lay long in bed today with my wife, merry and pleasant.

Until the eighteenth century sleeping remained a communal activity, and it was common for several people, not just two, to share a bed. This explains why many beds from this time were so enormous. Perhaps the most famous of these gargantuan beds was the Great Bed of Ware, as mentioned by William Shakespeare in *Twelfth Night* and Ben Jonson in *The Silent Woman*. This bed, which is now in the Victoria and Albert Museum in London, is 3.3 metres wide and 3.4 metres long, which is about average for a Tudor state bed of the time. It bears the carved date '1460', but experts believe it was actually made in the late sixteenth century. By 1612 it had been moved to an inn at Ware in Hertfordshire. The Great Bed of Ware could comfortably accommodate a dozen people and was reputedly used for orgies. According to one history, six couples from London once slept together in the Great Bed of Ware 'for a frolick'.

Even the uncontroversial practice of one man and one woman sharing a monogamous marriage bed has occasionally met with disapproval, however. Doctor James Graham, an eighteenth-century expert on almost everything, was unwavering in his condemnation:

> Gentlemen, there is not, in my opinion, anything in nature which is more immediately calculated totally to subvert health, strength, love, esteem, and indeed every thing that is desirable in the married state, than that odious, most indelicate, and most hurtful custom of man and wife continually *pigging* together, in one and the same bed. Nothing is more unwise – nothing more indecent – nothing more unnatural, than for a man and woman to sleep, and snore, and steam, and do every thing else that's indelicate together, three hundred and sixty-five times – every year.

(*Steam*?) To assess how bed partners actually influence each other's sleep, scientists monitored 46 obliging couples using automatic movement recorders and sleep diaries. They discovered that one third of all movements during the night involved both partners changing position more or less simultaneously. Most of the subjects were unaware that their movements during sleep were synchronised. The synchrony between partners' movements was strongest in younger couples. Men moved around more than women, and women were more disturbed by their partner's movements than men. (No surprises there.) People moved more when sleeping with their partner than when sleeping alone. Nonetheless, most said they slept better when their partner was present.

Adults are not the only ones to share beds, of course. Throughout almost the whole of human evolutionary history, babies have normally slept with one or both parents. Co-sleeping was, and still is, the norm in non-industrialised cultures. The practice of putting infants to sleep on their own is historically recent and largely a product of industrialised societies. In the UK and other European countries, babies slept with their mothers (or nurses) as a matter of course until the eighteenth century. This made good sense in poorly heated homes. But attitudes shifted during the nineteenth century, as mothers were increasingly encouraged to put their babies to sleep in a separate room (assuming they had one). By the 1920s, a popular baby-care book was able to assert, without any risk of controversy, the absurd advice that:

> The place where baby most likes to sleep is where he must not be.

Making infants sleep alone came to be regarded as normal, and the practice is still widely believed to help the developing infant attain independence. But this is a dubious belief. There are no solid grounds for supposing that infants who sleep with their parents for the first year or two have any greater problems becoming independent. In reality, many parents in industrialised societies 'break the rules' and co-sleep with their infants, albeit sometimes with vague feelings of guilt. If anything, the practice of co-sleeping has become more widespread again, as busy working parents seek ways of spending more time in contact with their offspring.

The important influence of co-sleeping on infants' sleep patterns has been largely neglected in laboratory studies, nearly all of which are based on measurements of solitary infants. The little research that has been done with co-sleeping mothers and infants has revealed a remarkable synchrony between their sleep patterns. The shared bed provides the developing infant with a much richer sensory environment than the solitary cot.

Co-sleeping can be pleasurable, both for baby and parents. It also makes breast-feeding easier, which is highly desirable. Babies who routinely sleep with their mothers spend three times longer breast-feeding during the night than breast-fed infants who sleep separately. Systematic observations have also found no support for the belief that mothers who share their beds get less sleep; co-sleeping mothers actually spend just as much time asleep as mothers whose babies sleep alone. All in all, co-sleeping, breast-feeding mothers generally have an easier time of it than non co-sleeping, bottle-feeding mothers, especially during the early months when babies want to feed frequently during the night. Even with a slightly older child, the experience can be mutually enjoyable. In *Cider with Rosie*, Laurie Lee's memoir of his Gloucestershire childhood, Lee describes the delights of sleeping in his mother's bed as a young boy:

> So in the ample night and the thickness of her hair I consumed my fattened sleep, drowsed and muzzling to her warmth of flesh, blessed by her bed and safety . . . My mother, freed from her noisy day, would sleep like a happy child, humped in her nightdress, breathing innocently and making soft drinking sounds in the pillow. In her flights of dreams she held me close, like a parachute to her back; or rolled and enclosed me with her great tired body so that I was snug as a mouse in a hayrick.

There are some drawbacks to co-sleeping, however – not least the detumescing effect on the parents' sex life. Most couples find that having a baby in the bed with them is erotically discouraging, even if the baby is far too young to know, let alone care, what its parents are doing with each other. More seriously, it has been suggested that infants who sleep in the same bed as their parents are at greater risk of succumbing to Sudden Infant Death Syndrome (SIDS), also known as cot death.

The putative link between co-sleeping and SIDS remains a controversial area. In the early 1990s researchers in New Zealand reported that infants who slept in their parents' bed were at greater risk of SIDS, especially if their mother smoked. Infants who shared the parental bed and had a mother who smoked were four times more likely to die of SIDS than infants with neither risk factor. In 1999 the US Consumer Product Safety Commission published even more alarming (some would say alarmist) conclusions. The Commission carried out a retrospective analysis of US statistics on young children who had died while co-sleeping. Between 1990 and 1997, 515 children aged two or younger had died while sleeping in adult beds or waterbeds. More than three quarters of these deaths were reportedly caused by strangulation or suffocation after the child became entangled in the structure of the bed; for example, between bed railings or in the folds of a waterbed. The other deaths were attributed to an adult or sibling lying on the infant. The Commission advised that children under the age of two should not sleep in adult beds.

The evidence that co-sleeping is generally hazardous is far from conclusive, however. Other respectable scientific studies have found that infants who share their parents' bed are at no greater risk of SIDS. For example, British investigators who assessed hundreds of SIDS cases found no evidence that sharing the parental bed was risky – provided the parents did not smoke. Infants younger than four months *were* at much greater risk of SIDS if their parents smoked and if they spent the whole night in the parental bed. Passive smoking, rather than smothering or overheating, seems to be the key link between co-sleeping and SIDS. Other research has indicated that co-sleeping might even have some protective effect against SIDS. Overnight observations show that co-sleeping mothers and their infants tend to sleep on their sides, facing each other, for most of the night. It has been suggested that the increased levels of carbon dioxide in the mother's exhaled air might help to stimulate and regulate the baby's breathing.

The scientific evidence increasingly suggests that it is safe for parents to share their bed with their baby – but not if they smoke, drink heavily or suffer from sleep deprivation. A baby who shares a bed with a mother or father who smokes is at greater risk of SIDS, as we have seen. Co-sleeping may also be inadvisable if one or both parents have been drinking heavily or if they are extremely tired. Statistics on cot deaths show that babies are at greater risk of SIDS at weekends and on public holidays. The likely explanation is that parents are more likely to drink heavily, sleep heavily and expose their babies to tobacco smoke on weekends and public holidays. No biologist would be surprised by the conclusion that co-sleeping – which has been the normal pattern of human behaviour throughout almost all of our species' evolutionary history – is generally safe, aside from the caveats about drinking and smoking.

Another aspect of the sleeping arrangements that has a clear influence on SIDS is the infant's posture. Infants who sleep on their fronts (prone) are at greater risk of SIDS. In 1992 the US authorities recommended that parents should always place their infants to sleep on their backs or sides, and a national 'Back To Sleep' campaign was launched to encourage this change in parenting. Between 1992 and

1996 the practice of placing infants to sleep on their fronts fell by two thirds in the US and the incidence of SIDS fell by 38 per cent. By 1996 fewer than one in four parents were still placing their infants to sleep on their fronts. The campaign was highly successful. Similar education programmes succeeded in many other countries. In the UK, for example, the number of SIDS cases dropped by a third between 1993 and 1998. Swedish research found that mothers who ignored the advice were also more likely to smoke, bottle-feed their infant and ignore other advice on childcare. The minority of infants who were still sleeping on their fronts were therefore more exposed to other risk factors such as passive smoking.

The bed, then, is a supremely important place. Do not neglect yours. Appreciate it for what it is. Buy the most comfortable one you can afford, spend plenty of time there, and enjoy sharing it.

17

An Excellent Thing

For do but consider what an excellent thing sleep is.
Thomas Dekker, *The Guls Horne Book* (1609)

Getting enough sleep of the right quality and at the right times is essential for your health, happiness, success and sanity. If you disbelieve that by now I have failed miserably. But sleeping and dreaming are far more than just basic maintenance activities or the neural equivalents of vitamins. They are also potential sources of great pleasure and satisfaction. We would all do well to enjoy them more.

So, cast aside any lingering puritanical concerns you might harbour about unproductive downtime, and focus instead on the sensual pleasures to be harvested from luxuriating in bed, snoozing, napping, sleeping and dreaming. Some warm words of encouragement will follow shortly. But first, a few words about the nay-sayers and sleep-slaggers who have tried, against all reason, to give sleep a bad name.

Puritans and hypocrites

I must confess that I, too, am a sleeper and until quite recently was riddled with guilt because of it.
Fran Lebowitz, *Metropolitan Life* (1974)

The puritans and dull, workaholic sleep-deniers of this world would have us believe that sleep squanders our precious time that should instead be spent in fruitful labour. This misguided perspective has a long and inglorious pedigree. Many annoying aphorisms, proverbs and sayings have been coined with the naked intention of making us feel guilty about sleeping. 'Up, sluggards, and waste no life; in the grave will be sleeping enough,' roared Benjamin Franklin. 'Love not sleep, lest thou come to poverty,' says the Bible. 'A little more Sleep, and a little more Slumber, thus he wastes half his Days, and his Hours without Number,' preached Isaac Watts in his sanctimonious *Divine Songs for Children*.

Someone once said that no human being believes that any other human being has a right to be in bed when he himself is up. The priggish disapproval we are all inclined to feel when awake in the presence of a sleeper is laid bare in Jerome K. Jerome's *Three Men in a Boat*:

> I don't know why it should be, I am sure, but the sight of another man asleep in bed when I am up maddens me. It seems to me so shocking to see the precious hours of a man's life – the priceless moments that will never come back to him again – being wasted in mere brutish sleep.
>
> There was George, throwing away in hideous sloth the inestimable gift of time; his valuable life, every second of which he would have to account for hereafter, passing away from him, unused. He might have been up stuffing himself with eggs and bacon, irritating the dog, or flirting with the slavey, instead of sprawling there, sunk in soul-clogging oblivion.
>
> It was a terrible thought. Harris and I appeared to be struck by it at the same instant. We determined to save him, and, in this noble resolve, our own dispute was forgotten. We flew across and slung the clothes off him, and Harris landed him one with a slipper, and I shouted in his ear, and he awoke.

Sleep is the great leveller. Every day it reduces even the most powerful and revered individuals – media celebrities, sporting heroes, prime ministers, presidents, mullahs, archbishops and monarchs – to the same inert and vulnerable state as the humblest and most under-achieving specimens of humanity. Samson had his hair cut off by Delilah while he slept, Jack took advantage of the giant's slumber to steal his golden hen, and Jason absconded with the Golden Fleece while the dragon guarding it slept. Gulliver was taken prisoner by the tiny inhabitants of Lilliput while he slumbered on the grass, and Dracula sucked the blood of his victims as they slept. Odysseus and his men escaped from the cave of the Cyclops by sending the monster to sleep with strong wine, then sneakily plunging a sharpened tree trunk into his solitary eye. In Jonathan Coe's novel *The House of Sleep*, an unhinged sleep-denier and admirer of Margaret Thatcher muses on the disturbingly democratic nature of sleep:

Sleep puts even the strongest people at the mercy of the weakest and most feeble. Can you imagine what it must be like for a woman of Mrs Thatcher's fibre, her moral character, to be obliged to prostrate herself every day in that posture of abject submission? The brain disabled, the muscles inert and flaccid? It must be insupportable.

G. K. Chesterton was a staunch defender of lying in bed for its own sake, and felt it should be done without any justification whatsoever. 'If a healthy man lies in bed,' he wrote, 'let him do it without a rag of excuse.' Nonetheless, Chesterton was conscious of swimming against the moral tide. In 1909 he wrote that:

The tone now commonly taken towards the practice of lying in bed is hypocritical and unhealthy. Instead of being regarded, as it ought to be, as a matter of personal convenience and adjustment, it has come to be regarded by many as if it were a part of essential morals to get up early in the morning.

But here comes the twist. Much of what has been said, written and believed about successful people not wasting their time on sleep turns out to be myth, bunkum or downright lies. Most of the finger-wagging scourges of sleep have actually slept or napped more than they or their admirers would have us believe.

Take, for example, the American inventor Thomas Alva Edison. He was famed for, and frequently boasted about, his capacity to work very long hours with little or no sleep. Edison once made the astonishingly stupid remark that 'there is really no reason why men should go to bed at all'. On another occasion he asserted that 'sleep is an absurdity, a bad habit'. Understandable, perhaps, from the man who invented the electric light bulb.

Edison, however, was the subject of much mythology. He stoked up his own cult of personality by encouraging admiring journalists to visit him at work. The stories they wrote created Edison's reputation for 'ceaseless activity', 'indomitable perseverance' and 60-hour sleep-less marathons. In 1888, for example, Edison and his workers perfected their version of the phonograph after what the newspapers claimed was a 'sleepless, five-day orgy of toil'. In fact, it took them three days. Nonetheless, there was some truth behind the mythology. It was common for Edison and his colleagues at his Menlo Park laboratory to work until at least midnight and sometimes straight through to the next morning. Edison often worked for several days at a time. But he did not do it without sleep.

Edison's secret was taking frequent catnaps on a folding bed near his experimental room. These naps more than compensated for his lack of scheduled sleep. As one of his biographers put it: 'he could sleep soundly on anything, anywhere, anytime'. Edison dreamt up many of his best ideas while drifting in and out of naps on his folding bed – another example of creative hypnagogia. Throughout history, tactical napping has allowed busy people to cope with unconventional sleep routines. It is commonly acknowledged that abstraction and theorising were not Edison's strengths. He worked best by charging headlong into intellectual obstacles and testing his ideas by experimentation. One biographer has argued that in harping on about his capacity to keep working without sleep, Edison might have been compensating for his inability to organise his thoughts, so that 'heroic quantities of effort would overcome deficiencies in the quality of method'.

Samuel Johnson at least had the honesty to admit his own brazen double standards. 'Preserve me from unseasonable and immoderate sleep,' preached Doctor Johnson in his *Prayers and Meditations*. But the truth came out when he spoke to James Boswell:

> People are influenced more by what a man says, if his practice is suitable to it, because they are blockheads . . . I have, all my life long, been lying till noon; yet I tell all young men, and tell them with great sincerity, that nobody who does not rise early will ever do any good.

Napoleon was famed for his ability to get by on very little sleep. But, as is so often the case, the reality was somewhat different. Despite his reputation, Napoleon actually relished sleep and slept for between seven and eight hours whenever circumstances permitted. Moreover, like many people celebrated for not sleeping, Napoleon compensated for temporary restrictions on his sleep, which usually resulted from some war he had started, by napping at every suitable opportunity. He had the knack, fortunate in a leader, of being able to nap at will, even when the guns were thundering around him. He also enjoyed long, relaxing soaks in his bath, where he would remain for an hour or more. Napoleon maintained that a hot bath was worth four hours' sleep. He probably napped there as well.

If Tolstoy's fictional account is authentic, then one of Napoleon's most illustrious military opponents was also a napper. In *War and Peace*, the Russian generals are considering their tactics on the night before their impending battle with Napoleon's troops. During this discussion General Kutúzov, the Commander of the Russian forces, closes his eyes and, to the surprise of his generals, goes to sleep. The generals continue their debate. Kutúzov eventually awakes from his nap and explains his behaviour:

> 'Gentlemen, the dispositions for tomorrow – or rather for today, for it is past midnight – cannot now be altered,' said he. 'You have heard them, and we shall all do our duty. But before a battle there is nothing more important . . .' he paused, 'than to have a good sleep.'

Alexander the Great is said to have slept so soundly on the day appointed for his desperate battle against the Persians that he had to be woken up before the fighting could commence.

Naps, nappers and napping

> Enjoy the honey-heavy dew of slumber.
> William Shakespeare, *Julius Caesar* (1599)

Napping works. It makes you sharper, smarter, happier and more energetic. It is safe, costs nothing and you need no special equipment or drugs to do it. And it feels good. Besides being a valuable skill, napping is one of the great, neglected pleasures of life.

More and more people are rediscovering the secret of napping, even if few of them brag about doing it. A millennium survey, which polled 1.2 million people in 251 countries in the year 2000, found that almost one in four people in the world nap on a regular basis. Many of them do so, of course, as part of the traditional afternoon nap, or siesta, that is still practised in many hot climates.

Winston Churchill is widely but falsely perceived to have been a man who worked ferociously long hours on little sleep. In fact, he was a consummate and enthusiastic napper. Churchill's biographer Walter Graebner relates how, on one occasion, Churchill quizzed an American guest about his work habits. The American replied that he was at his desk every morning by 8 o'clock and left at 5:30, with only a short break in between for lunch. He confessed to being so tired by 10 p.m. that he would fall into bed and be asleep within two minutes. Churchill was horrified, and told the man that this was the most perfect prescription for a short life he had ever heard. Churchill then expressed his firm view on how a life should be lived:

> You must sleep some time between lunch and dinner, and no half-way measures. Take off your clothes and get into bed. That's what I always do. Don't think you will be doing less work because you sleep during the day. That's a foolish notion held by people who have no imagination. You will be able to accomplish more.

Churchill continued that during the war he had been forced to sleep during the day because that was the only way he could cope with all his responsibilities. He had often been obliged to work late into the night, taking crucial decisions that could not wait until the following day. Experience had taught him that sleeping between lunch and dinner enabled him to squeeze at least one and a half days out of each day, if not two.

Pausing to relight his cigar and pour himself another large brandy, Churchill added that there was another good reason why he slept during the day. This was because it enabled him to be at his best in the evening when he dined with his wife, family and friends. Dinner was the most important part of the day for Churchill, and a focus for his personal relationships. The consumption of brandy and cigars after dinner was one of Churchill's greatest pleasures, second only to sleep. By sleeping during the day, he gained the extra energy he needed to sparkle in the evening. After dinner, when his guests and family had dispersed, Churchill would settle down to two or three hours of hard work, though he was always awake by eight.

Churchill took his naps before dinner, sleeping soundly for between one and two hours. His valet would wake him about ten minutes before dinner, giving Churchill just enough time to shave, bathe and dress before joining his family and guests. Even so, he would sometimes nod off in front of guests, out of tiredness rather than boredom. He would wake after these briefest of naps feeling much better.

The Victorian novelist Anthony Trollope was another great napper. When his first biographer once encountered Trollope on a rail journey, he noted that the great man slept for the first half and spent the second half of the journey writing several chapters of a novel. Another of his contemporaries described an encounter with Trollope at a literary dinner in London:

> Anthony Trollope was very much to the fore, contradicting everybody; afterwards saying kind things to everybody, and occasionally going to sleep on sofas or chairs; or leaning against sideboards, and even somnolent while standing erect on the hearthrug. I never knew a man who could take so many spells of 'forty winks' at unexpected moments, and then turn up quite wakeful, alert and pugnacious.

A nap has no precise definition, but it is usually taken to mean an intentional episode of sleep during the day. A nap can last anywhere from a few minutes to a few hours, although anything longer than about four hours is best regarded as plain sleep. Anything shorter than about five minutes is usually called a microsleep.

An afternoon nap, otherwise known as a siesta, is a natural reflection of the human circadian rhythm. A normal, healthy person who sleeps at night and works by day will go through a distinct dip in alertness and physiological rhythms during the afternoon. The desire to doze after lunch is entirely natural and has little to do with eating. A large lunch accompanied by lashings of alcohol may enhance the desire to sleep in the afternoon, but the underlying tendency will be there anyway. The siesta is an expression of human biology.

Naps really are restorative. Even healthy people who do not feel particularly fatigued will experience a lift following a nap. Research has repeatedly demonstrated that daytime naps boost alertness, mood and performance. One experiment, for example, assessed the effects of a 20-minute nap on young adults who were not habitual nappers. The subjects napped at around midday, before the afternoon circadian dip. Compared with control subjects, who rested in a semi-reclining chair but did not sleep, the nappers felt less sleepy in the afternoon and were more alert. Longer naps boost our objective performance even more than our subjective sense of wellbeing. In other words, they make more difference to how we do than how we feel. This is because the boost in alertness and performance that results from a longer nap is usually offset by the grogginess of sleep inertia for the first 15–30 minutes after waking. In an experiment where healthy volunteers took a two-hour nap during a long period of sleep deprivation, their reactions improved significantly after the nap but their subjective sense of tiredness was unchanged.

You can avoid the transient grogginess of sleep inertia by limiting a nap to about 20 minutes or less. With a little practice, almost anyone can learn to fall asleep quickly and wake up 10–20 minutes later. Fortunately, a nap does not have to last very long to be effective. A brief 'power nap' will leave you feeling refreshed and reinvigorated. Experiments have shown that even a nap of ten minutes can still boost alertness and performance. However, the longer a nap lasts, the greater the performance benefits, so 20 minutes will be better still. Do not overdo it though. If you sleep solidly for three hours every afternoon, you are unlikely to enjoy eight hours of unbroken slumber when you go to bed at night. Lengthy daytime napping is not a good idea if you are an insomniac.

You do not have to be particularly fatigued to benefit from a nap. Napping helps even if you have slept normally the night before. But it is even more beneficial if you are tired. Research has demonstrated that a 30-minute nap during the day substantially reduces the impact of sleep deprivation. Scientists at the NASA Ames Research Center in California showed that napping also helps in operational settings where people are often deprived of sleep, such as long-haul aeroplane flights. Suitably scheduled naps boosted the alertness and performance of flight crews during the critical phases of descent and landing. Crews that did not nap during long-haul flights exhibited an alarming number of microsleeps during the last few hours of the flight. The NASA research convinced several airlines to introduce planned rest periods for their long-haul flight crews, reversing the traditional practice of regarding it as a heinous offence for crew to sleep during a flight.

A useful distinction can be drawn between three basic types of nap, based on their purpose. The prophylactic nap is a nap taken ahead of a period of anticipated sleep deprivation, such as a late night or a long journey. The relief nap is a nap taken to make up for lost sleep and alleviate tiredness. And the pleasure nap is taken just because it feels good. Some people routinely use relief napping as a tactic for coping with insufficient sleep. A study of shift workers uncovered a higher proportion of nappers among those who worked at night than among day workers. The nappers spent less time asleep during their main sleep period, but napping lifted their total sleep to roughly the same duration as non-nappers. If you are going to take a relief nap during a long period of sleep deprivation, then the earlier you do it the greater its benefits. When experimental volunteers were allowed to nap at various points during a 56-hour period of sleep deprivation, naps taken during the first 24 hours of the vigil produced bigger and longer-lasting improvements in performance than naps taken near the end.

Naps can also be used prophylactically to help prepare for an impending period of sleep deprivation. Staying up very late will impair your performance and make you feel sleepier the next day, but these symptoms can be partially offset by taking a prophylactic nap beforehand. In one experiment, volunteers who took a four-hour afternoon nap before working all night were much more alert and performed better on complex tasks. A study of night-shift workers found that a short nap in the early hours of the morning improved alertness during the second half of the shift. In another experiment, young men took naps of various lengths, ranging from two to eight hours, before being deprived of sleep for two consecutive nights. The longer their prophylactic nap, the smaller the deterioration in their alertness and performance.

Pleasure naps are heavily constrained by lifestyle: busy, working people seldom find the time to nap just for the sheer enjoyment of it. (But then, having no free time used to be a standard excuse for not taking physical exercise.) Older people nap more than young people, on average. This may be partly because older people are more prone to insomnia, making them sleepier during the day. But their extra napping may also reflect a lifestyle which allows them more free time to nap. Younger people do nap when they have the opportunity – for example, on long flights and in boring lectures.

Will you dream during a nap? The answer depends on when you nap and for how long. REM sleep, which is when most ordinary dreams occur, comes at the end of the sleep cycle, so you are unlikely to experience REM sleep if you only nap for 10–15 minutes. A nap of an hour or more is a different matter, however. Your propensity to enter REM sleep also varies according to a circadian rhythm, so your likelihood of dreaming will depend to some extent on the time of day. A study of young medical students who habitually napped in the afternoons found that some of them usually dreamed during their naps whereas others rarely if ever dreamed. The likelihood of dreaming depended partly on when the nap occurred in relation to their circadian sleep–wake cycle. If you want to dream during your naps, try napping at different times of the day. Other things being equal, napping in the early morning is more likely to include REM sleep, whereas a late afternoon nap is more likely to comprise NREM sleep.

Napping is pleasurable, it makes us feel better and it enhances our objective performance at work. If napping were a commercial product, its manufacturer would not need a big advertising budget: it would just walk out of the shops by itself. Sadly, though, the prevailing culture in northern Europe and the USA remains essentially anti-napping. There is a deep-seated prejudice in most industrialised societies that anyone (apart from shift workers) who sleeps during the day must be unemployed, sick, drunk or idle. Even in countries such as Spain, where the siesta once reigned supreme, the tradition of afternoon napping is dying out, as more people commute long distances to work and globalisation irons out the national differences in work schedules. In the great rat race of life, there is no longer enough time to nap every afternoon.

There are some glimmers of hope, however. A few enlightened organisations have discovered that their employees are happier and work better when they have had some sleep. A few large American corporations, including one that is better known for hard-edged business methods than cuddliness, have introduced nap rooms for their employees. Far more important than the physical facilities themselves is the signal they send – that napping during the working day is now acceptable. Many large employers nowadays provide their staff with gyms and fitness centres where they can exercise their bodies, so why not somewhere to nap? Spending 20 minutes a day in a nap room is just as worthy and rational as spending 20 minutes in a gym. You should do both if you really want that sound mind to go with that sound body.

The healthy trend towards the decriminalisation of napping has its epicentre in the USA. However, the predominant workplace culture in the USA, UK and elsewhere remains one that views napping as a sign of drunkenness, weakness, illness or sloth. This anti-napping prejudice is reflected in our language, which still talks of people being 'caught napping' and imbues terms such as 'snooze' and 'forty winks' with distinct overtones of furtiveness. More employers should wake up to reality and follow the American example, if only because of the measurable benefits it would bring them in improved performance and morale. And why not do the same in schools? Napping makes sound practical sense, as one enlightened writer on health matters pointed out in the early nineteenth century:

> Is it not better economy of time to go to sleep for half an hour, than to go on noodling all day in a nerveless and semi-superannuated state – if not asleep, certainly not effectively awake?

The seeds of napping culture have started to drift into northern Europe, even if they have yet to take root in the UK. In Berlin, for example, one enlightened gentleman has opened a *dormitorium*, where members of the public can nap in a quiet, airy room on payment of a small fee. The would-be napper is guided to his or her bed by uniformed *maîtres morpheés* (masters of sleep) while a quiet recording of the distant sounds of playing children helps to induce slumber. In the UK, that bastion of the Protestant work ethic, a few pioneering members of the sleep-deprived chattering classes have been experimenting with an adult version of the slumber party. Guests assemble for Sunday lunch and then, having consumed their meal and done their chattering, they disperse around the house to beds, mattresses and armchairs to indulge in an hour or two of serious (solitary) napping. Most participants find that they adore the experience, once they are granted social permission to nap instead of pretending to enjoy supposedly worthy activities such as walking in the rain.

Sweet dreams

> It would be hard to conceive of anything in normal conscious life that could equal the freedom, the abandon, the sheer bliss I experienced in that dream.
>
> Arthur Schnitzler, *Dream Story* (1926)

One of the greatest of life's pleasures, and also one of the most neglected, is dreaming. Delights galore await those who explore the dream state. The proposition was elegantly put by Gérard de Nerval:

Sleep occupies a third of our life ... After several minutes of drowsiness, a new life begins, freed from the bounds of time and space.

Some fortunate individuals find that the remembrance and enjoyment of dreams come easily. They are usually enthusiasts. Simone de Beauvoir, for example, relished her dreams:

I anticipate my nocturnal adventures with pleasure as I go to sleep; and it is with regret that I say good-bye to them in the morning ... They are one of the pleasures I like best. I love their total unexpectedness and above all their gratuity ... That is why in the morning I often try to bring them together again, to reform them from the shreds that float behind my eyelids, glittering but still fading fast.

The film-maker Luis Buñuel also wrote about his passion for dreaming:

If someone were to tell me I had twenty years left, and ask me how I'd like to spend them, I'd reply: 'Give me two hours a day of activity, and I'll take the other twenty-two in dreams ... provided I can remember them.' I love dreams, even when they're nightmares, which is usually the case.

The writer Karen Blixen (alias Isak Dinesen) was another dream enthusiast:

People who dream when they sleep at night know of a special kind of happiness which the world of the day holds not, a placid ecstasy, and ease of heart, that are like honey on the tongue.

Well, it may have been lovely for Simone de Beauvoir, Karen Blixen and other dream aficionados, but most people seldom retain more than the dimmest recollection of their dreaming. Consequently they have little scope for savouring its many pleasures. They do dream, of course, every night. But the vast majority of dreams are not remembered, and without memory they might as well not exist. Individuals differ enormously in the ease and frequency with which they remember their dreams on waking. Some remember none, while others can recall dreams in considerable detail most days. A typical Western adult – who is chronically sleep-deprived, wakes abruptly to the clatter of an alarm clock and has various anxieties weighing on his or her mind – may be lucky to remember a fragment of a dream once every few days.

Fortunately, the capacity to remember dreams can be developed fairly easily, as we saw in chapter 10. Some simple, practical measures can help you to recall more of your dreams and milk their pleasures. The keys are desire and practice. The starting point is deciding you want to remember your dreams. Simply thinking about the subject of dreaming also helps. Volunteers who take part in scientific studies of dreaming often find that their general ability to remember their dreams improves as a consequence. When you awake in the mornings, lie still with your eyes closed and let the images flow into your mind. Try not to start fretting immediately about the day ahead: it can wait another five minutes. You are much more likely to remember dreaming if you wake from REM sleep. And you are more likely to wake from REM sleep if you wake naturally after a full night's sleep. Chronic sleep deprivation is one of the largest obstacles to recalling dreams. If you are still having no success, try resetting your alarm clock to half an hour earlier or later than usual; with luck, this might wake you during an episode of REM sleep. You should avoid using a radio-alarm, however, because the sounds from the radio may be incorporated into your thoughts as you wake up, expelling the remaining fragments of dream memories.

Once you do start to recall more of your dreams, the way is then open to extract maximum pleasure. We looked at some basic techniques for facilitating lucid dreaming in chapter 10. Many other methods have been devised for enhancing ordinary dreams.

One such method is known as the Senoi strategy. Though exotically named, this technique is at heart both simple and straightforward. The Senoi strategy is named after the Senoi people, a semi-nomadic tribe living in the highland jungles of Malaysia. In the 1930s a British anthropologist named Herbert Noone observed that the Senoi people placed great weight on their dreams. Each member of a household would describe their dreams each morning, and everyone was expected to remember them. An American anthropologist and psychoanalyst named Kilton Stewart later wrote about how the Senoi used special techniques for controlling their dreams. In essence, these techniques involve creating a positive expectation that you will remember your dreams and that they will be pleasant. In more recent years, doubts have been raised about whether the Senoi people actually practise the dream-control methods ascribed to them by Kilton Stewart. Whatever the case, the basic strategy of consciously adopting a positive attitude towards your dreams does seem to work. Research has confirmed that dreams do become more pleasurable after adopting this approach, with effects that are still detectable six months later.

If you wake tomorrow morning having enjoyed an hour or two of pleasant dreams, you can at least return to your other life without feeling you have wasted time that might have been better spent on more passive forms of entertainment.

Blessed oblivion

> What do we here
> In this Land of unbelief & fear?
> The Land of Dreams is better far
> William Blake, *The Land of Dreams* (c. 1803)

Sleep is always there to offer refuge in the bad times as well as the good. Throughout history, people have found relief from sorrows and tribulation by escaping into sleep. The eighteenth-century French playwright Nicolas Chamfort wrote that 'living is an illness to which sleep provides relief every sixteen hours'. In Zoroastrian creation mythology, Gayomart, the progenitor of humanity, was given the capacity to sleep so that he could have some respite from the repeated onslaughts of the evil spirit. And in Shakespeare's *A Midsummer Night's Dream*, the miserable Helena lies down, hoping that sleep will blot out her unhappiness:

> sleep, that sometimes shuts up sorrow's eye,
> Steal me a while from mine own company.

Soldiers struggling through the horrors of the First World War cherished the oblivion of sleep. Wilfred Owen's poems about the war were composed between January 1917, when he was sent to the Western Front, and November 1918, when he was killed. In 'Strange Meeting' Owen describes soldiers, fatigued beyond reason after long spells in the trenches, turning to sleep as their only way of blanking out the living nightmares that surrounded them:

> It seemed that out of battle I escaped
> Down some profound dull tunnel

Sleep can offer an escape from less horrendous pressures as well, including the boredom and frustrations of ordinary life. The novelist Iris Murdoch put it like this:

> The division of one day from the next must be one of the most profound peculiarities of life on this planet. It is, on the whole, a merciful arrangement. We are not condemned to sustained flights of being, but are constantly refreshed by little holidays from ourselves.

In Victorian England it was conventional for ladies to retire for a 'rest' in the afternoon. This allowed them to withdraw from their role of ministering angel, in which they were constantly subject to the needs and wishes of others.

Give sleep a chance

> The sweetest sleep, and fairest-boding dreams
> That ever entered in a drowsy head.
> William Shakespeare, *Richard III* (1591)

This is not a self-help book, and it is not for me to tell you how best to sleep. Individuals differ considerably in their sleep patterns, requirements and problems, so there is no stock advice that will suit everyone. Anyone who suspects that they might have a sleep disorder should certainly consult a qualified physician rather than rely on a book. That said, the scientific evidence outlined in this book does lead naturally to some simple, practical conclusions. And if one were to compile a list of those simple, practical conclusions, it would look something like this:

- Give sleep the high priority that it deserves in your life (which is probably a higher priority than it currently receives).

- Listen to your body: if you feel tired then you probably *are* tired and need more sleep.
- Try using a sleep diary to find out how much sleep you are actually getting (which will probably turn out to be less than you think). Be honest with yourself.
- If you have a suitable opportunity, use the method described in chapter 2 to measure your own preferred sleep duration (which will probably turn out to be longer than your current average).
- Pay off your accumulated sleep deficit by going to bed half an hour earlier for a few weeks. Half an hour isn't much.
- A regular sleep routine helps, so try to go to bed and get up at about the same time every day, even at weekends. (You should find it easier to do without your weekend lie-in once you have paid off your existing sleep deficit.)
- Learn to nap efficiently and without feeling guilty. (But don't overdo the daytime napping, especially if you find it hard to sleep at night.)
- Your bed is arguably the most important item of furniture in your home. You will spend longer there than anywhere else, and its comfort will affect how you feel throughout the day. Invest in a good bed with a comfortable mattress and decent pillows.
- Make sure your bedroom is at a sensible temperature, and especially not too hot. If in doubt, go for cooler.
- A gently declining body temperature helps to trigger the onset of sleep, so take a bath an hour or two before you go to bed.
- Your bedroom should be as peaceful as you can make it. If there is too much noise outside, try earplugs or use a detuned FM radio to generate masking white noise.
- Avoid excessive caffeine consumption. Aim to drink no more than three cups of coffee or tea a day and none after mid-afternoon. And watch out for those other sources of caffeine such as soft drinks.
- Avoid drinking large amounts of alcohol in the evenings, because it will disrupt your sleep.
- Do not smoke in the evening, because nicotine is a stimulant.
- Aim to avoid eating large meals just before you go to bed.

- Exercise as much as you like during the day, but avoid strenuous activity (other than sex) just before you go to bed, unless you are physically fit and used to it.
- Try to avoid doing anything mentally stimulating or anxiety-provoking just before bedtime, especially if you are prone to insomnia.
- Develop a relaxing pre-bed ritual. The aim should be unconsciously to associate bedtime with the sensations of relaxing and drifting off asleep. (Think Pavlov's dogs, but substituting sleep for salivation.)
- Ideally, your bedroom should be associated with only two activities: sleep and sex. If you can avoid it, do not use your bedroom as an office or for leisure activities such as watching TV.
- Start deriving more pleasure from your sleep by learning to remember your dreams. See if you can experience lucid dreaming.
- If you snore, try losing weight, avoid alcohol at night and sleep on your side. (But if you suspect you might have sleep apnoea, go and see a doctor.)
- Finally, remember Fran Lebowitz's dictum that sleep is both pleasant and safe to use.

Basic advice in a similar vein can be found in many sensible self-help books. None of it, of course, will make a scrap of difference to people who think sleep is unimportant.

In praise of horizontalism

> Sleep in the name of Morpheus your belly full.
> Thomas Dekker, *The Guls Horne Book* (1609)

We are the great unslept. Human sleep patterns have changed radically since the advent of industrialisation, electric lighting, electronic entertainment, shift work, the 24/7 society and other massive intrusions into the once dark hours. The blurring of the boundary between night and day marks a revolution in our environment. We are left with the depressing vision of a caffeine-assisted world in which people want to do little else besides work, commute, shop and be entertained. Sleep is a biological imperative that is now widely undervalued. Those who stumble through life on only five or six hours of poor-quality sleep a night are admired for their stamina, in the way that people were once admired for their capacity to drive cars while drunk. When it comes to sleep, we are still doing the equivalent of encouraging boozed-up drivers to have one more for the road.

Perhaps it is not too naively optimistic to believe that current social attitudes to sleep are so wildly out of kilter with the biological reality that they must be ripe for a change. If so, then future generations may look back on this present era of self-imposed sleep deprivation with the same sort of condescending amusement that we feel at our forebears' foibles. Perhaps regarding sleep as unimportant will one day seem as ludicrous as, say, believing that women are all feeble creatures or that smoking is good for you. Sleep is not a universal panacea. If you are disgruntled and unhappy with your life, better sleep will not magically transform you into a happy, smiley person. But it will certainly help.

Sleep deserves more attention in science, medicine, education, social policy and, most of all, in our everyday lives. A good night's sleep makes you feel more alive and helps you perform better at whatever you choose to do. It heightens the senses, sharpens the mind and mellows the spirit. Tasks no longer seem like too much trouble if you are not tired all the time. Sleeping, napping and dreaming are also untapped sources of enjoyment. We should give them a higher priority in our lives and learn to revel in their neglected pleasures.

Acknowledgements

I am very grateful to these kind people for their help: Gillon Aitken, Patrick Bateson, Rupert de Borchgrave, Malcolm Burrows, Bryn Caless, Nic Coombs, Yvonne Harrison, Stephen Hayes, Philip Gwyn Jones, Randal Keynes, Georgina Laycock, Harriet Martin, Frank McCourt, Phil Murphy, Geoffrey Prentice, Alison Reid, John Sants, David Thompson, Patsy Wilkinson, the staff of the University Library, Cambridge, the President and Fellows of Wolfson College, Cambridge, and the many hundreds of scientists whose work I have described. As always, the mistakes are mine.

The author and publisher are grateful to the following for permission to reproduce copyrighted material: The Minnesota Historical Society for *The Spirit of St Louis* by Charles A. Lindbergh (1953; reprint, St Paul, Minnesota, Minnesota Historical Society Press, 1993); Extract of 'Stopping By Woods on a Snowy Evening' from *The Poetry of Robert Frost*, edited by Edward Connery Lathem, the Estate of Robert Frost and Jonathan Cape as publisher, used by permission of The Random House Group Limited; David Higham Associates Ltd for *Under Milk Wood* by Dylan Thomas, published by J. M. Dent; Penguin Books Ltd for *The Canterbury Tales* by Geoffrey Chaucer, translated by Nevill Coghill (Penguin Classics 1951, fourth revised edition 1977), © 1951 by Nevill Coghill, © Nevill Coghill, 1958, 1960, 1975, 1977; Penguin Books Ltd for *The Physiology of Taste* by Jean-Anthelme Brillat-Savarin, translated by Anne Drayton (Penguin Classics 1994, first published by Penguin Books, 1970, as *The Philosopher in the Kitchen*), © Anne Drayton, 1970; and Penguin Books Ltd for *Selected Writings* by Gérard de Nerval, translated by Richard Sieburth (Penguin Classics, 1999), translation © Richard Sieburth, 1999.

References

References are listed here in approximately the same order as they are cited in the text. I have taken the liberty of shortening the titles of some journal articles. Despite its length, the list is not exhaustive.

General Reading

Ancoli-Israel, Sonia. *All I Want Is a Good Night's Sleep*. (St Louis, Mosby, 1996)

Caldwell, J. Paul. *Sleep*. (Willowdale, Firefly, 1997)

Coren, Stanley. *Sleep Thieves*. (NY, Free Press, 1996)

Dement, William C. *Some Must Watch While Some Must Sleep*. (San Francisco, W. H. Freeman, 1972)

Dement, William C. & Vaughan, Christopher. *The Promise of Sleep*. (NY, Delacorte, 1999)

Empson, Jacob. *Sleep and Dreaming*. (London, Faber & Faber, 1989)

Flanagan, Owen. *Dreaming Souls*. (Oxford, Oxford University Press, 2000)

Harrison, Yvonne. *Sleep Talking*. (London, Blandford, 1999)

Hobson, J. Allan. *Sleep*. (NY, Scientific American, 1995)

Hobson, J. Allan. *The Dreaming Brain*. (NY, Basic Books, 1988)

Horne, James. *Why We Sleep*. (Oxford, Oxford University Press, 1988)

Jouvet, Michel. *The Paradox of Sleep*. (Cambridge MA, MIT Press, 1999)

Lavie, Peretz. *The Enchanted World of Sleep*. (New Haven CT, Yale University Press, 1996)

Moor-Ede, Martin. *The 24 Hour Society*. (London, Piatkus, 1993)

General 1 – A Third of Life

Oswald I. Falling asleep open-eyed during intense rhythmic stimulation. *BMJ*, 1, 1450 (1960)

Johnson, Samuel. *The Idler*, No. 32, 25 November 1758

A sleep-sick society?

Dement, William C. & Vaughan, Christopher. *The Promise of Sleep*. (NY, Delacorte, 1999) p. 2

Rosekind M. R. et al. Alertness management in 24/7 settings. *Occup. Med.*, 17, 247 (2002)

Stores G. & Crawford C. Medical student education in sleep and its disorders. *J. R. Coll. Physicians Lond.*, 32, 149 (1998)

The universal imperative

Dickens, Charles. 'Lying Awake.' *Selected Journalism 1850–1879*. (London, Penguin, 1997)

McFarland, David. *Problems of Animal Behaviour*. (Harlow, Longman, 1989) p. 87

Tauber E. S. et al. Eye movements and EEG during sleep in diurnal lizards. *Nature*, 212, 1612 (1966)

Coren, Stanley. *Sleep Thieves*. (NY, Free Press, 1996) p. 159

Tauber E. S. et al. Behavioural inactivity in Bermuda reef fish. *Commun. Behav. Biol. A*, 3, 131 (1969)

Meddis R. On the function of sleep, *Anim. Behav.*, 23, 676 (1975)

Shaw P. J. et al. Correlates of sleep and waking in Drosophila melanogaster. *Science*, 287, 1834 (2000)

Greenspan R. J. et al. Sleep and the fruit fly. *Trends Neurosci.*, 24, 142 (2001)

Ayala-Guerrero F. et al. Sleep patterns of the volcano mouse. *Physiol. Behav.*, 64, 577 (1998)

Tobler I. & Deboer T. Sleep in the blind mole rat. *Sleep*, 24, 147 (2001)

De Moura Filho A. G. et al. Sleep in the three-toed sloth. *Comp. Biochem. Physiol. A*, 76, 345 (1983)

Ayala-Guerrero F. & Huitron-Resendiz S. Sleep in the lizard Ctenosaura pectinata. *Physiol. Behav.*, 49, 1305 (1991)

Meddis, Ray. *The Sleep Instinct*. (London, Routledge & Kegan Paul, 1977)

Szymczak J. T. Sleep pattern in the starling. *Acta Physiol. Pol.*, 36, 323 (1985)

Allan, Mea. *Darwin and his Flowers*. (London, Faber & Faber, 1977) p. 285

Half asleep

Mukhametov L. M. et al. Electrocorticographic sleep stages in bottle-nosed dolphins. *Neirofiziologiia*, 20, 532 (1988)

Mukhametov L. M. Unihemispheric sleep in the Amazonian dolphin. *Neurosci. Lett.*, 79, 128 (1987)

Lyamin O. I. et al. Unihemispheric slow wave sleep in a white whale. *Behav. Brain Res.*, 129, 125 (2002)

Rattenborg N. C. et al. Perspectives on unihemispheric sleep. *Neurosci. Biobehav. Rev.*, 24, 817 (2000)

Mascetti G. G. & Vallortigara G. Why do birds sleep with one eye open? *Curr. Biol.*, 11, 971 (2001)

Rattenborg N. C. et al. Facultative control of avian unihemispheric sleep under the risk of predation. *Behav. Brain Res.*, 105, 163 (1999)

Runyon, Damon. 'Bed-Warmers.' *From First to Last.* (London, Penguin, 1990)

Chapter 2 – Sleepy People

Are we sleep-deprived?

Ohayon M. M. et al. How sleep and mental disorders are related to complaints of daytime sleepiness. *Arch. Intern. Med.*, 157, 2645 (1997)

Martikainen K. et al. Natural evolution of sleepiness. *Eur. J. Neurol.*, 5, 355 (1998)

Haraldsson P. O. et al. Sleep apnea and automobile driving. *J. Clin. Epidemiol.*, 45, 821 (1992)

Zielinski J. et al. Excessive daytime somnolence in Warsaw. *Pol. Arch. Med. Wewn.*, 99, 407 (1998)

Johns M. & Hocking B. Daytime sleepiness of Australian workers. *Sleep*, 20, 844 (1997)

Billiard M. et al. Excessive daytime somnolence in young men. *Sleep*, 10, 297 (1987)

Roth T. & Roehrs T. A. Excessive daytime sleepiness. *Clin. Ther.*, 18, 562 (1996)

National Commission on Sleep Disorders Research. *Wake up America: A National Sleep Alert.* (Washington DC, Dept of Health & Human Services, 1993)

Hossain J. L. & Shapiro C. M. The prevalence, cost implications, and management of sleep disorders. *Sleep Breath.*, 6, 85 (2002)

Bliwise D. L. Historical change in the report of daytime fatigue. *Sleep*, 19, 462 (1996)

Hicks R. A. et al. The changing sleep habits of university students. *Percept. Mot. Skills*, 93, 648 (2001)

Wessely S. Chronic fatigue. *Ann. Intern. Med.*, 134, 838 (2001)

Le Bon O. et al. Primary sleep disorders in chronic fatigue syndrome. *Sleep Res. Online*, 3, 43 (2000)

Martin, Paul. *The Sickening Mind.* (London, Flamingo, 1998) p. 19

Bonnet M. H. & Arand D. L. We are chronically sleep-deprived. *Sleep*, 18, 908 (1995)

Breslau N. et al. Daytime sleepiness. *Am. J. Public Health*, 87, 1649 (1997)

Jean-Louis G. et al. Sleep duration, illumination, and activity patterns. *Biol. Psychiatry*, 47, 921 (2000)

Zielinski J. et al. Excessive daytime somnolence in Warsaw. *Pol. Arch. Med. Wewn.*, 99, 407 (1998)

Ban D. J. & Lee T. J. Sleep duration among students in Korea. *Korean J. Med. Sci.*, 16, 475 (2001)

Ohida T. et al. Sleep loss among the Japanese general population. *Sleep*, 24, 333 (2001)

Takemura T. et al. Sleep habits of students in Akita. *Psychiatry Clin. Neurosci.*, 56, 241 (2002)

Hublin C. et al. Insufficient sleep. *Sleep*, 24, 392 (2001)

Bonnet M. H. & Arand D. L. We are chronically sleep-deprived. *Sleep*, 18, 908 (1995)

Harrison Y. & Horne J. A. Should we be taking more sleep? *Sleep*, 18, 901 (1995)

Harrison Y. & Horne J. A. Are we really chronically sleep-deprived? *Psychophysiology*, 33, 22 (1996)

Are *you* sleep-deprived?
Dement, William C. & Vaughan, Christopher. *The Promise of Sleep.* (NY, Delacorte, 1999) p. 341

Reasons for not sleeping
Dement, William C. & Vaughan, Christopher. *The Promise of Sleep.* (NY, Delacorte, 1999) p. 218

de Grazia, Sebastian. *Of Time, Work, and Leisure.* (NY, Twentieth Century Fund, 1962) p. 341

Ancient and modern
Kealey, Terence. Slaves to the status. *New Scientist*, 10 July 1999

Wehr T. A. In short photoperiods, human sleep is biphasic. *J. Sleep Res.*, 1, 103 (1992)

Barbato G. et al. Extended sleep in humans in 14 hour nights. *Electroencephalogr. Clin. Neurophysiol.*, 90, 291 (1994)

Akerstedt T. et al. Fatal occupational accidents – relationship to sleeping difficulties. *J. Sleep Res.*, 11, 69 (2002)

Melamed S. & Oksenberg A. Daytime sleepiness and occupational injuries in non-shift daytime workers. *Sleep*, 25, 315 (2002)

Sleepy drivers
Maycock G. Sleepiness and driving. *J. Sleep Res.*, 5, 229 (1996)

Connor J. et al. Driver sleepiness and risk of serious injury. *BMJ*, 324, 1125 (2002)

Horne J. & Reyner L. Vehicle accidents related to sleep. *Occup. Environ. Med.*, 56, 289 (1999)

Martikainen K. et al. Daytime sleepiness. *Acta Neurol. Scand.*, 86, 337 (1992)

Ohayon M. M. et al. Sleep, mental disorders and daytime sleepiness. *Arch. Intern. Med.*, 157, 2645 (1997)

Laube I. et al. Accidents related to sleepiness. *Schweiz. Med. Wochenschr.*, 128, 1487 (1998)

Garbarino S. et al. Sleepiness in highway vehicle accidents. *Sleep*, 24, 203 (2001)

Horne J. A. & Reyner L. A. Sleep related vehicle accidents. *BMJ*, 310, 565 (1995)

Akerstedt T. & Kecklund G. Age, gender and early morning highway accidents. *J. Sleep Res.*, 10, 105 (2001)

REFERENCES

Torsvall L. & Akerstedt T. EEG changes in train drivers. *Electroencephalogr. Clin. Neurophysiol.*, 66, 502 (1987)

Philip T. et al. Self-induced sleep deprivation among automobile drivers. *Sleep*, 22, 475 (1999)

Mitler M. M. et al. The sleep of long-haul truck drivers. *N. Engl. J. Med.*, 337, 755 (1997)

Hicks G. J. et al. Fatal alcohol-related traffic crashes increase subsequent to changes to and from daylight savings time. *Percept. Mot. Skills*, 86, 879 (1998)

Varughese J. & Allen R. P. Fatal accidents following changes in daylight-saving time. *Sleep Med.*, 2, 31 (2001)

Rosenthal L. et al. The detection of brief daytime sleep episodes. *Sleep*, 22, 211 (1999)

Reyner L. A. & Horne J. A. Falling asleep whilst driving. *Int. J. Legal Med.*, 111, 120 (1998)

Pack A. I. et al. Characteristics of crashes attributed to the driver having fallen asleep. *Accid. Anal. Prev.*, 27, 769 (1995)

Sleepy pilots

Caldwell J. A. & Gilreath S. R. Aircrew fatigue in US Army aviation personnel. *Aviat. Space Environ. Med.*, 73, 472 (2002)

Wright N. & McGown A. Vigilance on the civil flight deck. *Ergonomics*, 44, 82 (2001)

Gander P. H. et al. Flight crew fatigue. *Aviat. Space Environ. Med.*, 69, B37 (1998)

Cho K. Chronic 'jet lag' produces temporal lobe atrophy and spatial cognitive deficits. *Nat. Neurosci.*, 4, 567 (2001)

Kecklund G. et al. Effects of early rising on sleep and alertness. *Sleep*, 20, 215 (1997)

Caldwell J. A. Fatigue in the aviation environment. *Aviat. Space Environ. Med.*, 68, 932 (1997)

Gundel A. et al. Sleep and circadian rhythm during a short space mission. *Clin. Investig.*, 71, 718 (1993)

Putcha L. et al. Pharmaceutical use by US astronauts on space shuttle missions. *Aviat. Space Environ. Med.*, 70, 705 (1999)

Czeisler, C. Reported in *New Scientist*, 24 April 1999

Sleepy doctors

Owens J. A. Sleep loss and fatigue in medical training. *Curr. Opin. Pulm. Med.*, 7, 411 (2001)

Lewis K. E. et al. Sleep deprivation and junior doctors' performance and confidence. *Postgrad. Med. J.*, 78, 85 (2002)

Millner P. A. Tired surgical trainees. *BMJ*, 324, 1154 (2002)

Patton D. V. et al. Sleep deprivation among resident physicians. *J. Health Law*, 34, 377 (2001)

Richardson G. S. et al. Sleep and alertness in medical house staff. *Sleep*, 19, 718 (1996)

Akerstedt T. et al. Physicians during and following night call duty. *Electroencephalogr. Clin. Neurophysiol.*, 76, 193 (1990)

Hawkins M. R. et al. Sleep and performance of house officers. *J. Med. Educ.*, 60, 530 (1985)

Defoe D. M. et al. Work schedules of residents in obstetrics and gynecology. *Obstet. Gynecol.*, 97, 1015 (2001)

Smith-Coggins R. et al. Rotating shiftwork schedules. *Acad. Emerg. Med.*, 4, 951 (1997)

Smith-Coggins R. et al. Relationship of day versus night sleep to physician performance and mood. *Ann. Emerg. Med.*, 24, 928 (1994)

Samkoff J. S. & Jacques C. H. Effects of sleep deprivation on residents' performance. *Acad. Med.*, 66, 687 (1991)

Leung L. & Becker C. E. Sleep deprivation and house staff performance. *J. Occup. Med.*, 34, 1153 (1992)

Deary I. J. & Tait R. Effects of sleep disruption on cognitive performance and mood in medical house officers. *Br. Med. J. (Clin. Res. Ed.)*, 295, 1513 (1987)

Nelson C. S. et al. Residents' performance before and after night call. *J. Am. Osteopath. Assoc.*, 95, 600 (1995)

Marcus C. L. & Loughlin G. M. Effect of sleep deprivation on driving safety in house staff. *Sleep*, 19, 763 (1996)

The madness of politicians

Foster, Christopher D. & Plowden, Francis J. *The State Under Stress.* (Buckingham, Open University Press, 1996) p. 220

Hennessy, Peter. *Cabinet.* (Oxford, Blackwell, 1986) p. 184

Cradock, Percy. *In Pursuit of British Interests.* (London, John Murray, 1997) p. 34

Clark, Alan. *Diaries.* (London, Weidenfeld & Nicolson, 1993) p. 72

Howe, Geoffrey. *Conflict of Loyalty.* (London, Pan, 1995) p. 568

Coles, John. *Making Foreign Policy.* (London, John Murray, 2000) p. 200

Truly, madly, sleepily

Leger D. The cost of sleep-related accidents. *Sleep*, 17, 84 (1994)

Dinges D. F. An overview of sleepiness and accidents. *J. Sleep Res.*, 4(S2), 4 (1995)

Rajaratnam S. M. W. & Arendt J. Health in a 24-h society. *Lancet*, 358, 999 (2001)

Mitler M. M. et al. Catastrophes, sleep, and public policy. *Sleep*, 11, 100 (1988)

Harrison, Yvonne. *Sleep Talking.* (London, Blandford, 1999)

Presidential Commission. *Report of the Presidential Commission on the Space Shuttle Challenger Accident.* Vol II, Appendix G. (Washington DC, US Government Printing Office, 1986)

Dalbokova D. et al. Stress states in nuclear operators under conditions of shift work. *Work Stress*, 9, 305 (1995)

Horne J. Sleep. *Psychologist*, 14, 302 (2001)

Akerstedt T. Work hours, sleepiness and the underlying mechanisms. *J. Sleep Res.*, 4S2, 15 (1995)

Chapter 3 – Dead Tired

Sleepiness
Devoto A. et al. Effects of different sleep reductions on daytime sleepiness. *Sleep*, 22, 336 (1999)
Oswald, Ian. *Sleep.* 3rd edn. (Harmondsworth, Penguin, 1974) p. 52
Harrison, Yvonne. *Sleep Talking.* (London, Blandford, 1999) p. 48
Harrison Y. & Horne J. A. 'High sleepability without sleepiness.' *Neurophysiol. Clin.*, 26, 15 (1996)
Dement, William C. & Vaughan, Christopher. *The Promise of Sleep.* (NY, Delacorte, 1999) p. 335
Mitler M. M. & Miller J. C. Methods of testing for sleepiness. *Behav. Med.*, 21, 171 (1996)
Mikulincer M. et al. The effects of 72 hours of sleep loss. *Br. J. Psychol.*, 80, 145 (1989)
Bonnet M. H. & Arand D. L. Sleepiness as a function of preceding activity. *Sleep*, 21, 477 (1998)

Fighting the beast
Lindbergh, Charles A. *The Spirit of St Louis.* (St Paul, Minnesota, Minnesota Historical Society Press, 1993; originally publ. 1953)
Forester, C. S. *Flying Colours.* (London, Michael Joseph, 1996)
Hibbert, Christopher. *Nelson.* (London, Viking, 1994) p. 330

A soil for peevishness
Reynolds C. F. et al. Sleep deprivation in healthy elderly men and women. *Sleep*, 9, 492 (1986)
Harrison Y. & Horne J. A. Impact of sleep deprivation on decision making. *J. Exp. Psychol. Appl.*, 6, 236 (2000)
Andersen M. L. & Tufik S. Effects of paradoxical sleep deprivation and cocaine on sexual behaviour in male rats. *Addict. Biol.*, 7, 251 (2002)
Ferraz M. R. et al. How REM sleep deprivation affects male rat sexual behaviour. *Pharmacol. Biochem. Behav.*, 69, 325 (2001)
Wright J. B. Mania following sleep deprivation. *Br. J. Psychiatry*, 163, 679 (1993)
Pilcher J. J. & Huffcutt A. I. Effects of sleep deprivation on performance. *Sleep*, 19, 318 (1996)
Seabrook, John. 'Sleeping with the Baby', *The New Yorker*, 8 November 1999
Totterdell P. et al. Sleep, everyday mood, minor symptoms and social interaction. *Sleep*, 17, 466 (1994)

Tired people are stupid and reckless

Glenville M. et al. Effects of sleep deprivation on short duration performance. *Sleep*, 1, 169 (1978)

Bonnett M. H. & Rosa R. R. Sleep and performance during acute sleep loss. *Biol. Psychol.*, 25, 153 (1987)

Carskadon M. A. et al. Sleep loss in young adolescents. *Sleep*, 4, 299 (1981)

Brendel D. H. et al. Responses to total sleep deprivation in healthy 80-year-olds and 20-year-olds. *Psychophysiology*, 27, 677 (1990)

Dinges D. F. et al. Sleepiness, mood disturbance, and performance decrements during a week of restricted sleep. *Sleep*, 20, 267 (1997)

Opstad P. K. et al. Performance, mood, and clinical symptoms in men exposed to prolonged, severe physical work and sleep deprivation. *Aviat. Space Environ. Med.*, 49, 1065 (1978)

Williams H. L. et al. Impaired performance with acute sleep loss. *Psychol. Mongr.*, 73, No. 14 (1959)

Doran S. M. et al. Sustained attention performance during sleep deprivation. *Arch. Ital. Biol.*, 139, 253 (2001)

Harrison Y. & Horne J. A. The impact of sleep deprivation on decision making. *J. Exp. Psychol. Appl.*, 6, 236 (2000)

Horne J. A. Sleep loss and 'divergent' thinking ability. *Sleep*, 11, 528 (1988)

Harrison Y. & Horne J. A. One night of sleep loss impairs innovative thinking and flexible decision making. *Organ. Behav. Hum. Decis. Process.*, 78, 128 (1999)

Harrison Y. & Horne J. A. Sleep loss impairs short and novel language tasks having a prefrontal focus. *J. Sleep Res.*, 7, 95 (1998)

Randazzo A. C. et al. Cognitive function following acute sleep restriction in children ages 10–14. *Sleep*, 21, 861 (1998)

Hoeksma-van Orden C. Y. et al. Social loafing under fatigue. *J. Pers. Soc. Psychol.*, 75, 1179 (1998)

Harrison Y. & Horne J. A. Sleep loss affects risk-taking. *J. Sleep Res.*, 7S2, 113 (1998)

Sicard B. et al. Risk propensity assessment in military special operations. *Mil. Med.*, 166, 871 (2001)

Alcohol, beauty and old age

Fairclough S. H. & Graham R. Impairment of driving performance caused by sleep deprivation or alcohol. *Hum. Factors*, 41, 118 (1999)

Arnedt J. T. et al. How do prolonged wakefulness and alcohol compare on a simulated driving task? *Accid. Anal. Prev.*, 33, 337 (2001)

Powell N. B. et al. The comparative risks of driving while sleepy. *Laryngoscope*, 111, 887 (2001)

Pilcher J. J. & Walters A. S. How sleep deprivation affects students' cognitive performance. *J. Am. Coll. Health*, 46, 121 (1997)

Zwyghuizen-Doorenbos A. et al. Increased daytime sleepiness enhances alcohol's sedative effects. *Neuropsychopharmacology*, 1, 279 (1988)

Roehrs T. et al. Sleep extension, enhanced alertness and the sedating effects of ethanol. *Pharmacol. Biochem. Behav.*, 34, 321 (1989)

Horne J. A. & Baumber C. J. Time-of-day effects of alcohol intake on simulated driving performance. *Ergonomics*, 34, 1377 (1991)

Roehrs T. et al. Sedating effects of ethanol and time of drinking. *Alcohol Clin. Exp. Res.*, 16, 553 (1992)

Harrison Y. et al. Prefrontal neuropsychological effects of sleep deprivation – a model for healthy aging? *Sleep*, 23, 1067 (2000)

Brower K. J. & Hall J. M. Effects of age and alcoholism on sleep. *J. Stud. Alcohol*, 62, 335 (2001)

Reimund E. Sleep deprivation-induced dermatitis. *Med. Hypotheses*, 36, 371 (1991)

Altemus M. et al. Stress-induced changes in skin barrier function. *J. Invest. Dermatol.*, 117, 309 (2001)

Van Cauter E. et al. Age-related changes in sleep and relationship with growth hormone and cortisol. *JAMA*, 284, 861 (2000)

Champion wakers

Patrick G. T. W. & Gilbert J. A. On the effects of loss of sleep. *Psychol. Rev.*, 3, 469 (1896)

Dement, William C. & Vaughan, Christopher. *The Promise of Sleep*. (NY, Delacorte, 1999) p. 45

Coren, Stanley. *Sleep Thieves*. (NY, Free Press, 1996) p. 50

Uses and abuses

de Montaigne, Michel. 'On Sleep', *Essays* (1580)

Hayes, Bill. *Sleep Demons*. (NY, Pocket Books, 2001) p. 238

Coren, Stanley. *Sleep Thieves*. (NY, Free Press, 1996) p. 58

Oswald, Ian. *Sleep*. 3rd edn (Harmondsworth UK, Penguin, 1974) p. 59

Chapter 4 – The Golden Chain

Kapur V. K. et al. The relationship between chronically disrupted sleep and healthcare use. *Sleep*, 25, 289 (2002)

Schwartz S. W. et al. Are sleep complaints an independent risk factor for myocardial infarction? *Ann. Epidemiol.*, 8, 384 (1998)

Philip P. et al. Daytime somnolence and sickness absenteeism. *J. Sleep Res.*, 10, 111 (2001)

Eriksen W. et al. Sleep problems: a predictor of long-term work disability? *Scand. J. Public Health*, 29, 23 (2001)

Imaki M. et al. Relationship between sleep and life style factors. *J. Physiol. Anthropol. Appl. Human Sci.*, 21, 115 (2002)

Wetzler H. P. & Ursano R. J. Physical health practices and psychological well-being. *J. Nerv. Ment. Dis.*, 176, 280 (1988)

Tsubono Y. et al. Health practices and mortality in rural Japanese. *Tohuku J. Exp. Med.*, 171, 339 (1993)

Seki N. Walking hours, sleeping hours, meaningfulness of life (*ikigai*) and mortality in the elderly. *Nippon Eiseigaku Zasshi*, 56, 535 (2001)

Stramba-Badiale M. et al. Sleep in the long-lived. *Minerva Med.*, 69, 1025 (1978)

Gale C. & Martyn C. Larks and owls and health, wealth, and wisdom. *BMJ*, 317, 1675 (1998)

Taub J. M. et al. Extended sleep and performance. *Nature*, 233, 142 (1971)

A waking death

Rechtschaffen A. et al. Prolonged sleep deprivation in rats. *Science*, 221, 182 (1983)

de Manacéïne, Marie. *Sleep.* (London, Walter Scott, 1897) p. 66

Bentivoglio M. & Grassi-Zucconi G. The pioneering experimental studies on sleep deprivation. *Sleep*, 20, 570 (1997)

Rechtschaffen A. & Bergmann B. M. Sleep deprivation in the rat by the disk-over-water method. *Behav. Brain Res.*, 69, 55 (1995)

Rechtschaffen A. et al. Sleep deprivation in the rat. *Sleep*, 12, 68 (1989)

Everson C. A. Sustained sleep deprivation in the rat. *Behav. Brain Res.*, 69, 43 (1995)

Everson C. A. et al. Effects of prolonged sleep deprivation on cerebral energy metabolism. *J. Neurosci.*, 14, 6769 (1994)

Body and soul

Leproult R. et al. Sleepiness, performance, and neuroendocrine function during sleep deprivation. *J. Biol. Rhythms*, 12, 245 (1997)

Symons J. D. et al. Physical performance and physiological responses following 60 hours of sleep deprivation. *Med. Sci. Sports Exerc.*, 20, 374 (1988)

Goodman J. et al. Maximal aerobic exercise following prolonged sleep deprivation. *Int. J. Sports Med.*, 10, 419 (1989)

Edinger J. D. et al. Daytime functioning and nighttime sleep before, during, and after a 146-hour tennis match. *Sleep*, 13, 526 (1990)

Cooper K. R. & Phillips B. A. Effect of short-term sleep loss on breathing. *J. Appl. Physiol.*, 53, 855 (1982)

White D. P. et al. Sleep deprivation and the control of ventilation. *Am. Rev. Respir. Dis.* 128, 984 (1983)

Chen H. I. & Tang Y. R. Sleep loss impairs inspiratory muscle endurance. *Am. Rev. Respir. Dis.*, 140, 907 (1989)

McMurray R. G. & Brown C. F. The effect of sleep loss on high intensity exercise and recovery. *Aviat. Space Environ. Med.*, 55, 1031 (1984)

Ilan Y. et al. Prolonged sleep-deprivation induced disturbed liver functions. *Eur. J. Clin. Invest.* 22, 740 (1992)

Gary K. A. et al. Total sleep deprivation and the thyroid axis. *Aviat. Space Environ. Med.*, 67, 513 (1996)

Toppila J. & Porkka-Heiskanen T. Transcriptional activity in the brain during sleep deprivation. *Ann. Med.*, 31, 146 (1999)

Spiegel K. et al. Impact of sleep debt on metabolic and endocrine function. *Lancet*, 354, 1435 (1999)

Sleep, immunity and health

Brown R. et al. Suppression of immunity to influenza virus infection following sleep disturbance. *Reg. Immunol.*, 2, 321 (1989)

Everson C. A. Sustained sleep deprivation impairs host defense. *Am. J. Physiol.*, 265, R1148 (1993)

Everson C. A. & Toth L. A. Systemic bacterial invasion induced by sleep deprivation. *Am. J. Physiol. Regul. Integr. Comp. Physiol.*, 278, R905 (2000)

Shephard R. J. et al. Immune deficits induced by strenuous exertion. *Crit. Rev. Immunol.*, 18, 545 (1998)

Born J. et al. Effects of sleep and circadian rhythm on immune cells. *J. Immunol.*, 158, 4454 (1997)

Irwin M. et al. Sleep deprivation reduces natural killer cell activity. *Psychosom. Med.*, 56, 493 (1994)

Irwin M. et al. Sleep deprivation reduces natural killer and cellular immune responses in humans. *FASEB J.*, 10, 643 (1996)

Palmblad J. et al. Lymphocyte and granulocyte reactions during sleep deprivation. *Psychosom. Med.*, 41, 273 (1979)

Moldofsky H. et al. Effects of sleep deprivation on human immune functions. *FASEB J.*, 3, 1972 (1989)

Rogers N. L. et al. Neuroimmunologic aspects of sleep loss. *Semin. Clin. Neuropsychiatry*, 6, 295 (2001)

Irwin M. Neuroimmunology of disordered sleep in depression and alcoholism. *Neuropsychopharmacology*, 25, S45 (2001)

Hall M. et al. Sleep as a mediator of the stress-immune relationship. *Psychosom. Med.*, 60, 48 (1998)

Benca R. M. & Quintas J. Sleep and host defenses. *Sleep*, 20, 1027 (1997)

Fang J. et al. Influenza viral infections enhance sleep. *Proc. Soc. Exp. Biol. Med.*, 210, 242 (1995)

Pollmacher T. et al. Experimental immunomodulation and sleepiness in humans. *Ann. NY Acad. Sci.*, 917, 488 (2000)

Martin, Paul. *The Sickening Mind*. (London, Flamingo, 1998) ch. 4

Leproult R. et al. Sleep loss results in an elevation of cortisol levels. *Sleep*, 20, 865 (1997)

Moldofsky H. Sleep and the immune system. *Int. J. Immunopharmacol.*, 17, 649 (1995)

The battle of Stalingrad

Beevor, Antony. *Stalingrad*. (London, Viking, 1998)

Sleepless in hospital

Cumming G. Sleep promotion and recovery from illness. *Med. Hypotheses*, 15, 31 (1984)

Yarrington A. & Mehta P. Does sleep promote recovery after bone marrow transplantation? *Pediatr. Transplant.*, 2, 51 (1998)

Lawrence J. W. et al. Sleep disturbance after burn injury. *J. Burn Care Rehabil.*, 19, 480 (1998)

Davidson J. R. et al. Sleep disturbance in cancer patients. *Soc. Sci. Med.*, 54, 1309 (2002)

Savard J. & Morin C. M. Insomnia in the context of cancer. *J. Clin. Oncol.*, 19, 895 (2001)

Winningham M. L. Managing cancer-related fatigue syndrome. *Cancer*, 92(S4), 988 (2001)

Hicks R. A. et al. Pain thresholds in rats during recovery from REM sleep deprivation. *Percept. Mot. Skills*, 48, 687 (1979)

Raymond I. et al. Quality of sleep and its daily relationship to pain intensity. *Pain*, 92, 381 (2001)

Krachman S. L. et al. Sleep in the intensive care unit. *Chest*, 107, 1713 (1995)

Olson D. M. et al. A nursing intervention to promote sleep in neurocritical care units. *Am. J. Crit. Care*, 10, 74 (2001)

Jarman H. et al. Allowing patients to sleep. *Int. J. Nurs. Pract.*, 8, 75 (2002)

Chapter 5 – The Shapes of Sleep

Measuring sleep

Maquet P. Functional neuroimaging of normal human sleep by PET. *J. Sleep Res.*, 9, 207 (2000)

Reynolds C. F. et al. Concordance between habitual sleep times and laboratory schedules. *Sleep*, 15, 571 (1992)

Falling asleep again, what am I to do?

Ogilvie R. D. The process of falling asleep. *Sleep Med. Rev.*, 5, 247 (2001)

Irving, Washington. 'The Adventure of My Uncle.' *Tales of a Traveller*. (London, John Murray, 1824)

Mavromatis, Andreas. *Hypnagogia*. (London, Routledge & Kegan Paul, 1987)

Kjaer T. W. et al. Regional cerebral blood flow during light sleep. *J. Sleep Res.*, 11, 201 (2002)

Rowley J. T. et al. Eyelid movements and mental activity at sleep onset. *Conscious. Cogn.*, 7, 67 (1998)

Dickens, Charles. 'Shy Neighbourhoods.' *Selected Journalism 1850–1870*. (London, Penguin, 1997)

Stickgold R. et al. Hypnagogic images in normals and amnesiacs. *Science*, 290, 350 (2000)

Palmer C. D. & Harrison G. A. Sleep latency and lifestyle in Oxfordshire villages. *Ann. Hum. Biol.*, 10, 417 (1983)

Harrison Y. et al. Can normal subjects be motivated to fall asleep faster? *Physiol. Behav.*, 60, 681 (1996)

Krauchi K. & Wirz-Justice A. Circadian clues to sleep onset mechanisms. *Neuropsychopharmacology*, 25, S92 (2001)

Murphy P. J. & Campbell S. S. Drop in body temperature: a trigger for sleep onset? *Sleep*, 20, 505 (1997)

Aubrey, John. *Brief Lives*. (London, Penguin, 2000) p. 145

Boswell, James. *The Life of Samuel Johnson, LL.D.* (London, 1791), entry for 19 Sept 1777

Horne J. A. & Reid A. J. Sleep EEG changes following a warm bath. *Electroencephalogr. Clin. Neurophysiol.*, 60, 154 (1985)

van den Heuvel C. J. et al. Changes in sleepiness and body temperature precede nocturnal sleep onset. *J. Sleep Res.*, 7, 159 (1998)

Krauchi K. et al. Warm feet promote the rapid onset of sleep. *Nature*, 401, 36 (1999)

Krauchi K. et al. Distal vasodilation and sleep-onset latency. *Am. J. Physiol. Regul. Integr. Comp. Physiol.*, 278, R741 (2000)

Pache M. et al. Cold feet and sleep-onset latency in vasospastic syndrome. *Lancet*, 358, 125 (2001)

Fletcher A. et al. Sleeping with an electric blanket. *Sleep*, 22, 313 (1999)

Brissette S. et al. Sexual activity and sleep in humans. *Biol. Psychiatry*, 20, 758 (1985)

The sleep cycle

Akerstedt T. & Gillberg M. Subjective and objective sleepiness. *Int. J. Neurosci.*, 52, 29 (1990)

Maquet P. Functional neuroimaging of normal human sleep by PET. *J. Sleep Res.*, 9, 207 (2000)

The paradoxical world of REM

Braun A. R. et al. Regional cerebral blood flow throughout the sleep–wake cycle. *Brain*, 120, 1173 (1997)

Maquet P. et al. Functional neuroanatomy of human REM sleep and dreaming. *Nature*, 383, 163 (1996)

Braun A. R. et al. Dissociated pattern of activity in visual cortices during REM sleep. *Science*, 279, 91 (1998)

Fontvieille A. M. et al. Sleep stages and metabolic rate in humans. *Am. J. Physiol.*, 267, E732 (1994)

Lucidi F. et al. Rapid eye movements density as a measure of sleep need. *Electroencephalogr. Clin. Neurophysiol.*, 99, 556 (1996)

Dement, William C. *Some Must Watch While Some Must Sleep*. (San Francisco, W. H. Freeman, 1972)

Zhou W. & King W. M. Binocular eye movements not coordinated during REM sleep. *Exp. Brain Res.*, 117, 153 (1997)

Aserinsky E. The discovery of REM sleep. *J. Hist. Neurosci.*, 5, 213 (1996)

Aserinsky E. & Kleitman N. Regularly occurring periods of eye motility, and concomitant phenomena during sleep. *Science*, 118, 273 (1953)

The sleep cycle continued

D'Olimpio F. & Renzi P. Ultradian rhythms in young and adult mice. *Physiol. Behav.*, 64, 697 (1998)

Halasz P. Micro-arousals and the microstructure of sleep. *Neurophysiol. Clin.*, 28, 461 (1998)

Bonnet M. H. et al. Metabolism during normal, fragmented, and recovery sleep. *J. Appl. Physiol.*, 71, 1112 (1991)

Tan K. O. et al. Changes in tear film composition during sleep. *Curr. Eye Res.*, 12, 1001 (1993)

Velluti R. A. Interactions between sleep and sensory physiology. *J. Sleep Res.*, 6, 61 (1997)

Oswald I. et al. Discriminative responses to stimulation during human sleep. *Brain*, 83, 440 (1960)

Perrin F. et al. Differential brain response to subject's own name persists during sleep. *Clin. Neurophysiol.*, 110, 2153 (1999)

Stickgold R. Sensory processing during sleep. *Trends Neurosci.*, 24, 307 (2001)

Portas C. M. et al. Auditory processing across the sleep–wake cycle. *Neuron*, 28, 991 (2000)

Badia P. et al. Responsiveness to olfactory stimuli presented in sleep. *Physiol. Behav.*, 48, 87 (1990)

Ayala-Guerrero F. et al. Sleep patterns of the volcano mouse. *Physiol. Behav.*, 64, 577 (1998)

Tobler I. & Deboer T. Sleep in the blind mole rat. *Sleep*, 24, 147 (2001)

Ayala-Guerrero F. & Huitron-Resendiz S. Sleep patterns in the lizard Ctenosaura pectinata. *Physiol. Behav.*, 49, 1305 (1991)

Mukhametov L. M. et al. Sleep in an Amazonian manatee. *Experientia*, 48, 417 (1992)

Szymczak J. T. Sleep pattern in the starling. *Acta Physiol. Pol.*, 36, 323 (1985)

Ayala-Guerrero F. Sleep patterns in the parakeet. *Physiol. Behav.*, 46, 787 (1989)

Buchet C. et al. Sleep in emperor penguins. *Physiol. Behav.*, 38, 331 (1986)

Waking up

Dickens, Charles. *Oliver Twist*. (London, 1838)

Lavie P. et al. 'It's time, you must wake up now.' *Percept. Mot. Skills*, 49, 447 (1979)

Born J et al. Timing the end of nocturnal sleep. *Nature*, 397, 29 (1999)

Achermann P. et al. Time course of sleep inertia. *Arch. Ital. Biol.*, 134, 109 (1995)

Bruck D. & Pisani D. L. Effects of sleep inertia on decision making. *J. Sleep Res.*, 8, 95 (1999)

Jewett M. E. et al. Time course of sleep inertia dissipation. *J. Sleep Res.*, 8, 1 (1999)

Balkin T. J. & Badia P. Sleep inertia and sleepiness. *Biol. Psychol.*, 27, 245 (1988)

The quality of sleep

Levine B. et al. Effects of sleep restriction and extension on sleep efficiency. *Int. J. Neurosci.*, 43, 139 (1988)

Akerstedt T. et al. The meaning of good sleep. *J. Sleep Res.*, 3, 152 (1994)

Keklund G. & Akerstedt T. Individual differences in subjective sleep quality. *J. Sleep Res.*, 6, 217 (1997)

Akerstedt T. et al. Good sleep – its timing and physiological characteristics. *J. Sleep Res.*, 6, 221 (1997)

Hobson J. A. et al. Postural immobility and sleep-cycle phase in humans. *Science*, 201, 1251 (1978)

Meditating – or only sleeping?

Jevning R. et al. The physiology of meditation. *Neurosci. Biobehav. Rev.*, 16, 415 (1992)

Mavromatis, Andreas. *Hypnagogia*. (London, Routledge & Kegan Paul, 1987)

Stigsby B. et al. EEG during TM. *Electroencephalogr. Clin. Neurophysiol.*, 51, 434 (1981)

Aftanas L. & Golocheikine S. Non-linear dynamic complexity of the human EEG during meditation. *Neurosci. Lett.*, 330, 143 (2002)

Younger J. et al. Sleep during TM. *Percept. Mot. Skills*, 40, 953 (1975)

Pagano R. R. et al. Sleep during TM. *Science*, 191, 308 (1976)

Mason L. I. et al. Electrophysiological correlates of higher states of consciousness during sleep in long-term practitioners of TM. *Sleep*, 20, 102 (1997)

Delmonte M. M. Electrocortical activity associated with meditation. *Int. J. Neurosci.*, 24, 217 (1984)

Throll D. A. Transcendental meditation and progressive relaxation. *J. Clin. Psychol.*, 38, 522 (1982)

Fenwick P. B. et al. Metabolic and EEG changes during TM. *Biol. Psychol.*, 5, 101 (1977)

Lazar S. W. et al. Brain mapping of the relaxation response and meditation. *Neuroreport*, 11, 1581 (2000)

Jevning R. et al. Effects on regional cerebral blood flow of TM. *Physiol. Behav.*, 59, 399 (1996)

Miles, Barry. *Paul McCartney: many years from now*. (London, Vintage, 1998) p. 413

Chapter 6 – Morpheus Undressed

The rhythms of life

Jazwinska E. C. & Adam K. Diurnal change in stature. *Experientia*, 41, 1533 (1985)

Panda S. et al. Circadian rhythms from flies to human. *Nature*, 417, 329 (2002)

Dijk D. J. & Lockley S. W. Integration of human sleep-wake regulation and circadian rhythmicity. *J. Appl. Physiol.*, 92, 852 (2002)

DeCoursey P. J. et al. Circadian performance of SCN-lesioned antelope ground squirrels in a desert enclosure. *Physiol. Behav.*, 62, 1099 (1997)

DeCoursey P. J. et al. A circadian pacemaker in free-living chipmunks: essential for survival? *J. Comp. Physiol. [A]*, 186, 169 (2000)

Boivin D. B. & Czeisler C. A. Resetting of circadian melatonin and cortisol rhythms in humans by ordinary room light. *Neuroreport*, 9, 779 (1998)

Czeisler C. A. et al. Stability, precision, and near-24-hour period of the human circadian pacemaker. *Science*, 284, 2177 (1999)

Miller A. M. et al. The superior colliculus-pretectum mediates the direct effects of light on sleep. *Proc. Natl. Acad. Sci. USA*, 95, 8957 (1998)

Berson D. M. et al. Phototransduction by retinal ganglion cells that set the circadian clock. *Science*, 295, 1070 (2002)

Campbell S. S. & Murphy P. J. Extraocular circadian phototransduction. *Science*, 279, 396 (1998)

Barrera-Mera B. & Barrera-Calva E. The Cartesian clock metaphor for pineal gland operation. *Neurosci. Biobehav. Rev.*, 23, 1 (1998)

Shochat T. et al. Melatonin – the key to the gate of sleep. *Ann. Med.*, 30, 109 (1998)

Kelly T. L. et al. Circadian rhythms of melatonin in submariners. *J. Biol. Rhythms*, 14, 190 (1999)

Van Cauter E. et al. Sleep and the somatotropic axis. *Sleep*, 21, 553 (1998)

Luboshitzky R. et al. REM sleep and testosterone secretion in normal men. *J. Androl.*, 20, 731 (1999)

Czeisler C. A. et al. Sleep–wake habits in older people and changes in circadian pacemaker. *Lancet*, 340, 933 (1992)

Miles L. E. et al. Blind man has circadian rhythms of 24.9 hours. *Science*, 198, 421 (1977)

Tabandeh H. et al. Disturbance of sleep in blindness. *Am. J. Ophthalmol.*, 126, 707 (1998)

Klein T. et al. Circadian sleep regulation in the absence of light perception. *Sleep*, 16, 333 (1993)

Arendt J. et al. Melatonin treatment in jet lag, shift work, and blindness. *J. Biol. Rhythms*, 12, 604 (1997)

Cardinali D. P. et al. Melatonin in sleep disorders and jet-lag. *Neuroendocrinol. Lett.*, 23S1, 9 (2002)

Skene D. J. et al. Melatonin in the treatment of phase shift and sleep disorders. *Adv. Exp. Med. Biol.*, 467, 79 (1999)

Zhdanova I. V. et al. Melatonin: a sleep-promoting hormone. *Sleep*, 20, 899 (1997)

Herxheimer A. & Petrie K. J. Melatonin for the prevention and treatment of jet lag. *Cochrane Database Syst. Rev.*, 2, CD001520 (2002)

Bergstrom W. H. & Hakanson D. O. Melatonin: the dark force. *Adv. Pediatr.*, 45, 91 (1998)

So SAD

Wehr T. A. et al. Suppression of responses to seasonal changes in day length by artificial lighting. *Am. J. Physiol.*, 269, R173 (1995)

Schwartz P. J. et al. Winter seasonal affective disorder. *Am. J. Psychiatry*, 153, 1028 (1996)

Guillemette J. et al. Natural bright light exposure and seasonal mood variations. *Biol. Psychiatry*, 44, 622 (1998)

Eastman C. I. et al. Bright light treatment of winter depression. *Arch. Gen. Psychiatry*, 55, 883 (1998)

Larks and owls

Shattock, Roger. *Marcel Proust.* (Princeton, Princeton University Press, 1974) p. 15

Boswell, James. *The Life of Samuel Johnson, LL.D.* (London, 1791) entry for 19 Sept 1777

Manber R. et al. Effects of regularizing sleep–wake schedules on daytime sleepiness. *Sleep*, 19, 432 (1996)

Kerkhof G. A. & Lancel M. EEG activity, REM sleep, and rectal temperature in morning-type and evening-type subjects. *Psychophysiology*, 28, 678 (1991)

Gale C. & Martyn C. Larks and owls and health, wealth, and wisdom. *BMJ*, 317, 1675 (1998)

Edwards S. et al. Time of awakening and diurnal cortisol secretory activity. *Psychoneuroendocrinology*, 26, 613 (2001)

Dekker, Thomas. *The Guls Horne Booke.* (Menston, Scolar Press, 1969; first publ. 1609)

Reyner L. A. & Horne J. A. Gender- and age-related differences in sleep. *Sleep*, 18, 127 (1995)

Aeschbach D. et al. Short sleepers live under higher sleep pressure than long sleepers. *Neuroscience*, 102, 493 (2001)

Hicks R. A. et al. Handedness and sleep duration. *Cortex*, 15, 327 (1979)

Martin, Paul. *The Sickening Mind.* (London, Flamingo, 1998) p. 106

O'Leary D. J. & Millodot M. Eyelid closure causes myopia in humans. *Experientia*, 35, 1478 (1979)

Genes and sleep

Schibler U. & Tafti M. The isolation of sleep-related genes. *J. Sleep Res.*, 8S1, 1 (1999)

Lowrey P. L. et al. The mammalian circadian mutation tau. *Science*, 288, 483 (2000)

Jones C. R. et al. Familial advanced sleep-phase syndrome. *Nat. Med.*, 5, 1062 (1999)

Toh K. L. et al. A mutation in familial advanced sleep-phase syndrome. *Science*, 291, 1040 (2001)

Katzenberg D. et al. CLOCK polymorphism associated with diurnal preference. *Sleep*, 21, 569 (1998)

Webb W. B. & Campbell S. S. Sleep characteristics of identical and fraternal twins. *Arch. Gen. Psychiatry*, 40, 1093 (1983)

Heath A. C. et al. Effects of lifestyle and genetic predisposition on sleep. *Twin Res.*, 1, 176 (1998)

Bateson, Patrick & Martin, Paul. *Design for a Life*. (London, Vintage, 2000) ch. 4

Sackett G. & Korner A. Sleep-waking states in conjoined twin neonates. *Sleep*, 16, 414 (1993)

Webb W. B. The sleep of conjoined twins. *Sleep*, 1, 205 (1978)

Willekens D. et al. Children with Smith-Magenis syndrome. *Genet. Couns.*, 11, 103 (2000)

Allanson J. E. et al. The face of Smith-Magenis syndrome. *J. Med. Genet.*, 36, 394 (1999)

Smith A. C. et al. Sleep disturbance in Smith-Magenis syndrome. *Am. J. Med. Genet.*, 81, 186 (1998)

De Leersnyder H. et al. Inversion of the circadian rhythm of melatonin in Smith-Magenis syndrome. *J. Pediatr.*, 139, 111 (2001)

Potocki L. et al. Circadian rhythm abnormalities in Smith-Magenis syndrome. *J. Med. Genet.*, 37, 428 (2000)

A sleeplessness that kills

Gallassi R. et al. Fatal familial insomnia. *Neurology*, 46, 935 (1996)

Montagna P. et al. Fatal familial insomnia. *Adv. Neuroimmunol.*, 5, 13 (1995)

Cortelli P. et al. Fatal familial insomnia. *J. Sleep Res.*, 8S1, 23 (1999)

Lugaresi E. et al. The pathophysiology of fatal familial insomnia. *Brain Pathol.*, 8, 521 (1998)

Sy M. S. et al. Human prion diseases. *Med. Clin. North Am.*, 86, 551 (2002)

Fiorino A. S. Sleep, genes and death. *Brain Res. Brain Res. Rev.*, 22, 258 (1996)

Collins S. et al. GSS, FFI, and kuru. *J. Clin. Neurosci.*, 8, 387 (2001)

Chapter 7 – Strange Tales of Erections and Yawning

Nocturnal erections

Hirshkowitz M. & Moore C. A. Sleep-related erectile activity. *Neural. Clin.*, 14, 721 (1996)

Rogers G. S. et al. Vaginal pulse amplitude response patterns during erotic conditions and sleep. *Arch. Sex. Behav.*, 14, 327 (1985)

Henton C. L. Nocturnal orgasm in college women. *J. Genet. Psychol.*, 129, 245 (1976)

Reynolds C. F. et al. Nocturnal penile tumescence in healthy 20- to 59-year-olds. *Sleep*, 12, 368 (1989)

Burris A. S. et al. Nocturnal penile tumescence and rigidity in normal men. *J. Androl.*, 10, 492 (1989)

Gordon C. M. & Carey M. P. Penile tumescence monitoring during morning naps. *Arch. Sex. Behav.*, 24, 291 (1995)

Oksenberg A. et al. Sleep-related erections in vegetative state patients. *Sleep*, 23, 953 (2000)

Schmidt M. H. et al. Penile erections during paradoxical sleep in the rat. *Neuroreport*, 5, 561 (1994)

Ohlmeyer P. et al. Periodische Vorgänge im Schlaf. *Pflügers Archiv*, 258, 559 (1944)

Ruspoli, Mario. *The Cave of Lascaux*. (London, Thames & Hudson, 1987)

Bataille, Georges. *Prehistoric Painting: Lascaux or the birth of art*. (London, Macmillan, 1980)

Jouvet, Michel. *The Paradox of Sleep*. (Cambridge MA, MIT Press, 1999) p. 170

Rosen R. C. et al. Nocturnal penile tumescence: *Psychosom. Med.*, 48, 423 (1986)

Hatzichristou D. G. et al. Nocturnal penile tumescence: effect of sexual intercourse. *J. Urol.*, 159, 1921 (1998)

Morlet A. et al. Effects of acute alcohol on penile tumescence. *Urology*, 35, 399 (1990)

Hirshkowitz M. et al. Nocturnal penile tumescence in smokers with erectile dysfunction. *Urology*, 39, 101 (1992)

Takahashi Y. & Hirata Y. Nocturnal penile tumescence monitoring with stamps. *Diabetes Res. Clin. Pract.*, 4, 197 (1988)

Ek A. et al. Nocturnal penile rigidity measured by the snap-gauge band. *J. Urol.*, 129, 964 (1983)

Ware J. C. Monitoring erections during sleep. In *Principles and Practice of Sleep Medicine*, ed. by Meir H. Kryger et al. (London, Saunders, 1989) p. 689

Campbell P. I. et al. Visual estimates of erectile fullness. *Sleep*, 10, 480 (1987)

Moreland R. B. Pathophysiology of erectile dysfunction. *Int. J. Impot. Res.*, 12S4, S39 (2000)

Thase M. E. et al. Diminished nocturnal penile tumescence in depression. *Biol. Psychiatry*, 31, 1136 (1992)

Nofzinger E. A. et al. Nocturnal penile tumescence in depressed men. *Psychiatry Res.*, 49, 139 (1993)

Wincze J. P. et al. Nocturnal penile tumescence and penile response to erotic stimulation. *Arch. Sex. Behav.*, 17, 333 (1988)

Malhotra H. K. & Wig N. N. Dhat syndrome. *Arch. Sex. Behav.*, 4, 519 (1975)

Dement, William C. & Vaughan, Christopher, *The Promise of Sleep*. (NY, Delacorte, 1999) p. 295

Montorsi F. et al. Sildenafil increases nocturnal erections. *Urology*, 56, 906 (2000)

Johansen L. V. et al. Prevention of erection after penile surgery. *Urol. Res.*, 17, 393 (1989)

The mystery of yawning
Darwin, Charles. Notebook M, 85. *Charles Darwin's Notebooks, 1836–1844*, ed. by P. H. Barrett et al. (Ithaca NY, Cornell University Press, 1987)

Luttenberger F. The problem of yawning in reptiles. *Z Tierpsychol.*, 37, 113 (1975)

Askenasy J. J. Is yawning an arousal defense reflex? *J. Psychol.*, 123, 609 (1989)

Hanning C. D. Yawning. *Sleep Med. Rev.*, 5, 411 (2001)

Koch P. et al. Behavioural and physiological variables in children attending kindergarten and primary school. *Chronobiol. Int.*, 4, 525 (1987)

Greco M. & Baeninger R. Effects of yawning on skin conductance and heart rate. *Physiol. Behav.*, 50, 1067 (1991)

Provine R. R. et al. Yawning: no effect of 3–5% CO_2, 100% O_2, and exercise. *Behav. Neural Biol.*, 48, 382 (1987)

Sepulveda W. & Mangiamarchi M. Fetal yawning. *Ultrasound Obstet. Gynecol.*, 5, 57 (1995)

Baenninger R. et al. Field observations of yawning and activity. *Physiol. Behav.*, 59, 421 (1996)

Giganti F. et al. Yawning and behavioral states in premature infants. *Dev. Psychobiol.*, 41, 289 (2002)

Holmgren B. et al. Food anticipatory yawning in the rat. *Acta Neurobiol. Exp. (Warsz.)*, 51, 97 (1991)

Adams A. The big yawn. *New Scientist*, 19/26 Dec 1998 – 2 Jan 1999

Holmgren B. et al. Dopaminergic-induced yawning and penile erections in the rat. *Pharmacol. Biochem. Behav.*, 22, 31 (1985)

Wessells H. et al. Synthetic melanotropic peptide initiates erections in men with psychogenic erectile dysfunction. *J. Urol.*, 160, 389 (1998)

Chapter 8 – Friends and Enemies of Sleep

Brother caffeine
Weinberg, Bennett & Bealer, Bonnie. *The World of Caffeine*. (London, Routledge, 2001) numerous

Pendergrast, Mark. *Uncommon Grounds*. (NY, Basic Books, 1999) numerous

Brillat-Savarin, Jean-Anthelme. *The Physiology of Taste*. (London, Penguin, 1994)

Hindmarch I. et al. Effects of tea, coffee and water on alertness and sleep. *Psychopharmacology (Berl.)*, 149, 203 (2000)

Brown, Guy. *The Energy of Life* (London, Flamingo, 2000) p. 209

Hishikawa Y. Stimulant drugs. In *The Pharmacology of Sleep*, ed. by Anthony Kales. (Berlin, Springer, 1995) p. 421

Leonard T. K. et al. Effects of caffeine on various body systems. *J. Am. Diet. Assoc.*, 87, 1048 (1987)

Nehlig A. et al. Caffeine and the central nervous system. *Brain Res. Brain Res. Rev.*, 17, 139 (1992)

Nehlig A. & Debry G. Caffeine and sports activity. *Int. J. Sports Med.*, 15, 215 (1994)

Bonnet M. H. & Arand D. L. Metabolic rate and the restorative function of sleep. *Physiol. Behav.*, 59, 777 (1996)

Karacan I. et al. Sleep disturbances induced by coffee and caffeine. *Clin. Pharmacol. Ther.*, 20, 682 (1976)

Kamimori G. H. et al. Effect of caffeine on plasma catecholamines and alertness. *Eur. J. Clin. Pharmacol.*, 56, 537 (2000)

Levy M. & Zylber-Katz E. Caffeine metabolism and sleep disturbances. *Clin. Pharmacol. Ther.*, 33, 770 (1983)

Griffiths R. R. & Vernotica E. M. Is caffeine a flavoring agent in cola soft drinks? *Arch. Fam. Med.*, 9, 727 (2000)

Nehlig A. Are we dependent upon coffee and caffeine? *Neurosci. Biobehav. Rev.*, 23, 563 (1999)

Hofer I. & Battig K. Cardiovascular, behavioral, and subjective effects of caffeine. *Pharmacol. Biochem. Behav.*, 48, 899 (1994)

Evans S. M. & Griffiths R. R. Caffeine withdrawal. *J. Pharmacol. Exp. Ther.*, 289, 285 (1999)

Schuh K. J. & Griffiths R. R. Caffeine reinforcement: the role of withdrawal. *Psychopharmacology (Berl.)*, 130, 320 (1997)

Griffiths R. R. & Chausmer A. L. Caffeine as a model drug of dependence. *Nihon Shinkei Seishin Yakurigaku Zasshi*, 20, 223 (2000)

Nehlig A. & Boyet S. Caffeine effects on cerebral functional activity. *Brain Res.*, 858, 71 (2000)

Sister alcohol

Aubrey, John. *Brief Lives*. (London, Penguin, 2000) p. 29

Johnson E. O. et al. Alcohol and medication as aids to sleep. *Sleep*, 21, 178 (1998)

Roehrs T. & Roth T. Sleep, sleepiness, and alcohol use. *Alcohol Res. Health*, 25, 101 (2001)

Van F. et al. Effects of alcohol on a nap in the afternoon. *Biol. Psychol.*, 41, 55 (1995)

Williams D. L. et al. Effects of ethanol on the sleep of young women. *J. Stud. Alcohol*, 44, 515 (1983)

Landholt H. P. et al. Late-afternoon ethanol intake affects nocturnal sleep. *J. Clin. Psychopharmacol.*, 16, 428 (1996)

Palmer C. D. et al. Smoking, drinking and sleep duration. *Ann. Hum. Biol*, 7, 103 (1980)

Mulder E. J. et al. Maternal alcohol consumption disrupts behavioral state in the fetus. *Pediatr. Res.*, 44, 774 (1998)

Block A. J. & Hellard D. W. Scotch or vodka induce equal effects on sleep. *Arch. Intern. Med.*, 147, 1145 (1987)

Brower K. J. et al. Insomnia, self-medication, and relapse to alcoholism. *Am. J. Psychiatry*, 158, 399 (2001)

Brower K. J. et al. Sleep predictors of alcoholic relapse. *Alcohol Clin. Exp. Res.*, 22, 1864 (1998)

Drummond S. P. et al. The sleep of abstinent alcoholic patients. *Alcohol Clin. Exp. Res.*, 22, 1796 (1998)

Eccles R. & Tolley N. S. Effect of alcohol upon nasal airway resistance. *Rhinology*, 25, 245 (1987)

Taasan V. C. et al. Alcohol increases sleep apnea and oxygen desaturation. *Am. J. Med.* 71, 240 (1981)

Mitler M. M. et al. Bedtime ethanol produces sleep apneas in asymptomatic snorers. *Alcohol Clin. Exp. Res.*, 12, 801 (1988)

Scanlan M. F. et al. Effect of moderate alcohol upon sleep apnoea. *Eur. Respir. J.*, 16, 909 (2000)

Castaneda R. et al. Effects of moderate alcohol intake on psychiatric and sleep disorders. *Recent Dev. Alcohol.*, 14, 197 (1998)

Tobacco

Palmer C. D. et al. Smoking, drinking and sleep duration. *Ann. Hum. Biol.*, 7, 103 (1980)

Phillips B. A. & Danner F. J. Smoking and sleep disturbance. *Arch. Intern. Med.*, 155, 734 (1995)

Parkin C. et al. Effects of smoking on overnight performance. *Psychopharmacology*, 136, 172 (1998)

Wetter D. W. et al. Tobacco withdrawal and nicotine replacement influence sleep. *J. Consult. Clin. Psychol.*, 63, 658 (1995)

Soldatos C. R. et al. Cigarette smoking associated with sleep difficulty. *Science*, 207, 551 (1980)

Food for sleep

Brillat-Savarin, Jean-Anthelme. *The Physiology of Taste*. (London, Penguin, 1994) p. 201

Wells A. S. et al. Effects of meals on daytime sleepiness. *J. Appl. Physiol.*, 84, 507 (1998)

Zammit G. K. et al. Postprandial sleep and thermogenesis. *Physiol, Behav.*, 52, 251 (1992)

Harnish M. J. et al. A comparison of feeding to cephalic stimulation on postprandial sleepiness. *Physiol, Behav.*, 64, 93 (1998)

Orr W. C. et al. Meal composition and postprandial sleepiness. *Physiol, Behav.*, 62, 709 (1997)

Neumann M. & Jacobs K. W. Dietary components and sleep. *Percept. Mot. Skills*, 75, 873 (1992)

Wells A. S. et al. Effects of meals on daytime sleepiness. *J. Appl. Physiol.*, 84, 507 (1998)

Edwards S. J. et al. Spicy meal disturbs sleep. *Int. J. Psychophysiol.*, 13, 97 (1992)

Exercise is bunk, isn't it?

Walker J. M. et al. Effects of exercise on sleep. *J. Appl. Physiol.*, 44, 945 (1978)

Shapiro C. M. et al. Slow-wave sleep: a recovery period after exercise. *Science*, 214, 1253 (1981)

Kern W. et al. Hormonal secretion during sleep indicating stress of exercise. *J. Appl. Physiol.*, 79, 1461 (1995)

O'Connor P. J. & Youngstedt S. D. Influence of exercise on sleep. *Exerc. Sport Sci. Rev.*, 23, 105 (1995)

Torsvall L. et al. Effects on sleep of different degrees of exercise. *Electroencephalogr. Clin. Neurophysiol.*, 57, 347 (1984)

Sherrill D. L. et al. Physical activity and human sleep disorders. *Arch. Intern. Med.*, 158, 1894 (1998)

Van Someren E. J. et al. Long-term fitness training improves circadian rhythm in healthy elderly males. *J. Biol. Rhythms*, 12, 146 (1997)

Shirota A. et al. Effects of daytime activity on sleep in the elderly. *Psychiatry Clin. Neurosci.*, 54, 309 (2000)

Youngstedt S. D. et al. Is sleep disturbed by vigorous late-night exercise? *Med. Sci. Sports Exerc.*, 31, 864 (1999)

Horne J. A. The effects of exercise upon sleep. *Biol. Psychol.*, 12,241 (1981)

Horne J. A. & Moore V. J. Sleep EEG effects of exercise with and without body cooling. *Electroencephalogr. Clin. Neurophysiol.*, 60, 33 (1985)

Things that go bump in the night

Horne J. A. et al. A field study of sleep disturbance: effects of aircraft noise. *Sleep*, 17, 146 (1994)

Fidell S. et al. Community response to a step change in aircraft noise. *J. Acoust. Soc. Am.*, 111, 200 (2002)

Thiessen G. J. Disturbance of sleep by noise. *J. Acoust. Soc. Am.*, 64, 216 (1978)

Kawabata T. Annoyance in school children caused by Shinkansen noise. *Nippon Koshu Eisei Zasshi*, 41, 1131 (1994)

Kawada T. Effects of traffic noise on sleep. *Nippon Eiseigaku Zasshi*, 50, 932 (1995)

Nivison M. E. & Endresen I. M. Environmental noise and the consequences for health and sleep. *J. Behav. Med.*, 16, 257 (1993)

Shift work

Garbarino S. et al. Brain function and effects of shift work. *Neuropsychobiology*, 45, 50 (2002)

Akerstedt T. Work hours and sleepiness. *Neurophysiol. Clin.*, 25, 367 (1995)

Rajaratnam S.M.W. & Arendt J. Health in a 24-h society. *Lancet*, 358, 999 (2001)

Imbernon E. et al. Effects on health of on-call shifts. *J. Occup. Med.*, 35, 1131 (1993)

Akerstedt T. & Torsvall L. Napping in shift work. *Sleep*, 8, 105 (1985)

Scott A. J. et al. Shift work as a risk factor for depression. *Int. J. Occup. Environ. Health*, 3S2, S2 (1997)

Penev P. D. et al. Chronic circadian desynchronization decreases survival of animals with heart disease. *Am. J. Physiol.*, 275, H2334 (1998)

Knutsson A. et al. Increased risk of heart disease in shift workers. *Lancet*, 2 (8498), 89 (1986)

Davis S. et al. Night shift work and risk of breast cancer. *J. Natl. Cancer Inst.*, 93, 1557 (2001)

Schernhammer E. S. et al. Rotating night shifts and risk of cancer. *J. Natl. Cancer Inst.*, 93, 1563 (2001)

Porcu S. et al. Sleepiness, alertness and performance during simulation of acute shift of wake–sleep cycle. *Ergonomics*, 41, 1192 (1998)

Akerstedt T. Shifted sleep hours. *Ann. Clin. Res.*, 17, 273 (1985)

Gordon N. P. et al. The prevalence and impact of shift work. *Am. J. Public Health*, 76, 1225 (1986)

Gold D. R. et al. Rotating shift work, sleep, and accidents in hospital nurses. *Am. J. Public Health*, 82, 1011 (1992)

Pilcher J. J. et al. Differential effects of permanent and rotating shifts on sleep. *Sleep*, 23, 155 (2000)

Arendt J. et al. Melatonin treatment in jet lag, shift work, and blindness. *J. Biol. Rhythms*, 12, 604 (1997)

Poppy, mandragora and drowsy syrups

Merritt S. L. et al. Herbal remedies. *Nurse Pract. Forum*, 11, 87 (2000)

Kitchiner, William. *The Art of Invigorating and Prolonging Life, Etc.* 4th edn (London, Hurst, Robinson, 1822) p. 97

Cauffield J. S. & Forbes H. J. Dietary supplements in the treatment of depression, anxiety, and sleep disorders. *Lippincotts Prim. Care Pract.*, 3, 290 (1999)

Burton, Robert. *The Anatomy of Melancholy*. (NY, New York Review Books, 2001) p. 252

Leathwood P. D. et al. Extract of valerian improves sleep quality. *Pharmacol. Biochem. Behav.*, 17, 65 (1982)

Donath F. et al. Effect of valerian on sleep structure and quality. *Pharmacopsychiatry*, 33, 47 (2000)

Sherry C. J. et al. Pharmacological effects of extract of nutmeg. *J. Ethnopharmacol.*, 6, 61 (1982)

Mills M. H. & Faunce T. A. Melatonin supplementation from early morning auto-urine drinking. *Med. Hypotheses*, 36, 195 (1991)

Hardy M. et al. Replacement of drug treatment for insomnia by ambient odour. *Lancet*, 346, 701 (1995)

Sproule B. A. et al. The use of non-prescription sleep products in the elderly. *Int. J. Geriatr. Psychiatry*, 14, 851 (1999)

Saitou K. et al. Sleep-inducing effects of H1-antagonists. *Biol. Pharm. Bull.*, 22, 1079 (1999)

Hypnotic exotica

Polyakov, Vladimir. *When Lovers Ruled Russia*. (NY, Appleton, 1928)

Limoge A. et al. Transcutaneous cranial electrical stimulation. *Neurosci. Biobehav. Rev.*, 23, 529 (1999)

Southworth S. Effects of cranial electrical stimulation. *Integr. Physiol. Behav. Sci.*, 34, 43 (1999)

Huber R. et al. Exposure to pulsed high-frequency electromagnetic field during waking affects human sleep EEG. *Neuroreport*, 11, 3321 (2000)

Preece A. W. et al. Effect of a simulated mobile phone signal on cognitive function. *Int. J. Radiat. Biol.*, 75, 447 (1999)

Jech R. et al. Electromagnetic field of mobile phones affects visual event related potential. *Bioelectromagnetics*, 22, 519 (2001)

Akerstedt T. et al. A 50-Hz electromagnetic field impairs sleep. *J. Sleep Res.*, 8, 77 (1999)

Vaughan, William. *Naturall and Artificial Directions for Health, etc.* (London, R. Bradocke, 1602)

Chapter 9 – The Children of an Idle Brain?

To sleep, perchance . . .

Kahn D. et al. Dreaming and waking consciousness. *J. Sleep Res.*, 9, 317 (2000)

Schredl M. Dream length and creativity. *Percept. Mot. Skills*, 78, 1297 (1994)

Katz M. & Shapiro C. M. Dreams and medical illness. *BMJ*, 306, 993 (1993)

Hall C. S. 'A ubiquitous sex difference in dreams' revisited. *J. Pers. Soc. Psychol.*, 46, 1109 (1984)

McNamara P. et al. Impact of attachment styles on dream recall and content. *J. Sleep Res.*, 10, 117 (2001)

Fosse R. et al. The mind in REM sleep: reports of emotional experience. *Sleep*, 24, 947 (2001)

Hobson J. A. & Pace-Schott E. F. The cognitive neuroscience of sleep. *Nat. Rev. Neurosci.*, 3, 679 (2002)

Weed S. & Hallam F. A study of the dream-consciousness. *Am. J. Psychol.*, April 1896.

Schredl M. et al. Dreaming and insomnia. *J. Sleep Res.*, 7, 191 (1998)

Hobson J. A. et al. The neuropsychology of REM sleep dreaming. *Neuroreport*, 16, R1 (1998)

Zadra A. L. et al. Auditory, olfactory, and gustatory experiences in dreams. *Percept. Mot. Skills*, 87, 819 (1998)

de Manacéïne, Marie. *Sleep*. (London, Walter Scott, 1897) p. 302

Flanagan, Owen. *Dreaming Souls*. (Oxford, Oxford University Press, 2000) p. 148

Calkins, M. W. Statistics of dreams. *Am. J. Psychol.*, April 1893.

Stevens, Anthony. *Private Myths: dreams and dreaming.* (London, Penguin, 1996) p. 90

Cavallero C. et al. Slow-wave sleep dreaming. *Sleep*, 15, 562 (1992)

Stickgold R. et al. Sleep/wake factors influencing mentation report length. *Sleep*, 24, 171 (2001)

Hobson J. A. et al. Dreaming and the brain. *Behav. Brain Sci.*, 23, 793 (2000)

Do flies dream?

Aristotle, *Historia Animalium* and *De Divinatione per Somnum*

Montaigne. 'An apology for Raymond Sebond.' *Essays.* (1580–1592) Book II

Darwin, Erasmus. *Zoonomia.* (London, Johnson, 1801) v1, s8

Darwin, Charles. *The Descent of Man.* (London, 1871)

A dream within a dream?

Schopenhauer, Arthur. *The World as Will and Idea.* (London, Everyman, 1995)

Kahan T. L. et al. Similarities and differences between dreaming and waking cognition. *Conscious. Cogn.*, 6, 132 (1997)

Llinás R. R. & Paré D. Of dreaming and wakefulness. *Neuroscience*, 44, 521 (1991)

Flanagan, Owen. *Dreaming Souls.* (Oxford, Oxford University Press, 2000) p. 171

Starker S. Daydreaming styles and nocturnal dreaming. *Percept. Mot. Skills*, 45, 411 (1977)

Reynolds C. F. et al. Sleep after spousal bereavement. *Biol. Psychiatry*, 34, 791 (1993)

Cartwright R. D. et al. Effects of divorce and depression on dream content. *Psychiatry*, 47, 251 (1984)

Garcia-Garcia F. et al. Manipulations during forced wakefulness have differential impact on sleep. *Brain Res. Bull.*, 47, 317 (1998)

Dement W. C. & Wolpert E. The relation of eye movement, body motility, and external stimuli to dream content. *J. Exp. Psychol.*, 55, 543 (1958)

de Manacéïne, Marie. *Sleep.* (London, Walter Scott, 1897) p. 259

Dreaming as madness

Dickens, Charles. 'Night Walks.' *Selected Journalism 1850–1870.* (London, Penguin, 1997)

Hobson J. A. Dreaming as delirium. *Semin. Neurol.*, 17, 121 (1997)

Schwartz S. & Maquet P. Sleep imaging and the neuro-psychological assessment of dreams. *Trends Cogn. Sci.*, 6, 23 (2002)

Mazzoni G. A. & Loftus E. F. When dreams become reality. *Conscious. Cogn.*, 5, 442 (1996)

Kelly P. H. Defective inhibition of dream event memory formation. *Brain Res. Bull.*, 46, 189 (1998)

Stevenson, Robert Louis. 'A Chapter on Dreams.' *Essays and Poems.* (London, Everyman, 1992)

Are dreams meaningful?

Stevens, Anthony. *Private Myths: dreams and dreaming*. (London, Penguin, 1996) pp. 14, 25

Descola, Philippe. *The Spears of Twilight*. (London, Flamingo, 1997)

Schiller F. The inveterate paradox of dreaming. *Arch. Neurol.*, 42, 903 (1985)

Horton P. C. Detecting cancer in dream content. *Bull. Menninger Clin.*, 62, 326 (1998)

Freud, Sigmund. *The Interpretation of Dreams*. (London, Penguin, 1976)

Lebzeltern G. Sigmund Freud and cocaine. *Wien Klin. Wochenschr.*, 95, 765 (1983)

Webster, Richard. *Why Freud Was Wrong*. (London, HarperCollins, 1995) p. 260

Dunlap K. The pragmatic advantage of Freudo-analysis. *Psyehoanal. Rev.*, 1, 151 (1913)

McCarley R. W. & Hobson J. A. The neurobiological origins of psychoanalytic dream theory. *Am. J. Psychiatry*, 134, 1211 (1977)

Like wine through water

Smellie, William. *The Philosophy of Natural History*. (Edinburgh, 1790–99) vII, ch5

Flanagan, Owen. *Dreaming Souls*. (Oxford, Oxford University Press, 2000) p. 148

Can dreams be sinful?

Saint Augustine, *The Confessions* (c. 397 AD), Book XX

Flanagan, Owen. *Dreaming Souls*. (Oxford, Oxford University Press, 2000) p. 180

Browne, Sir Thomas. *On Dreams*. (ca. 1650)

Chapter 10 – A Second Life

A creative state

Jouvet, Michel. *The Paradox of Sleep*. (Cambridge MA, MIT Press, 1999) p. 87

Koestler, Arthur. *The Act of Creation*. (London, Hutchinson, 1964)

Mueller E. T. *Daydreaming in Humans and Machines*. (London, Intellect Books, 1990)

Miles, Barry. *Paul McCartney: many years from now*. (London, Vintage, 1998)

Mavromatis, Andreas. *Hypnagogia*. (London, Routledge & Kegan Paul, 1987)

Holmes, Richard. *Coleridge: darker reflections*. (London, HarperCollins, 1998) p. 436

Flanagan, Owen. *Dreaming Souls*. (Oxford, Oxford University Press, 2000) p. 188

Buñuel, Luis. *My Last Breath*. (London, Vintage, 1994) p. 134

de Becker, Raymond. *The Understanding of Dreams or the Machinations of the Night*. (London, George Allen & Unwin, 1968) p. 84

Loewi, O. An autobiographic sketch. *Perspect. Biol. Med.*, 4, 17 (1960)

Stevens, Anthony. *Private Myths: dreams and dreaming*. (London, Penguin, 1996) p. 283

Barrett D. Dreams of scientific problem solving. *Dreaming*, 11, 93 (2001)

Stevenson's Brownies

Stevenson, Robert Louis. 'A Chapter on Dreams.' *Essays and Poems*. (London, Everyman, 1992)

Lucid dreams

Cicogna P. C. & Bosinelli M. Consciousness during dreams. *Conscious. Cogn.*, 10, 26 (2001)

Green, Celia & McCreery, Charles. *Lucid Dreaming*. (London, Routledge, 1994)

Hearne, Keith M. T. *The Dream Machine*. (Wellingborough, Aquarian Press, 1990)

Garfield, Patricia L. *Creative Dreaming*. 2nd edn (NY, Simon & Schuster, 1995)

Lequerica A. Lucid dreaming and the mind–body relationship. *Percept. Mot. Skills*, 83, 331 (1996)

LaBerge S. et al. Lucid dreaming verified by volitional communication during REM sleep. *Percept. Mot. Skills*, 52, 727 (1981)

LaBerge S. & Dement W. Voluntary control of respiration during REM sleep. *Sleep Res.*, 11, 107 (1982)

Brylowski A. et al. H-reflex suppression and autonomic activation during lucid REM sleep. *Sleep*, 12, 374 (1989)

Tholey P. Abilities of dream characters during lucid dreaming. *Percept. Mot. Skills*, 68, 567 (1989)

Purcell S. et al. Dream self-reflectiveness as a learned cognitive skill. *Sleep*, 9, 423 (1986)

LaBerge, Stephen & Rheingold, Howard. *Exploring the World of Lucid Dreaming*. (NY, Ballantine, 1990)

Abramovitch H. Nightmare disorder treated by lucid dreaming. *Isr. J. Psychiatry Relat. Sci.*, 32, 140 (1995)

The great dreamer

Schatzman, Morton. Introduction to *Dreams and How to Guide Them* by Hervey de Saint-Denys, transl. by Nicholas Fry. (London, Duckworth, 1982)

de Saint-Denys, Hervey. *Les Rêves et les Moyens de les Diriger*. (Paris, Amyot, 1867)

Van Eeden F. A study of dreams. *Proc. Soc. Psychical Res.*, 26, 431 (1913)

Chapter 11 – From Egg to Grave

Thorleifsdottir B. et al. Sleep habits from childhood to young adulthood over a 10-year period. *J. Psychosom. Res.*, 53, 529 (2002)

Screaming babies

Mauri M. et al. Sleep in the premenstrual phase. *Acta Psychiatr. Scand.*, 78, 82 (1988)

Baker F. C. et al. Oral contraceptives alter sleep. *Pflügers Arch.*, 442, 729 (2001)

Santiago J. R. et al. Sleep and sleep disorders in pregnancy. *Ann. Intern. Med.*, 134, 396 (2001)

Schorr S. J. et al. Sleep patterns in pregnancy. *J. Perinatol.*, 18, 427 (1998)

Franklin K. A. et al. Snoring, pregnancy-induced hypertension, and growth retardation of the fetus. *Chest*, 117, 137 (2000)

Edwards N. et al. Nasal continuous positive airway pressure reduces sleep-induced blood pressure increments in preeclampsia. *Am. J. Respir. Crit. Care Med.*, 162, 252 (2000)

Mirmiran M. & Lunshof S. Perinatal development of circadian rhythms. *Prog. Brain Res.*, 111, 217 (1996)

Scher A. Night waking in the first year. *Child Care Health Dev.*, 17, 295 (1991)

Scott G. & Richards M. P. Night waking in 1-year-old children. *Child Care Health Dev.*, 16, 283 (1990)

Louis J. et al. Sleep ontogenesis revisited. *Sleep*, 20, 323 (1997)

Coons S. & Guilleminault C. Development of sleep–wake patterns during the first six months. *Pediatrics*, 69, 793 (1982)

Bamford F. N. et al. Sleep in the first year of life. *Dev. Med. Child Neurol.*, 32, 718 (1990)

Butte N. F. et al. Sleep organization of breast-fed and formula-fed infants. *Pediatr. Res.*, 32, 514 (1992)

Hiscock H. & Wake M. Behavioural intervention to improve infant sleep and maternal mood. *BMJ*, 324, 1062 (2002)

Anders T. F. et al. Sleeping through the night. *Pediatrics*, 90, 554 (1992)

Schmitt B. D. The prevention of sleep problems and colic. *Pediatr. Clin. North Am.*, 33, 763 (1986)

Adams L. A. & Rickert V. I. Reducing bedtime tantrums: comparison between positive routines and graduated extinction. *Pediatrics*, 84, 756 (1989)

Nelson E. A. et al. Child care practices in nonindustrialized societies. *Pediatrics*, 105, E75 (2000)

Wilson C. E. Cree infant care practices. *Can. J. Public Health*, 91, 133 (2000)

Masataka N. On the function of swaddling practiced by Native South Americans. *Shinrigaku Kenkyu*, 67, 285 (1996)

Tronick E. Z. et al. The Quechua manta pouch: a caretaking practice for buffering the Peruvian infant against the stressors of high altitude. *Child Dev.*, 65, 1005 (1994)

Yoshida Y. et al. Childcare and home education in China. *Nippon Koshu Eisei Zasshi.*, 48, 470 (2001)

Li Y. et al. Child-rearing behaviors in rural Yunnan, China. *J. Dev. Behav. Pediatr.*, 21, 114 (2000)

Chisholm J. S. Swaddling, cradleboards and development of children. *Early Hum. Dev.*, 2, 255 (1978)

Caglayan S. et al. Swaddling above the waist. *Turk. J. Pediatr.*, 33, 117 (1991)

Kutlu A. et al. Congenital dislocation of the hip and swaddling. *J. Pediatr. Orthop.*, 12, 598 (1992)

Yurdakok K. et al. Swaddling and respiratory infections. *Am. J. Public Health*, 80, 873 (1990)

Ponsonby A. L. et al. Factors potentiating the risk of sudden infant death syndrome associated with the prone position. *N. Engl. J. Med.*, 329, 377 (1993)

Locard E. et al. Is it possible to improve sleep in children? *Arch. Pediatr.*, 4, 1247 (1997)

Bad children

Sadeh A. et al. Sleep, neurobehavioral functioning and behavior problems in school-age children. *Child Dev.*, 73, 405 (2002)

Stein M. A. et al. Sleep and behavior problems in school-aged children. *Pediatrics*, 107, E60 (2001)

Dahl R. E. The development and disorders of sleep. *Adv. Pediatr.*, 45, 73 (1998)

Lavigne J. V. et al. Sleep and behavior problems among preschoolers. *J. Dev. Behav. Pediatr.*, 20, 164 (1999)

Meijer A. M. et al. Time in bed and school functioning of children. *J. Sleep Res.*, 9, 145 (2000)

Schneider S. C. et al. ADHD. *Postgrad. Med.*, 101, 231 (1997)

Schweitzer J. B. et al. ADHD. *Med. Clin. North Am.*, 85, 757 (2001)

Patzold L. M. et al. Sleep characteristics of children with autism and Asperger's Disorder. *J. Paediatr. Child Health*, 34, 528 (1998)

Ring A. et al. Sleep disturbances in children with ADHD. *J. Learn. Disabil.*, 31, 572 (1998)

Brown T. E. & McMullen W. J. Attention deficit disorders and sleep/arousal disturbances. *Ann. NY Acad. Sci.*, 931, 271 (2001)

Gruber R. et al. Instability of sleep patterns in children with ADHD. *J. Am. Acad. Child Adolesc. Psychiatry*, 39, 495 (2000)

Corkum P. et al. Sleep in children with ADHD. *Sleep*, 24, 303 (2001)

Chervin R. D. et al. Sleep disorders, inattention, and hyperactivity in children. *Sleep*, 20, 1185 (1997)

Chervin R. D. et al. Inattention, hyperactivity, and sleep-disordered breathing. *Pediatrics*, 109, 449 (2002)

Liu X. et al. Sleep problems in Chinese schoolchildren. *Sleep*, 23, 1053 (2000)

Hishikawa Y. Stimulant drugs. In *The Pharmacology of Sleep*, ed. by Anthony Kales. (Berlin, Springer, 1995) p. 421

Yawning youth

Mercer P. W. et al. Reported sleep need among adolescents. *J. Adolesc. Health*, 23, 259 (1998)

Reid A. et al. Sleep behavior of South African adolescents. *Sleep*, 25, 423 (2002)

Carskadon M. A. Patterns of sleep and sleepiness in adolescents. *Pediatrician*, 17, 5 (1990)

Gillin J. C. Are sleep disturbances risk factors for anxiety, depressive and addictive disorders? *Acta Psychiatr. Scand. Suppl.*, 393, 39 (1998)

Jean-Louis G. et al. Mood states and sleepiness in college students. *Percept. Mot. Skills*, 87, 507 (1998)

Johnson E. O. & Breslau N. Sleep problems and substance abuse in adolescence. *Drug Alcohol Depend.*, 64, 1 (2001)

Carskadon M. A. et al. Adolescent sleep patterns. *Sleep*, 21, 871 (1998)

Wolfson A. R. & Carskadon M. A. Sleep and daytime functioning in adolescents. *Child Dev.*, 69, 875 (1998)

Epstein R. et al. Starting times of school: effects on daytime functioning. *Sleep*, 21, 250 (1998)

Strauch I. & Meier B. Sleep in adolescents. *Sleep*, 11, 378 (1988)

Vignau J. et al. Sleep quality and troubles in French adolescents. *J. Adolesc. Health*, 21, 343 (1997)

Macgregor I. D. & Balding J. W. Bedtimes and family size in schoolchildren. *Ann. Hum. Biol.*, 15, 435 (1988)

Andrade M. M. et al. Sleep characteristics of adolescents. *J. Adolesc. Health*, 14, 401 (1993)

Szymczak J. T. et al. Changes in the sleep–wake rhythm of school children. *Sleep*, 16, 433 (1993)

Hicks R. A. & Pellegrini R. J. The sleep habits of students. *Percept. Mot. Skills*, 72, 1106 (1991)

Old and grey and full of sleep

Hill, Dr *The Old Man's Guide to Health and Longer Life*. 5th edn (London, R. Baldwin & J. Ridley, 1764)

Ancoli-Israel S. et al. Sleep problems in older adults. *Geriatrics*, 52, 20 (1997)

Hume K. I. et al. Age and gender differences in habitual adult sleep. *J. Sleep Res.*, 7, 85 (1998)

Webb W. B. Sleep structures of 50- to 60-year-old men and women. *J. Gerontol.*, 37, 581 (1982)

Alvarez A. *Night.* (London, Jonathan Cape, 1995) p. 56

Webb W. B. Sleep stage responses of older and younger subjects after sleep deprivation. *Electroencephalogr. Clin. Neurophysiol.*, 52, 368 (1981)

Bixler E. O. et al. Sleep and wakefulness: effects of age and sex. *Int. J. Neurosci.*, 23, 33 (1984)

Carrier J. et al. Sleep and morningness–eveningness in the 'middle' years. *J. Sleep Res.*, 6, 230 (1997)

Reilly T. et al. Aging, rhythms of physical performance, and adjustment to changes in the sleep–activity cycle. *Occup. Environ. Med.*, 54, 812 (1997)

Czeisler C. A. et al. Sleep–wake habits in older people and changes in circadian pacemaker. *Lancet*, 340, 933 (1992)

Duffy J. F. & Czeisler C. A. Age-related change in the relationship between circadian period, circadian phase, and diurnal preference. *Neurosci. Lett.*, 318, 117 (2002)

Bundlie S. R. Sleep in aging. *Geriatrics*, 53 S1, S41 (1998)

Moran M. G. et al. Sleep disorders in the elderly. *Am. J. Psychiatry*, 145, 1369 (1988)

Owens J. F. & Matthews K. A. Sleep disturbance in middle-aged women. *Maturitas*, 30, 41 (1998)

Hoch C. C. et al. Stability of EEG and sleep quality in healthy seniors. *Sleep*, 11, 521 (1988)

Mallon L. & Hetta J. Sleep difficulties in an elderly Swedish population. *Ups. J. Med. Sci.*, 102, 185 (1997)

Reyner L. A. & Horne J. A. Gender- and age-related differences in sleep. *Sleep*, 18, 127 (1995)

Carskadon M. A. & Dement W. C. Respiration during sleep in the aged human. *J. Gerontol.*, 36, 420 (1981)

Kelly J. & Feigenbaum L. Z. Another cause of reversible dementia: sleep deprivation due to prostatism. *J. Am. Geriatr. Soc.*, 30, 645–6 (1982)

Zepelin H. et al. Effects of age on auditory awakening thresholds. *J. Gerontol.*, 39, 294 (1984)

Cricco M. et al. Insomnia and cognitive functioning in older adults. *J. Am. Geriatr. Soc.*, 49, 1185 (2001)

Ohayon M. M. & Vecchierini M. F. Daytime sleepiness and cognitive impairment in the elderly population. *Arch. Intern. Med.*, 162, 201 (2002)

Tamaki M. et al. Restorative effects of a short nap in the elderly. *Sleep Res. Online*, 3, 131 (2000)

Chapter 12 – The Reason of Sleep

Kleitman, Nathaniel. Obituary. *The Times*, 30 Sept 1999

Donne, John. 'Meditation 15.' *Devotions Upon Emergent Occasions*. (London, 1624)

The evolution of sleep
Nicol S. C. et al. The echidna manifests characteristics of REM sleep. *Neurosci. Lett.*, 283, 49 (2000)

Siegel J. M. et al. Monotremes and the evolution of REM sleep. *Philos. Trans. R. Soc. Lond. B Biol. Sci.*, 353, 1147 (1998)

Siegel J. M. et al. Sleep in the platypus. *Neuroscience*, 91, 391 (1999)

What is sleep for?
Horne J. Human slow-wave sleep. *Experientia*, 48, 941 (1992)

Maquet P. Sleep function(s) and cerebral metabolism. *Behav. Brain Res.*, 69, 75 (1995)

Berger R. J. & Phillips N. H. Energy conservation and sleep. *Behav. Brain Res.*, 69, 65 (1995)

Kilduff T. S. et al. Sleep and mammalian hibernation. *Sleep*, 16, 372 (1993)

Walker J. M. et al. Hibernation at moderate temperatures. *Experientia*, 37, 726 (1981)

Meddis R. On the function of sleep. *Anim. Behav.*, 23, 676 (1975)
Meddis, Ray. *The Sleep Instinct.* (London, Routledge & Kegan Paul, 1977)
Allison T. & Cicchetti D. V. Sleep in mammals: ecological and constitutional correlates. *Science*, 194, 732 (1976)

What is REM sleep for?
Maurice D. M. An ophthalmological explanation of REM sleep. *Exp. Eye Res.*, 66, 139 (1998)
Mirmiran M. The function of fetal/neonatal REM sleep. *Behav. Brain Res.*, 69, 13 (1995)
Marks G. A. et al. A functional role for REM sleep in brain maturation. *Behav. Brain Res.*, 69, 1 (1995)
Shaffery J. P. et al. REM sleep deprivation in monocularly occluded kittens. *Sleep*, 21, 837 (1998)
Shaffery J. P. et al. REM sleep deprivation modifies expression of long-term potentiation in visual cortex. *Neuroscience*, 110, 431 (2002)

Reverse learning
Crick F. & Mitchison G. The function of dream sleep. *Nature*, 304, 111 (1983)
Crick F. & Mitchison G. REM sleep and neural nets. *Behav. Brain Res.*, 69, 147 (1995)
Hopfield J. J. et al. 'Unlearning' has a stabilizing effect in collective memories. *Nature*, 304, 158 (1983)
Revonsuo A. An evolutionary hypothesis of the function of dreaming. *Behav. Brain Sci.*, 23, 877 (2000)

To sleep, perchance to learn
Storr, Anthony. *Solitude.* (London, HarperCollins, 1988) p. 22
Peigneux P. et al. Sleeping brain, learning brain. *Neuroreport*, 12, A111 (2001)
Sejnowski T. J. Sleep and memory. *Curr. Biol.*, 5, 832 (1995)
Laureys S. et al. Experience-dependent changes in cerebral functional connectivity during REM sleep. *Neuroscience*, 105, 521 (2001)
Laureys S. et al. Sleep and motor skill learning. *Neuron*, 35, 5 (2002)
Smith C. & Rose G. M. Posttraining paradoxical sleep in rats is increased after spatial learning. *Behav. Neurosci.*, 111, 1197 (1997)
Vescia S. et al. Sleep related to the learning ability of rats. *Physiol. Behav.*, 60, 1513 (1996)
Mandai O. et al. REM sleep modifications following Morse code learning. *Physiol. Behav.*, 46, 639 (1989)
De Koninck J. et al. Language learning and increases in REM sleep. *Int. J. Psychophysiol.*, 8, 43 (1989)
Buchegger J. & Meier-Koll A. An EEG study of trampoliners. *Percept. Mot. Skills*, 67, 635 (1988)
Smith C. Sleep states, memory processes and synaptic plasticity. *Behav. Brain Res.*, 78, 49 (1996)

Idzikowski C. Sleep and memory. *Br. J. Psychol.*, 75, 439 (1984)

Smith C. T. et al. Brief paradoxical sleep deprivation impairs reference, but not working, memory. *Neurobiol. Learn. Mem.*, 69, 211 (1998)

Tilley A. J. & Empson J. A. REM sleep and memory consolidation. *Biol. Psychol.*, 6, 293 (1978)

Karni A. et al. Dependence on REM sleep of overnight improvement of a perceptual skill. *Science*, 265, 679 (1994)

Wilson M. A. & McNaughton B. L. Reactivation of hippocampal ensemble memories during sleep. *Science*, 265, 676 (1994)

Louie K. & Wilson M. A. Temporally structured replay of awake hippocampal ensemble activity during REM sleep. *Neuron*, 29, 145 (2001)

Maquet P. et al. Experience-dependent changes in cerebral activation during REM sleep. *Nat. Neurosci.*, 3, 831 (2000)

Dave A. S. et al. Behavioral state modulation of auditory activity in a vocal motor system. *Science*, 282, 2250 (1998)

Dave A. S. & Margoliash D. Song replay during sleep. *Science*, 290, 812 (2000)

Siegel J. M. The REM sleep–memory consolidation hypothesis. *Science*, 294, 1058 (2001)

Stickgold R. et al. Visual discrimination learning requires sleep after training. *Nat. Neurosci.*, 3, 1237 (2000)

Plihal W. & Born J. Effects of early and late nocturnal sleep on memory. *Psychophysiology*, 36, 571 (1999)

Stickgold R. et al. Sleep, learning and dreams. *Science*, 294, 1052 (2001)

Maquet P. The role of sleep in learning and memory. *Science*, 294, 1048 (2001)

Chapter 13 – Bad Sleepers

Leger D. Public health and insomnia: economic impact. *Sleep*, 23S3, S69 (2000)

An intolerable lucidity

Rosekind M. R. The epidemiology and occurrence of insomnia. *J. Clin. Psychiatry*, 53S, 4 (1992)

Ancoli-Israel S. & Roth T. Characteristics of insomnia in the United States. *Sleep*, 22S2, S347 (1999)

Kupperman M. et al. Sleep problems and their correlates. *J. Gen. Intern. Med.*, 10, 25 (1995)

Hochstrasser B. Epidemiology of sleep disorders. *Ther. Umsch.*, 50, 679 (1993)

Hajak G. Insomnia in Germany. *Eur. Arch. Psychiatry Clin. Neurosci.*, 251, 49 (2001)

Mniszek D. H. Sleep in 20–45-year-olds. *J. Int. Med. Res.*, 16, 61 (1988)

Ohayon M. M. et al. How a general population perceives its sleep. *Sleep*, 20, 715 (1997)

Ohayon M. Insomnia in the general population. *Sleep*, 19, S7 (1996)

Tachibana H. et al. Insomnia in Japanese industrial workers. *Psychiatry Clin. Neurosci.*, 52, 397 (1998)

Janson C. et al. Sleep disturbances in three European countries. *Sleep*, 18, 589 (1995)

Althuis M. D. et al. Insomnia and mortality among older women. *J. Am. Geriatr. Soc.*, 46, 1270 (1998)

Kerr S. & Jowett S. Sleep problems in pre-school children. *Child Care Health Dev.*, 20, 379 (1994)

Kahn A. et al. Sleep problems in healthy preadolescents. *Pediatrics*, 84, 542 (1989)

Quine L. Sleep problems in children with mental handicap. *J. Ment. Defic. Res.*, 35, 269 (1991)

Ohayon M. M. et al. Problematic sleep among adolescents. *J. Am. Child Adolesc. Psychiatry*, 39, 1549 (2000)

Roth T. & Ancoli-Israel S. Daytime consequences of insomnia in the US. *Sleep*, 22S2, S354 (1999)

Crawford B. Clinical economics and sleep disorders. *Sleep*, 20, 829 (1997)

Leger D. et al. The direct costs of insomnia in France. *Sleep*, 22S2, S394 (1999)

Why can't you sleep?

Russell, Bertrand. *The Conquest of Happiness*. (London, George Allen & Unwin, 1930) ch. 1

Adam K. et al. Differences between good and poor sleepers. *J. Psychiatr. Res.*, 20, 310 (1986)

Mercer J. D. et al. Insomniacs' perception of wake instead of sleep. *Sleep*, 25, 564 (2002)

Waugh, Evelyn. *Decline and Fall* (1928) pt. 2, ch. 3

Perlis M. L. et al. The mesograde amnesia of sleep may be attenuated in primary insomnia. *Physiol. Behav.*, 74, 71 (2001)

Bonnet M. H. & Arand D. L. Diagnosis and treatment of insomnia. *Respir. Care Clin. N. Am.*, 5, 333 (1999)

Dodge R. et al. Insomnia and respiratory symptoms. *Arch. Intern. Med.*, 155, 1797 (1995)

Wittig R. M. et al. Disturbed sleep in patients complaining of chronic pain. *J. Nerv. Ment. Dis.*, 170, 429 (1982)

Bonnet M. H. & Arand D. L. The consequences of a week of insomnia. *Sleep*, 19, 453 (1996)

Vgontzas A. N. et al. Chronic insomnia and the stress system. *J. Psychosom. Res.*, 45, 21 (1998)

Mendelson W. B. Insomnia and related sleep disorders. *Psychiatr. Clin. North Am.*, 16, 841 (1993)

de Botton, Alain. *How Proust Can Change Your Life*. (London, Picador, 1997) p. 63

Vgontzas A. N. & Kales A. Sleep and its disorders. *Annu. Rev. Med.*, 50, 387 (1999)

Reite M. Sleep disorders presenting as psychiatric disorders. *Psychiatr. Clin. North Am.*, 21, 591 (1998)

Storm and stress and sleep

Bouyer J. J. et al. Reaction of sleep–wakefulness cycle to stress. *Brain Res.*, 804, 114 (1998)

Buguet A. et al. Sleep and stress in man. *Can. J. Physiol. Pharmacol.*, 76, 553 (1998)

Tachibana H. et al. Insomnia in Japanese industrial workers. *Psychiatry Clin. Neurosci.*, 52, 397 (1998)

Hyyppa M. T. et al. Quality of sleep during economic recession. *Soc. Sci. Med.*, 45, 731 (1997)

Ferrie J. E. et al. Health effects of organisational change and job insecurity. *Soc. Sci. Med.*, 46, 243 (1998)

Martin, Paul. *The Sickening Mind*. (London, Flamingo, 1998) ch. 7

Mattiasson I. et al. Threat of unemployment and cardiovascular risk factors. *BMJ*, 301, 461 (1990)

Cartwright R. D. & Wood E. The sleep effects of a stressful event. *Psychiatry Res.*, 39, 199 (1991)

Lavie P. et al. War-related environmental insomnia. *Isr. J. Med. Sci.*, 27, 681 (1991)

Mellman T. A. Sleep disturbances in posttraumatic stress disorder. *Ann. N. Y. Acad. Sci.*, 821, 142 (1997)

Astrom C. et al. Sleep disturbances in torture survivors. *Acta Neurol. Scand.*, 79, 150 (1989)

Neylan T. C. et al. Sleep disturbances in the Vietnam generation. *Am. J. Psychiatry*, 155, 929 (1998)

Sadavoy J. The late-life effects of prior psychological trauma. *Am. J. Geriatr. Psychiatry*, 5, 287 (1997)

Schlosberg A. & Benjamin M. Sleep in acute combat fatigue cases. *J. Clin. Psychiatry*, 39, 546 (1978)

Lavie P. et al. Elevated awaking thresholds during sleep. *Biol. Psychiatry*, 44, 1060 (1998)

What to do?

Idzikowski, Chris. *The Insomnia Book*. (NY, Penguin Studio, 1999)

Parker, Dorothy. 'The Little Hours.' *The Portable Dorothy Parker*. (NY, Viking Press, 1944)

Harvey A. G. & Payne S. The management of unwanted pre-sleep thoughts in insomnia. *Behav. Res. Ther.*, 40, 267 (2002)

Walsh J. K. et al. Non-nightly use of zolpidem for primary insomnia. *Sleep*, 23, 1087 (2000)

Lader M. H. Implications of hypnotic flexibility on patterns of clinical usage. *Int. J. Clin. Pract.* Suppl. 2001, Jan, 14 (2001)

Ancoli-Israel S. Insomnia in the elderly. *Sleep*, 23S1, S23 (2000)

Schneider D. L. Safe and effective therapy for sleep problems in the older patient. *Geriatrics*, 57, 24 (2002)

Lacks P. & Morin C. M. Assessment and treatment of insomnia. *J. Consult. Clin. Psychol.*, 60, 586 (1992)

Morin C. M. et al. Nonpharmacological interventions for insomnia. *Am. J. Psychiatry*, 151, 1172 (1994)

Perlis M. L. et al. Behavioral treatment of insomnia. *J. Behav. Med.*, 24, 281 (2001)

Smith M. T. et al. Pharmacotherapy and behavior therapy for persistent insomnia. *Am. J. Psychiatry*, 159, 5 (2002)

Levin Y. 'Brain music' in the treatment of insomnia. *Neurosci. Behav. Physiol*, 28, 330 (1998)

Staying awake

Penetar D. et al. Caffeine reversal of sleep deprivation. *Psychopharmacology*, 112, 359 (1993)

Reyner L. A. & Horne J. A. Early morning driver sleepiness: effectiveness of 200 mg caffeine. *Psychophysiology*, 37, 251 (2000)

Wright K. P. et al. Bright light and caffeine during sleep deprivation. *J. Sleep Res.*, 6, 26 (1997)

Horne J. A. & Reyner L. A. Beneficial effects of an 'energy drink' given to sleepy drivers. *Amino Acids*, 20, 83 (2001)

Bonnet M. H. & Arand D. L. The use of prophylactic naps and caffeine to maintain performance. *Ergonomics*, 37, 1009 (1994)

Reyner L. A. & Horne J. A. Suppression of sleepiness in drivers: combination of caffeine with a short nap. *Psychophysiology*, 34, 721 (1997)

Bonnet M. H. & Arand D. L. The impact of music upon sleep tendency. *Physiol. Behav.*, 71, 485 (2000)

Engber T. M. et al. The novel wake-promoting agent modafinil. *Neuroscience*, 87, 905 (1998)

Batejat D. M. & Lagarde D. P. Naps and modafinil as countermeasures for sleep deprivation. *Aviat. Space Environ. Med.*, 70, 493 (1999)

Matsumoto Y. et. al. Physical activity increases dissociation between sleepiness and performance during extended wakefulness. *Neurosci. Lett.*, 326, 133 (2002)

Chapter 14 – Dark Night

Walking and talking

Vgontzas A. N. & Kales A. Sleep and its disorders. *Annu. Rev. Med.*, 50, 387 (1999)

Hublin C. et al. Prevalence and genetics of sleepwalking. *Neurology*, 48, 177 (1997)

Joncas S. et al. Sleep deprivation and sleepwalking. *Neurology*, 58, 936 (2002)

Ohayon M. M. et al. Night terrors, sleepwalking, and confusional arousals. *J. Clin. Psychiatry*, 60, 268 (1999)

Guilleminault C. et al. Sleep and wakefulness in somnambulism. *J. Psychosom. Res.*, 51, 411 (2001)

Dement, William C. *Some Must Watch While Some Must Sleep.* (San Francisco, W. H. Freeman, 1972) p. 79

Hare, Augustus J. C. *The Story of My Life.* (London, George Allen, 1896–1900)

Crisp A. H. The sleepwalking/night terrors syndrome in adults. *Postgrad. Med. J.*, 72, 599 (1996)

Ohayon M. M. et al. Violent behavior during sleep. *J. Clin. Psychiatry*, 58, 369 (1997)

Guilleminault C. et al. Nocturnal wandering and violence. *Sleep*, 18, 740 (1995)

Oswald I. & Evans J. On serious violence during sleep-walking. *Br. J. Psychiatry*, 147, 688 (1985)

Guilleminault C. et al. Atypical sexual behavior during sleep. *Psychosom. Med.*, 64, 328 (2002)

Hublin C. et al. Sleeptalking in twins. *Behav. Genet.*, 28, 289 (1998)

Pareja J. A. et al. Native language shifts in bilingual sleeptalkers. *Sleep*, 22, 243 (1999)

Nightmares, night terrors, sleep paralysis and the Old Hag

Darwin, Erasmus. *Zoonomia.* (London, J. Johnson, 1801) vl, s8

Llorente M. D. et al. Night terrors in adults. *J. Clin. Psychiatry*, 53, 392 (1992)

Cheyne J. A. Situational factors affecting sleep paralysis and associated hallucinations. *J. Sleep Res.*, 11, 169 (2002)

Ohayon M. M. et al. Sleep paralysis. *Neurology*, 52, 1194 (1999)

Wing Y. K. et al. Sleep paralysis in Chinese. *Sleep*, 17, 609 (1994)

Fukuda K. et al. Kanashibari phenomenon in Japan. *Sleep*, 10, 279 (1987)

Ness R. C. The Old Hag phenomenon as sleep paralysis. *Cult. Med. Psychiatry*, 2, 15 (1978)

Fukuda K. et al. Recognition of sleep paralysis among adults in Canada and Japan. *Psychiatry Clin. Neurosci.*, 54, 292 (2000)

Cheyne J. A. et al. Hypnagogic and hypnopompic hallucinations during sleep paralysis. *Conscious. Cogn.*, 8, 319 (1999)

Stores G. Sleep paralysis and hallucinosis. *Behav. Neurol.*, 11, 109 (1998)

Spanos N. P. et al. An examination of UFO experiences. *J. Abnorm. Psychol.*, 102, 624 (1993)

Holmes, Richard. *Coleridge: darker reflections.* (London, HarperCollins, 1998) p. 229

Holmes, Richard. *Coleridge: early visions* (London, Hodder & Stoughton, 1989) pp. 354, 293

Nashe, Thomas. *The Terrors of the Night.* (London, 1594)

Claridge G. et al. Nightmares, dreams, and schizotypy. *Br. J. Clin. Psychol.*, 36, 377 (1997)

Belicki K. & Belicki D. Predisposition for nightmares. *J. Clin. Psychol.*, 42, 714 (1986)

Hublin C. et al. Nightmares. *Am. J. Med. Genet.*, 88, 329 (1999)

Krakow B. et al. Imagery rehearsal treatment for chronic nightmares. *Behav. Res. Ther.*, 33, 837 (1995)

Moving sleep
Montagna P. et al. Motor disorders in sleep. *Eur. Neurol.*, 38, 190 (1997)
Youngstedt S. D. et al. Periodic leg movements during sleep. *J. Gerontol. A. Biol. Sci. Med. Sci.*, 53, M391 (1998)
Montagna P. et al. Paroxysmal arousals during sleep. *Neurology*, 40, 1063 (1990)
Sheldon S. H. & Jacobsen J. REM-sleep motor disorder in children. *J. Child Neurol.*, 13, 257 (1998)
Chiu H. F. & Wing Y. K. REM sleep behaviour disorder. *Int. J. Clin. Pract.*, 51, 451 (1997)
Schenck C. H. et al. A parasomnia overlap disorder. *Sleep*, 20, 972 (1997)

Midnight feasting
Spaggiari M. C. et al. Nocturnal eating syndrome. *Sleep*, 17, 339 (1994)
Manni R. et al. Nocturnal eating. *Sleep*, 20, 734 (1997)
Birketvedt G. S. et al. The night-eating syndrome. *JAMA*, 282, 689 (1999)

Soggy sheets
Hublin C. et al. Nocturnal enuresis in a nationwide twin cohort. *Sleep*, 21, 579 (1998)
Kalo B. B. & Bella H. Enuresis in Saudi Arabia. *Acta Paediatr.*, 85, 1217 (1996)
Oge O. et al. Enuresis among Turkish children. *Turk. J. Pediatr.*, 43, 38 (2001)
Liu X. et al. Sleep problems in Chinese schoolchildren. *Sleep*, 23, 1053 (2000)
Laberge L. et al. Development of parasomnias. *Pediatrics*, 106, 67 (2000)
Ferrara P. et al. Nocturnal enuresis and left-handedness. *Scand. J. Urol. Nephrol.*, 35, 184 (2001)
Rona R. J. et al. Determinants of nocturnal enuresis. *Dev. Med. Child Neurol.*, 39, 677 (1997)
Bailey J. N. et al. Primary nocturnal enuresis and ADHD. *Acta Paediatr.*, 88, 1364 (1999)
Hjalmas K. Nocturnal enuresis. *Eur. Urol.*, 33 S3, 53 (1998)

Aching heads
Paiva T. et al. Headaches and sleep disturbances. *Headache*, 35, 590 (1995)
Spierings E. L. et al. Precipitating factors of migraine versus tension-type headaches. *Headache*, 41, 554 (2001)
Inamorato E. et al. The role of sleep in migraine attacks. *Arq. Neuropsiquiatr.*, 51, 429 (1993)
Sacks, Oliver. *Migraine*. (London, Picador, 1995)
Wilkinson M. Migraine treatment. *Headache*, 34, S13 (1994)
Ohayon M. M. et al. Risk factors for sleep bruxism. *Chest*, 119, 53 (2001)

Nishigawa K. et al. Bite force during sleep bruxism. *J. Oral Rehabil.*, 28, 485 (2001)

Tan E. K. & Jankovic J. Treating severe bruxism with botulinum toxin. *J. Am. Dent. Assoc.*, 131, 211 (2000)

Troubled guts

Semple J. I. et al. Dramatic diurnal variation in human trefoil peptide TFF2. *Gut*, 48, 648 (2001)

Furukawa Y. et al. Sleep and human colonic motor patterns. *Gastroenterology*, 107, 1372 (1994)

Fukudo S. et al. Brain–gut interactions in irritable bowel syndrome. *Nippon Rinsho.*, 50, 2703 (1992)

Mayer E. A. et al. Intestinal and extraintestinal symptoms in gastrointestinal disorders. *Eur. J. Surg. Suppl. 1998*, 29 (1998)

Orr W. C. et al. Sleep and gastric function in irritable bowel syndrome. *Gut*, 41, 390 (1997)

Kumar D. et al. Abnormal REM sleep in irritable bowel syndrome. *Gastroenterology*, 103, 12 (1992)

Jarrett M. et al. Sleep disturbance influences gastrointestinal symptoms in IBS. *Dig. Dis. Sci.*, 45, 952 (2000)

Goldsmith G. & Levin J. S. Effect of sleep quality on IBS. *Dig. Dis. Sci.*, 38, 1809 (1993)

Troubled minds

Sateia M. J. et al. Sleep in neuropsychiatric disorders. *Semin. Clin. Neuropsychiatry*, 5, 227 (2000)

Nestler E. J. et al. Neurobiology of depression. *Neuron*, 34, 13 (2002)

Gillin J. C. Are sleep disturbances risk factors for anxiety, depressive and addictive disorders? *Acta Psychiatr. Scand. Suppl.*, 393, 39 (1998)

Roth T. Psychiatric diseases and insomnia. *Int. J. Clin. Pract. Suppl.*, Jan, 3–8 (2001)

Livingston G. et al. Does sleep disturbance predict depression in elderly people? *Br. J. Gen. Pract.*, 43, 445 (1993)

Roberts R. E. et al. Sleep complaints and depression. *Am. J. Psychiatry*, 157, 81 (2000)

Benca R. M. et al. Sleep and psychiatric disorders. *Arch. Gen. Psychiatry*, 49, 651 (1992)

Stefos G. et al. Shortened REM latency as a marker for psychotic depression? *Biol. Psychiatry*, 44, 1314 (1998)

Beauchemin K. M. & Hays P. REM sleep and dreaming in affective disorders. *J. Affect. Disord.*, 41, 125 (1996)

Nofzinger E. A. et al. Changes in forebrain function in depression. *Psychiatry Res.*, 91, 59 (1999)

Wu J. C. & Bunney W. E. Antidepressant response to sleep deprivation and relapse. *Am J. Psychiatry*, 147, 14 (1990)

Riemann D. et al. Sleep and sleep-wake manipulations in bipolar depression. *Neuropsychobiology*, 45S1, 7 (2002)

Kuhs H. & Tolle R. Sleep deprivation therapy. *Biol. Psychiatry*, 29, 1129 (1991)

Gillin J. C. et al. Sleep deprivation as a model antidepressant treatment. *Depress. Anxiety*, 14, 37 (2001)

Sudden nocturnal death

Melles R. B. & Katz B. Night terrors and sudden unexplained nocturnal death. *Med. Hypotheses*, 26, 149 (1988)

Munger R. G. & Booton E. A. Bangungut in Manila. *Int. J. Epidemiol.*, 27, 677 (1998)

Tatsanavivat P. et al. Sudden and unexplained deaths in sleep (Laitai) of young men in rural northeastern Thailand. *Int. J. Epidemiol.*, 21, 904 (1992)

Baron R. C. et al. Sudden death among Southeast Asian refugees. *JAMA*, 250, 2947 (1983)

Parmar M. S. & Luque-Coqui A. F. Killer dreams. *Can. J. Cardiol.*, 14, 1389 (1998)

Narcolepsy

Ohayon M. M. et al. Prevalence of narcolepsy in the European general population. *Neurology*, 58, 1826 (2002)

Hublin C. et al. Epidemiology of narcolepsy. *Sleep*, 17, S7 (1994)

Kryger M. H. et al. Diagnoses received by narcolepsy patients in the year prior to diagnosis by a sleep specialist. *Sleep*, 25, 36 (2002)

Thorpy M. Current concepts in narcolepsy. *Sleep Med.*, 2, 5 (2001)

Hungs M. & Mignot E. Hypocretin/orexin, sleep and narcolepsy. *Bioessays*, 23, 397 (2001)

Faraco J. et al. Genetic studies in narcolepsy. *J. Hered.*, 90, 129 (1999)

Chemelli R. M. et al. Narcolepsy in orexin knockout mice. *Cell*, 98, 437 (1999)

Nishino S. et al. Hypocretin deficiency in human narcolepsy. *Lancet*, 355, 39 (2000)

Green P. M. & Stillman M. J. Narcolepsy. *Arch. Fam. Med.*, 7, 472 (1998)

Mignot E. et al. Complex HLA-DR and -DQ interactions confer risk of narcolepsy-cataplexy. *Am. J. Hum. Genet.*, 68, 686 (2001)

Honda M. et al. Monozygotic twins incompletely concordant for narcolepsy. *Biol. Psychol.*, 49, 943 (2001)

Lin L. et al. Narcolepsy and the HLA region. *J. Neuroimmunol.*, 117, 9 (2001)

Thannickal T. C. et al. Reduced hypocretin neurons in human narcolepsy. *Neuron*, 27, 469 (2000)

Taheri S. et al. The role of hypocretins (orexins) in sleep regulation and narcolepsy. *Annu. Rev. Neurosci.*, 25, 283 (2002)

Chapter 15 – Pickwickian Problems

The wonderful world of snoring

Agrawal S. et al. Sound frequency analysis and the site of snoring. *Clin. Otolaryngol.*, 27, 162 (2002)

Lipman, Derek S. *Snoring From A to ZZZZ*. (Portland OR, Spencer Press, 1998) p. 7

Elliott A. R. et al. Microgravity reduces sleep-disordered breathing. *Am. J. Respir. Crit. Care Med.*, 164, 478 (2001)

Hoffstein V. et al. Comparing perceptions and measurements of snoring. *Sleep*, 19, 783 (1996)

Phillips B. et al. Sleep apnea: prevalence of risk factors. *South. Med. J.*, 82, 1090 (1989)

Teculescu D. et al. Habitual snoring in French males. *Respiration*, 68, 365 (2001)

Lindberg E. et al. Snoring in men. *Chest*, 114, 1048 (1998)

Ipsiroglu O. S. et al. Sleep disorders in school children. *Wien. Klin. Wochenschr.*, 113, 235 (2001)

Ferreira A. M. et al. Snoring in Portuguese primary school children. *Pediatrics*, 106, E64 (2000)

Martikainen K. et al. Natural evolution of snoring. *Acta Neurol. Scand.*, 90, 437 (1994)

Ohayon M. M. et al. Snoring and breathing pauses during sleep. *BMJ*, 314, 860 (1997)

Whittle A. T. et al. Neck soft tissue and fat distribution. *Thorax*, 54, 323 (1999)

Norton P. G. & Dunn E. V. Snoring as a risk factor for disease. *Br. Med. J. (Clin. Res. Ed.)*, 291, 630 (1985)

Hoffstein V. Is snoring dangerous to your health? *Sleep*, 19, 506 (1996)

Seppala T. et al. Sudden death and sleeping history among Finnish men. *J. Intern. Med.*, 229, 23 (1991)

Hoffstein V. Blood pressure, snoring, obesity, and nocturnal hypoxaemia. *Lancet*, 344, 643 (1994)

Stradling J. R. et al. Self reported snoring and daytime sleepiness. *Thorax*, 46, 807 (1991)

Verstraeten E. et al. Performance in nonapneic snorers. *Percept. Mot. Skills*, 84, 1211 (1997)

Ficker J. H. et al. Are snoring medical students at risk of failing their exams? *Sleep*, 22, 205 (1999)

Jennum P. et al. Headache and cognitive dysfunctions in snorers. *Arch. Neurol.*, 51, 937 (1994)

Ferreira A. M. et al. Snoring in Portuguese primary school children. *Pediatrics*, 106, E64 (2000)

390

Silence is golden

Stopes, Marie Carmichael. *Sleep*. (London, Chatto & Windus, 1956) p. 120

Braver H. M. et al. Treatment for snoring. *Chest*, 107, 1283 (1995)

Ulfberg J. & Fenton G. Effect of Breathe Right nasal strip on snoring. *Rhinology*, 35, 50 (1997)

Todorova A. et al. Effect of the nasal dilator Breathe Right on snoring. *Eur. J. Med. Res.*, 3, 367 (1998)

Loth S. et al. Better quality of life when nasal breathing of snoring men is improved at night. *Arch. Otolaryngol. Head Neck Surg.*, 125, 64 (1999)

Lipman, Derek S. *Snoring from A to ZZZZ*. (Portland OR, Spencer Press, 1998) p. 138

Breathless in bed

Malhotra A. & White D. P. Obstructive sleep apnoea. *Lancet*, 360, 237 (2002)

Lugaresi E. & Plazzi G. Heavy snorer disease. *Respiration*, 64 S1, 11 (1997)

Lavie P. Rediscovering the importance of nasal breathing in sleep. *J. Laryngol. Otol.*, 101, 558 (1987)

Kryger M. H. Fat, sleep, and Charles Dickens. *Clin. Chest Med.*, 6, 555 (1985)

Lavie P. Historical accounts of sleep apnea syndrome. *Arch. Intern. Med.*, 144, 2025 (1984)

Stradling, John R. *Handbook of Sleep-Related Breathing Disorders*. (Oxford, Oxford University Press, 1993) p. 2

Commare M. C. et al. Ondine's curse. *Neuropediatrics*, 24, 313 (1993)

Kyzer S. & Charuzi I. Obstructive sleep apnea in the obese. *World J. Surg.*, 22, 998 (1998)

Royal College of Physicians of London. *Sleep Apnoea and Related Conditions*. (London, Royal College of Physicians, 1993) p. 5

Partinen M. & Telakivi T. Epidemiology of obstructive sleep apnea syndrome. *Sleep*, 15, S1 (1992)

Enright P. L. et al. Snoring and observed apneas in older adults. *Sleep*, 19, 531 (1996)

Partinen M. Epidemiology of obstructive sleep apnea syndrome. *Curr. Opin. Pulm. Med.*, 1, 482 (1995)

Gibson G. J. et al. Sleep apnoea. *J. R. Coll. Physicians Lond.*, 32, 540 (1998)

Viner S. et al. Are history and physical examination a good screening test for sleep apnea? *Ann. Intern. Med.*, 115, 356 (1991)

Young T. et al. Estimation of the clinically diagnosed proportion of sleep apnea in middle-aged men and women. *Sleep*, 20, 705 (1997)

Ohayon M. M. et al. Snoring and breathing pauses during sleep. *BMJ*, 314, 860 (1997)

Royal College of Physicians of London. *Sleep Apnoea and Related Conditions*. (London, Royal College of Physicians, 1993) pp. 2, 30

O'Brien L. M. & Gozal D. Behavioural and neurocognitive implications of snoring and obstructive sleep apnoea in children. *Paediatr. Respir. Rev.*, 3, 3 (2002)

Gislason T. & Benediktsdottir B. Snoring, apneic episodes, and nocturnal hypoxemia among children. *Chest*, 107, 963 (1995)

Guilleminault C. & Pelayo R. Sleep-disordered breathing in children. *Ann. Med.*, 30, 350 (1998)

Rosen C. L. Obstructive sleep apnea syndrome in children. *Sleep*, 19, S274 (1996)

Young T. et al. Predictors of sleep-disordered breathing. *Arch. Intern. Med.*, 162, 893 (2002)

Itasaka Y. et al. Influence of sleep position and obesity on sleep apnea. *Psychiatry Clin. Neurosci.*, 54, 340 (2000)

Miyazaki S. et al. Influence of nasal obstruction on sleep apnea. *Acta Otolaryngol. Suppl. (Stockh.)*, 537, 43 (1998)

Katz I. et al. Do patients with obstructive sleep apnea have thick necks? *Am. Rev. Respir. Dis.*, 141 1228 (1990)

Grunstein R. R. Metabolic aspects of sleep apnea. *Sleep*, 19, S218 (1996)

Kikuchi M. et al. Facial patterns of obstructive sleep apnea patients. *Psychiatry Clin. Neurosci.*, 54, 336 (2000)

el Bayadi S. et al. A family study of sleep apnea. *Chest*, 98, 554 (1990)

Hendricks J. C. et al. The English bulldog: a natural model of sleep-disordered breathing. *J. Appl. Physiol.*, 63, 1344 (1987)

Lavie P. et al. Breathing disorders in sleep associated with 'microarousals' in patients with allergic rhinitis. *Acta Otolaryngol. (Stockh.)*, 92, 529 (1981)

McNicholas W. T. et al. Obstructive apneas during sleep in patients with seasonal allergic rhinitis. *Am. Rev. Respir. Dis.*, 126, 625 (1982)

Hoffstein V. et al. Handedness and sleep apnea. *Chest*, 103, 1860 (1993)

Consequences – mostly dire

Colt H. G. et al. Hypoxemia vs sleep fragmentation as cause of daytime sleepiness in sleep apnea. *Chest*, 100, 1542 (1991)

Martin S. E. et al. Microarousals in sleep apnoea/hypopnoea sydrome. *J. Sleep Res.*, 6, 276 (1997)

Mitler M. M. Daytime sleepiness and cognitive functioning in sleep apnea. *Sleep*, 16, S68 (1993)

Ulfberg J. et al. Daytime sleepiness and work performance among heavy snorers and patients with sleep apnea. *Chest*, 110, 659 (1996)

Powell N. B. et al. Reaction time in sleep-disordered breathing versus alcohol-impaired controls. *Laryngoscope*, 109, 1648 (1999)

Findley L. J. et al. Automobile accidents involving patients with sleep apnea. *Am. Rev. Respir. Dis.*, 138, 337 (1988)

Mitler M. M. Sleepiness and human behavior. *Curr. Opin. Pulm. Med.*, 2, 488 (1996)

Stoohs R. A. et al. Traffic accidents in commercial long-haul truck drivers. *Sleep*, 17, 619 (1994)

Teran-Santos J. et al. Sleep apnea and the risk of traffic accidents. *N. Engl. J. Med.*, 340, 847 (1999)

Findley L. J. et al. Automobile accidents in patients with sleep apnea. *Chest*, 108, 619 (1995)

Findley L. J. et al. Driving simulator performance in patients with sleep apnea. *Am. Rev. Respir. Dis.*, 140, 529 (1989)

Young T. et al. Sleep-disordered breathing and vehicle accidents. *Sleep*, 20, 608 (1997)

Haraldsson P. O. et al. Sleep apnea and automobile driving. *J. Clin. Epidemiol*, 45, 821 (1992)

Findley L. J. et al. Treatment with CPAP decreases automobile accidents. *Am. J. Respir. Crit. Care Med.* 161, 857 (2000)

George C. F. Reduction in vehicle collisions following treatment of sleep apnoea. *Thorax*, 56, 508 (2001)

Lindberg E. et al. Increased mortality among sleepy snorers. *Thorax*, 53, 631 (1998)

Lavie P. et al. Mortality in sleep apnea patients. *Sleep*, 18, 149 (1995)

Pankow W. et al. Influence of sleep apnea on 24-hour blood pressure. *Chest*, 112, 1253 (1997)

Lavie P. et al. Obstructive sleep apnoea as a risk factor for hypertension. *BMJ*, 320, 479 (2000)

Hung J. et al. Association of sleep apnoea with myocardial infarction. *Lancet*, 336, 261 (1990)

Shahar E. et al. Sleep-disordered breathing and cardiovascular disease. *Am. J. Respir. Crit. Care Med.*, 163, 19 (2001)

Marcus C. L. et al. Growth in children with obstructive sleep apnea. *J. Pediatr.*, 125, 556 (1994)

Gozal D. Sleep-disordered breathing and school performance in children. *Pediatrics*, 102, 616 (1998)

Naseem S. et al. ADHD in adults and obstructive sleep apnea. *Chest*, 119, 294 (2001)

Millman R. P. et al. Depression as a manifestation of sleep apnea. *J. Clin. Psychiatry*, 50, 348 (1989)

Robb M. P. et al. Vocal tract resonance characteristics of adults with obstructive sleep apnea. *Acta Otolaryngol. (Stockh.)*, 117, 760 (1997)

Ulfberg J. et al. Headache, snoring and sleep apnoea. *J. Neurol.*, 243, 621 (1996)

Karacan I. & Karatas M. Erectile dysfunction in sleep apnea. *J. Sex. Marital Ther.*, 21, 239 (1995)

Fischer J. & Raschke F. Economic and medical significance of sleep-related breathing disorders. *Respiration*, 64 Sl, 39 (1997)

Unblocking those tubes

Jan M. A. et al. Effect of posture on upper airway dimensions. *Am. J. Respir. Crit. Care Med.*, 149, 145 (1994)

Penzel T. et al. Effect of sleep position on the collapsibility of the upper airways. *Sleep*, 24, 90 (2001)

Chaudhary B. A. et al. Therapeutic effect of posture in sleep apnea. *South. Med. J.*, 79, 1061 (1986)

Itasaka Y. et al. Influence of sleep position and obesity on sleep apnea. *Psychiatry Clin. Neurosci.*, 54, 340 (2000)

Cartwright R. D. et al. Sleep position training as a treatment for sleep apnea. *Sleep*, 8, 87 (1985)

Phillips B. A. et al. Effect of sleep position on sleep apnea. *Chest*, 90, 424 (1986)

Cartwright R. et al. Treatments for positional sleep apnea. *Sleep*, 14, 546 (1991)

Loube D. I. et al. Weight loss for obstructive sleep apnea. *J. Am. Diet. Assoc.*, 94, 1291 (1994)

Sampol G. et al. Efficacy of weight loss in sleep apnoea/hypopnoea. *Eur. Respir. J.*, 12, 1156 (1998)

Lorino A. M. et al. Effects of mandibular advancement on respiratory resistance. *Eur. Respir. J.*, 16, 928 (2000)

Schmidt-Nowara W. et al. Oral appliances for the treatment of sleep apnea. *Sleep*, 18, 501 (1995)

Sullivan C. E. et al. Reversal of sleep apnoea by CPAP. *Lancet*, 1, 862 (1981)

McArdle N. et al. Partners of patients with sleep apnoea/hypopnoea. *Thorax*, 56, 513 (2001)

Liistro G. et al. Management of sleep apnoea. *Eur. Respir. J.*, 8, 1751 (1995)

Chervin R. D. & Guilleminault C. Obstructive sleep apnea. *Neurol. Clin.*, 14, 583 (1996)

Rollheim J. et al. Obstructive sleep apnoea and body mass index. *Clin. Otolaryngol.*, 22, 419 (1997)

Lee W. C. et al. Complications of palatoplasty. *J. Laryngol. Otol.*, 111, 1151 (1997)

Back L. et al. Sleep-disordered breathing: radiofrequency thermal ablation. *Laryngoscope*, 111, 464 (2001)

Berger G. et al. Laser-assisted uvulopalatoplasty for snoring. *Arch. Otolaryngol. Head Neck Surg.*, 127, 412 (2001)

Barthel S. W. & Strome M. Snoring, sleep apnea, and surgery. *Med. Clin. North Am.*, 83, 85 (1999)

Chapter 16 – And So to Bed

Wright, Lawrence. *Warm and Snug*. (London, Routledge & Kegan Paul, 1962) p. 177

Burgess, Anthony. *On Going to Bed*. (London, André Deutsch, 1982) p. 59

Chesterton, G. K. 'On lying in bed.' *Tremendous Trifles*. (Beaconsfield, Darwen Finlayson, 1968)

Shattock, Roger. *Marcel Proust*. (Princeton, Princeton University Press, 1974) p. 21

Mann, Thomas. *Sleep, Sweet Sleep*. (Privately printed, 1934) p. 4

Graebner, Walter. *My Dear Mister Churchill*. (London, Michael Joseph, 1965) p. 48

A brief history of beds
Runyon, Damon. 'Sweet Dreams.' *From First to Last*. (London, Penguin, 1990)
Wright, Lawrence. *Warm and Snug*. (London, Routledge & Kegan Paul, 1962) numerous
Beldegreen, Alecia. *The Bed*. (NY, Stewart, Tabori & Chang, 1991) numerous
Eden, Mary & Carrington, Richard. *The Philosophy of the Bed*. (London, Hutchinson, 1961) numerous
Reynolds, Reginald. *Beds*. (London, André Deutsch, 1952) numerous
Gray, Cecil & Gray, Margery. *The Bed*. (London, Nicholson & Watson, 1946) numerous
Stopes, Marie Carmichael. *Sleep*. (London, Chatto & Windus, 1956)
Sparkes, Ivan G. *Four-poster and Tester Beds*. (Princes Risborough, Shire, 1990) numerous
Burgess, Anthony. *On Going to Bed*. (London, André Deutsch, 1982) numerous
Picard, Liza. *Restoration London*. (London, Weidenfeld & Nicolson, 1997) p. 50
Descola, Philippe. *The Spears of Twilight*. (London, Flamingo, 1997) p. 43
Brownlow A. R. et al. The nesting behavior of chimpanzees. *Am. J. Primatol.*, 55, 49 (2001)
Anderson J. R. Sleep, sleeping sites, and sleep-related activities. *Am. J. Primatol.*, 46, 63 (1998)
Reichard U. Sleeping sites and presleep behavior of gibbons. *Am. J. Primatol.*, 46, 35 (1998)
von Hippel F. A. Use of sleeping trees by colobus monkeys. *Am. J. Primatol.*, 45, 281 (1998)

Sleeping partners
De Waterman A. L. & Kerkhof G. Sleep–wake patterns of partners. *Percept. Mot. Skills*, 86, 1141 (1998)
Graham, James. *A Lecture on the Generation, Increase and Improvement of the Human Species!* (London, 1783)
Pankhurst F. P & Horne J. A. Influence of bed partners on movement during sleep. *Sleep*, 17, 308 (1994)
Hardyment, Christina. *Dream Babies*. (Oxford, Oxford University Press, 1984) p. 53
The Glaxo Baby Book, 14th edn (London, Joseph Nathan, 1921)
Mosko S. et al. Infant arousals during mother–infant bed sharing. *Pediatrics*, 100, 841 (1997)
McKenna J. J. & Mosko S. S. Sleep and arousal, synchrony and independence, among mothers and infants sleeping apart and together. *Acta Paediatr. Suppl.*, 397, 94 (1994)
McKenna J. J. et al. Bedsharing promotes breastfeeding. *Pediatrics*, 100, 214 (1997)

Mosko S. et al. Maternal sleep and arousals during bedsharing. *Sleep*, 20, 142 (1997)

Scragg R. et al. Bed sharing, smoking, and alcohol in SIDS. *BMJ*, 307, 1312 (1993)

Nakamura S. et al. Hazards associated with children placed in adult beds. *Arch. Pediatr. Adolesc. Med.*, 153, 1019 (1999)

Blair P. S. et al. Babies sleeping with parents. *BMJ*, 319, 1457 (1999)

Mosko S. et al. Maternal proximity and infant CO_2 environment during bedsharing. *Am. J. Phys. Anthropol.*, 103, 315 (1997)

Mitchell E. Cot death – the story so far. *BMJ*, 319, 1461 (1999)

Willinger M. et al. Transition to nonprone sleep positions of infants in the US. *JAMA*, 280, 329 (1998)

Douglas A. S. et al. Seasonality of sudden infant death syndrome. *Arch. Dis. Child.*, 79, 269 (1998)

Lindgren C. et al. Sleeping position, bedsharing and passive smoking in Swedish infants. *Acta Paediatr.*, 87, 1028 (1998)

Chapter 17 – An Excellent Thing

Puritans and hypocrites
Chesterton, G. K. 'On lying in bed.' *Tremendous Trifles*. (Beaconsfield, Darwen Finlayson, 1968)

Wachhorst, Wyn. *Thomas Alva Edison*. (Cambridge MA, MIT Press, 1981)

Israel, Paul. *Edison: a life of invention*. (New York, Wiley, 1998)

Boswell, James. *The Journal of a Tour to the Hebrides with Samuel Johnson, LL.D.* (1786) entry for 14 Sept 1773

Cronin, Vincent. *Napoleon*. (London, Collins, 1971)

Naps, nappers and napping
Graebner, Walter. *My Dear Mister Churchill*. (London, Michael Joseph, 1965)

Mullen, Richard. *Anthony Trollope*. (London, Duckworth, 1990) p. 164

Sala, George A. *Things I Have Seen and People I Have Known*. (London, Cassell, 1894) Vol. I, p. 30

Campbell S. S. et al. When the human circadian rhythm is caught napping. *Sleep*, 16, 638 (1993)

Mednick S. C. et al. The restorative effect of naps on perceptual deterioration. *Nat. Neurosci.*, 5, 677 (2002)

Taub J. M. et al. Effects of afternoon naps on physiological variables, performance and activation. *Biol. Psychol.*, 5, 191 (1977)

Hayashi M. et al. Effects of a 20-min nap on sleepiness, performance and EEG. *Int. J. Psychophysiol.*, 32, 173 (1999)

Dinges D. F. et al. Temporal placement of a nap for alertness. *Sleep*, 10, 313 (1987)

Takahashi M. et al. Brief naps during post-lunch rest. *Eur. J. Appl. Physiol.*, 78, 93 (1998)

Tietzel A. J. & Lack L. C. Benefits of naps following sleep restriction. *Sleep*, 24, 293 (2001)

Gillberg M. et al. Effects of a short daytime nap after restricted sleep. *Sleep*, 19, 570 (1996)

Tanaka H. et al. Short naps and exercise improve sleep quality and mental health in the elderly. *Psychiatry Clin. Neurosci.*, 56, 233 (2002)

Rosekind M. R. et al. Strategic naps in operational settings. *J. Sleep Res.*, 4(S2), 62 (1995)

Chan O. Y. et al. Sleep quality of day and night workers. *Sleep*, 12, 439 (1989)

Dinges D. F. et al. Temporal placement of a nap for alertness. *Sleep*, 10, 313 (1987)

Bonnet M. H. & Arand D. L. Impact of naps and caffeine on extended nocturnal performance. *Physiol. Behav.*, 56, 103 (1994)

Sallinen M. et al. Promoting alertness with a short nap during a night shift. *J. Sleep Res.*, 7, 240 (1998)

Bonnet M. H. Effect of prophylactic naps on performance, alertness and mood. *Sleep*, 14, 307 (1991)

Buysse D. J. et al. Napping and 24-hour sleep/wake patterns. *J. Am. Geriatr. Soc.*, 40, 779 (1992)

Dinges D. F. Differential effects of prior wakefulness and circadian phase on nap sleep. *Electroencephalogr. Clin. Neurophysiol.*, 64, 224 (1986)

Islas-Marroquin J. & Delgado-Brambila H. A. Nap sleep in students. *Arch. Med. Res.*, 29, 149 (1998)

Kitchiner, William. *The Art of Invigorating and Prolonging Life, Etc.* 4th edn, (London, Hurst, Robinson, 1822) p. 74

Sweet dreams

de Nerval, Gérard. 'Aurélia.' *Selected Writings*. (London, Penguin, 1999)

de Beauvoir, Simone. *All Said and Done*. (London, Penguin, 1977)

Buñuel, Luis. *My Last Breath*. (London, Vintage, 1994) p. 92

Blixen, Karen. *Out of Africa*. (London, Penguin, 1954) p. 82

Garfield, Patricia L. *Creative Dreaming*. 2nd edn, (NY, Simon & Schuster, 1995)

Doyle M. C. Enhancing dream pleasure with Senoi strategy. *J. Clin. Psychol.*, 40, 467 (1984)

Blessed oblivion

Murdoch, Iris. *The Black Prince*. (London, Penguin, 1975) p. 231

Storr, Anthony. *Solitude*. (London, HarperCollins, 1988) p. 94

Give sleep a chance

O'Hanlon, Brenda. *Sleep: the common sense approach*. (Dublin, Newleaf, 1998)

Kavey, Neil B. *50 Ways to Sleep Better*. (Lincolnwood ILL, Publications International, 1996)

Ball, Nigel & Hough, Nick. *The Sleep Solution*. (London, Vermilion, 1998)

Chokroverty, Sudhansu. *100 Questions About Sleep and Sleep Disorders*. (Oxford, Blackwell Science, 2001)

Idzikowski, Christopher. *Learn to Sleep Well*. (San Francisco, Chronicle Books, 2000)

Index

Paula

Isabel Allende

*'Listen Paula, I am going to tell you a story,
so that when you wake up, you will not feel lost.'*

In December 1991, Isabel Allende's daughter Paula, aged 28, fell gravely ill and sank into a coma. This book was written during the interminable hours the novelist spent in the corridors of a Madrid hospital, in the hotel room where she lived for several months, and beside Paula's bed at home in California, during the summer and autumn of 1992.

Faced with the loss of her child, Isabel Allende turned to the thing she does best, to her art, to storytelling to sustain her own spirit, but also to convey to her daughter the will to wake up, to survive. The story she tells is that of her own life, her family history and the tragedy of her nation, Chile, in the years leading up to Pinochet's brutal military coup.

Paula is a testament to the ties which bind mothers and children, a brave, enlightening, inspiring book filled with the insight, passion, humour and magic which characterises the work of one of the world's greatest stroytellers.

Naomi Klein

No Logo

'The *Das Kapital* of the growing anti-corporate movement'
Guardian

'Don't miss this visionary book, perfectly pitched for our suspect, over-corporate times.' *i-D*

'If the world really is just one big global village, then the logo is its common language understood by – if not accessible to – everyone. In *No Logo*, Klein undertakes an arduous journey to the centre of a post-national planet. Starting with the brand's birth, as a means of bringing soul to mass marketing, she follows in the logo's wake and notes its increasing capacity for making the product subservient. Beyond this she reaches her core argument – the now uneasy struggle between corporate power and anti-corporate activism – via sweatshop labour, submerged identity and subversive action. Part sociological thesis, part design history, *No Logo* is entirely engrossing and emphatic.' *GQ*

'A riveting, conscientious piece of journalism and a strident call to arms. Packed with enlightening statistics and extraordinary anecdotal evidence, *No Logo* is fluent, undogmatically alive to its contradictions and omissions and positively seethes with intelligent anger.' SAM LEITH, *Observer*

0 00 653040 0

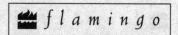

Gwyneth Lewis

Sunbathing in the Rain

'Undoubtedly the best book I have ever read about one person's experience of depression.'
DOROTHY ROWE

'*Sunbathing In The Rain* is such a lovely title for a book about beating depression. Gwyneth Lewis says she wants this book to be easy to pick up and read at any page. It is . . . A delicate, vulnerable triumph.'
SIMON HATTENSTONE, *Guardian*

Drawing on her own experience, Lewis re-embarks on a journey that nearly killed her first time round and returns from it with this, perhaps the first truly undogmatic, undemanding, down-right useful companion to depression. With an overall structure that moves from darkness to light, from the profound to the frivolous, it looks to offer comfort and encouragement to others going through this agonising and perplexing experience. Alongside a paragraph about the proper relationship between the ego, the mind and the emotion, nestles a passage on the therapeutic value of nail varnish. Practical hints (diet, read *Hello!*) sit alongside striking quotations.

'What gives the book its edge is her determination that the illness must be seen as an early warning system, to be welcomed as a timely indication that something needs addressing . . . this upbeat, very readable and engaging view of depression as a temporary retrenchment, a breathing space in which to adjust better to life, makes encouraging reading.'
CAROLINE MOOREHEAD, *Spectator*

'A flush of hard-won knowledge and hope for a brighter future suffuses these pages.'
TLS

0 00 712062 1

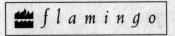